TRACE ELEMENTS IN COAL

Volume II

Author

Vlado Valković

Institute Ruder Bošković
Zagreb, Yugoslavia

CRC Press, Inc.
Boca Raton, Florida

Library of Congress Cataloging in Publication Data
Valkovič, V.
　Trace elements in coal

　Includes bibliographical references.
　1. Coal — Analysis.　2. Trace elements—
Analysis. I. Title.
TP325.V34　　662.6'22　　82-4386
ISBN 0-8493-5491-9 (v. 1)　　AACR2
ISBN 0-8493-5492-7 (v. 2)

　This book represents information obtained from authentic and highly regarded sources. Reprinted material is quoted with permission, and sources are indicated. A wide variety of references are listed. Every reasonable effort has been made to give reliable data and information, but the author and the publisher cannot assume responsibility for the validity of all materials or for the consequences of their use.

　All rights reserved. This book, or any parts thereof, may not be reproduced in any form without written consent from the publisher.

　Direct all inquiries to CRC Press, Inc., 2000 Corporate Blvd., N.W., Boca Raton, Florida, 33431.

© 1983 by CRC Press, Inc.

International Standard Book Number 0-8493 (Volume I) 5491-9
International Standard Book Number 0-8493 (Volume II) 5492-7
Library of Congress Card Number 82-4386

Printed in the United States.

PREFACE

Coal has been widely used as a principal energy source since the 18th century. It is believed that coal was the main driving force behind the Industrial Revolution; for the first time in history, a truly cheap metal, iron, was made available. By the end of the 19th century coal surpassed wood as the main energy source in the U.S. (see Figure 1, Chapter 1, showing projections of the U.S. Energy Information Administration). Since then it has been used as the principal fuel for heating of buildings, manufacturing processes, and electrical power generation. For a long time coal was king.

Then, at the beginning of the 1950s it was dethroned by petroleum with the help of attacks by environmentalists. The problems of energy and environment were brought together. In order to maintain or increase the standard of living, an increase in the development of energy sources is required. So, to preserve the quality of life, the production of energy in an environmentally acceptable manner is required. So far, energy has been extracted from coal by direct combustion in steam boilers and generation of electricity. This results in damage to the environment, with the worst offender probably being sulfur in the coal.

With the predictable exhaustion of oil reserves and the finite amounts of deposits of some elements, attention is again being focused on coal. There have been some very important developments in coal utilization technology in the areas of coal liquefaction, coal gasification, and advanced methods of coal combustion to generate electric power. It may turn out to be equally important that coal combustion produces an ash which may offer an economical raw material for resource recovery. It has been generally accepted that coal technologies (unlike nuclear technologies, for example) can proceed in a largely empirical way. However, recently the need to accelerate the transition from experience-based technology to science-based technology has been recognized. This new approach can produce more efficient and cleaner technologies.

Coal has been described as a "bridge to the future". It is the most plentiful fossil fuel in the world, and it has a potential for filling a growing proportion of the demand for energy. It is found around the globe, but three countries (the U.S. the U.S.S.R., and China) own nearly two thirds of all known coal reserves. At present rates of consumption, these reserves would last the world more than 200 years, according to conservative estimates. Furthermore, geologists think the world probably has 15 times this much coal.

Developments in the U.S. are of great interest. The goal of the U.S. government to reduce dependency on imported fossil fuels, as well as some other factors, is moving the U.S. toward an economy based on coal as the primary fossil fuel. The use of coal as fuel is expected to triple in the period 1975 to 2000, from approximately 10 quads (1 quad = 10^{15} Btu) to 30 quads (see Figure 1, Chapter 1). Electric power utilities are the major users of coal, and most of the projected increase will also be for the generation of power. Minor increases in coal use as a chemical feedstock to replace petroleum sources is anticipated, but not in the next few years. The use of coal in coking will probably remain at about the present level.

Because of the way it was formed, coal may contain every naturally occurring element. Early chemical analyses of coal and its ashes were limited to the major elements (C, H, O, N, S, Si, Fe, Al, K, Ca, and Mg). Trace elements is often defined as any element whose concentration in coal is 1000 ppm (0.1%) or less. However, elemental composition is very variable, and for the sake of completeness, major elements should be included in discussions on elemental composition of coals.

The first trace element analyses of coal were performed in the 1930s. In the U.S., an ongoing program for the sampling and analysis of coal for trace elements began in 1948. At that time the lack of trace element analysis resulted from the difficulty of applying the classical wet chemistry methods of analysis to trace constituents. However, the advent of newer techniques of instrumental analysis has made analysis of trace elements more feasible. In recent years, general interest in chemical analysis of coal and coal related materials has significantly increased to the state where adequate analytic procedures and suitable standards are currently available for many trace elements. The number of complete modern analyses for trace elements in coal has increased greatly in recent years. As of 1979, the largest accumulation of these data, on 3700 samples of U.S. coal, is publicly available from the data bank of the U.S. Geological Survey in its National Coal Resources Data System.

The main emphasis of this book is on the occurrence and distribution of trace elements in coal. Nature and mode of occurrence of trace elements, variations in distributions, and concentration levels in ash are discussed in detail. The accumulated knowledge about all elements detected in coal is summarized. Special attention is paid to rare and uncommon elements and to radionuclides. Trace elements distribution in different phases of coal utilization processes is discussed next. Possibilities of trace element recovery from coal and coal ash are elaborated.

Environmental considerations are of great importance for the acceptance of coal-utilization techniques. Trace element contaminants from coal-fired power plants are studied by many researchers. The effects of radionuclides released in the environment are critically summarized.

Methods for the measurements of trace elements concentration in coal and its ash are described in the last chapter of Volume II. Optical methods, X-ray analysis, and nuclear methods are presented, in addition to classical chemical methods.

THE AUTHOR

Vlado Valkovic̀, Ph.D., is presently Scientific Advisor at the Institute Ruder Bos̆kovic̀, Zagreb, Croatia, Yugoslavia.

Dr. Valkovic̀ graduated in 1961 from the University of Zagreb, with a B.A. degree in experimental physics and obtained his M.A. degree in nuclear physics in 1963 and his Ph.D. degree in 1964. The thesis title was "Nuclear Reactions with 14.4 MeV Neutrons on Light Elements".

Dr. Valkovic̀ has been associated with two institutions during all his professional career — one in the U.S., one in Yugoslavia. The institutions are the Institute Ruder Bŏskovic̀ in Zagreb and the T. W. Bonner Nuclear Laboratories at Rice University, Houston, Texas.

Dr. Valkovic̀ is a fellow of the American Physical Society, a member of the European Physical Society, Yugoslav Physical Society (Croatian section), Society for Environmental Geochemistry and Health, and other professional and honorary organizations.

His research has been done on both sides of the Atlantic: in the U.S. as well as in a number of laboratories in Europe.

Dr. Valkovic̀ has published more than 100 research papers on nuclear physics and applications of nuclear techniques to the problems in biology, medicine, environmental research, and trace element analysis. He is the author of five additional books: *Trace Element Analysis* (Taylor and Francis, London, 1975), *Nuclear Microanalysis* (Garland, New York, 1977), *Trace Elements in Human Hair* (Garland, New York, 1977), *Trace Elements in Petroleum* (Petroleum Publishing, Tulsa, 1978), and *Analysis of Biological Material for Trace Elements Using X-ray Spectroscopy* (CRC Press, Boca Raton, 1980). His current major research interests include the study of the role and movements of the elements in nature.

To Georgia

TABLE OF CONTENTS

Volume I

Chapter 1
COAL — ORIGIN, CLASSIFICATION, PHYSICAL AND CHEMICAL PROPERTIES 1
I. Coal Origin 1
II. Coal Deposits and Reserves 4
III. Characterization of Coal 5
IV. Physical Properties 17
    A. Optical Properties of Coal 17
    B. Mechanical properties of Coal 19
        1. Pore Structure 20
        2. Coal Fracture 22
    C. Electrical Conductivity 24
    D. Magnetic Properties of Coal 26
    E. Thermal Properties of Coal 28
V. Chemical Properties 28
VI. Chemical Composition 30
    A. Coal Structure 31
    B. Macerals 36
    C. Mineral Matter 38
VII. Radioactivity 43

References 48

Chapter 2
TRACE ELEMENTS IN COAL — OCCURRENCE AND DISTRIBUTION 57
I. General Consideration 57
II. Nature and Mode of Occurrence of Trace Elements 70
III. Ash 77
IV. Summary of Data on Element Concentrations in Coal 83

References 177

Volume II

Chapter 1
COAL UTILIZATION 1
I. Combustion 1

II.	Material Balance in Power Plant	5
III.	Ash	14
	A. Ash Fusion	18
	B. Ash Determination	21
	C. Bottom Ash	25
	D. Fly Ash	27
	E. Ash Utilization	31
IV.	Desulfurization	35
V.	Trace Element Recovery	46
	A. Aluminum Recovery	49
	B. Germanium Recovery	52
	C. Titanium Recovery	54
	D. Uranium Recovery	56
VI.	New Technologies of Coal Utilization	64
	A. Coal Conversion	67
	B. Coal Liquefaction	69
	C. Coal Gasification	74
References		80

Chapter 2
ENVIRONMENTAL CONSIDERATIONS 103

I.	Trace Contaminants from Coal-Fired Power Plants	103
	A. Air Pollution	105
	B. Atmospheric Deposition and Soil Pollution	116
II.	Coal Cleaning	120
III.	Waste Management	129
IV.	Radioactivity	133
V.	Environmental Impacts of New Technologies	146
VI.	Biological and Health Aspects of Coal Use	156
References		164

Chapter 3
METHODS OF ANALYSIS 173

I.	General Consideration	173
	A. Introduction	173
	B. Sample Preparation	177
	C. Standards	179
II.	Classical Chemical Methods	181

III.	Optical Methods.		185
	A.	Emission Spectroscopy	185
	B.	Atomic Absorption Spectroscopy	191
IV.	Electron Microscopy		197
V.	X-ray Analysis		202
	A.	X-ray Diffraction	202
	B.	X-ray Emission Spectroscopy	203
		1. Tube Excited X-ray Emission	204
		2. Radioactive Source Excitation	219
		3. Proton Induced X-ray Emission (PIXE) Spectroscopy	220
VI.	Neutron Activation		228
	A.	Neutron Activation With Thermal Neutrons	228
	B.	Neutron Activation Using ^{252}Cf Source	237
	C.	Fast Neutron Activation Analysis	243
VII.	Other Nuclear Methods		247
	A.	Charged Particle Activation	248
	B.	Alpha, Beta and Gamma Counting	250
	C.	Mössbauer Spectroscopy	253
	D.	Nuclear Magnetic Resonance (nmr) Spectroscopy	253
	E.	Electron Spin Resonance (esr) Spectroscopy	257
VIII.	Mass Spectroscopy		258
IX.	Comparision of Different Methods		262
References			266
Index			275

Chapter 1

COAL UTILIZATION

I. COMBUSTION

Combustion converts the organic part of coal to carbon dioxide and water and the nonvolatile inorganic consituents to coal ash residues. A variable amount (up to 17%) of the coal burned is recoverable as coal ash residues. Loevblad[1] has presented data for ash content, volatile compounds content, sulfur content, and heat of combustion for coals of different origin. His findings are shown in Table 1. Coal ash residues consist of approximately 30% bottom ash collected from the bottom of the combustion chamber and approximately 70% fly ash collected from the stack gases by the air pollution control equipment. In the absence of efficient pollution control equipment, fly ash is released to the atmosphere with the stack gases and ultimately settles to the earth.

A more general list of residues from coal-fired power-generation plants contains the following categories:

1. Boiler bottom ash (boiler slag)
2. Fly ash, captured in emission-control devices and/or emitted to the atmosphere through stacks
3. Flue-gas desulfurization (FGD) sludge resulting from the removal of SO_2 from the combustion gases
4. Atmospheric gases
5. Wastewaters

Three categories of boiler are commonly used in power-generation plants. These are stoker-fired boilers, cyclone furnaces, and pulverized-coal-fired (PC-fired) furnaces. In the cyclone unit, 70 to 80% of the ash is melted and is continuously tapped from the bottom of the furnace as slag. The remaining 20 to 30% is collected from the stack gas by techniques similar to those used in the PC-fired units. The PC-fired boiler unit is of two types — wet and dry bottom — with newer units being almost exclusively the latter. In both, the bottom of the fire box contains one or more hopper-shaped openings through which 20 to 30% of the ash in the form of large fused pieces drops into a water-filled hopper, whence they are subsequently sluiced to a settling pond. The wet-bottom furnace, whether PC or cyclone, operates at higher temperatures in the fire box so that more of the ash is melted and flows through a slag tap into a water-filled hopper below. This material is referred to as boiler slag. Small ash particles entrained with the combustion products from the boiler are collected at the stack by means such as electrostatic precipitators and bag filters (fly ash) and wet scrubbers (FGD sludge), with collection efficiencies as high as 95 to 99.7% by weight. The percentage of bottom ash, boiler slag, and fly ash in relation to the furnace configuration is shown in Table 2.[2]

The knowledge of the process of coal burning is rather limited. Recently computed tomography has been used to supply first inside views of lumps of coal in the process of burning. Results of the first such investigation were reported during the March 1981 meeting of American Physical Society in Phoenix by Maylotte.[3] Maylotte's team, working at the General Electric Research and Development Center in Schenectady, N.Y., built a special coal-burning furnace that fits neatly into a computer tomography (CT) scanner. The cylindrical furnace is about 30.5 cm in diameter and 66 cm long. It can heat its cargo of coal to 980°C.

Table 1
ASH, VOLATILE COMPOUNDS, SULFUR CONTENT, AND HEAT OF COMBUSTION OF VARIOUS COALS[1]

Origin of the coals:			Ash (%)	Volatile compounds (%)	Sulfur (%)	Heat of combustion (kWh/kg)
Norway		Spitsbergen	7	39	0.78	8.4
England			9	30	1.12	7.4
U.S.	A	Massey	—	35	1.02	8.1
U.S.	B	Massey	—	32	0.61	8.6
West Germany	HA	Ruhr	5	28	0.85	9.7
West Germany	W		11	35	0.79	9.4
Poland	A		—	—	0.60	7.1
Poland	B		—	—	0.76	7.1
Soviet Union	A	Petchora	15—17	34—35	0.70	7.8—8.0
Soviet Union	B	Petchora	15	36	0.50	9.4
Soviet Union	C	Kuznetsk	12	28	0.40	9.3
Australia		50% Lithgow 50% Katoomba (New South Wales)	15	27	0.47	8.3

Table 2
PERCENTAGE OF COAL ASH FROM VARIOUS FURNACE CONFIGURATIONS[2]

Furnace	Bottom ash (%)	Boiler slag (%)	Fly ash (%)
Stoker fired, traveling	100	—	—
Stoker fired, spreader	—	45—85	15—55
Cyclone fired	—	80—85	15—20
Pulverized coal, wet bottom	50	—	50
Pulverized coal, dry bottom	20—25	—	75—80

Inside the furnace are thermocouples to monitor the temperature of the coal and its surroundings and exit passages to pipe away gases produced as the coal burns. The gases are then channeled to a mass spectrometer that detects the amounts of carbon monoxide, oxygen, nitrogen, hydrogen, and other constituents of the combustion process.

Coal was introduced into this furnace one lump at a time; the optimum size for a lump was found to be 5 to 6 cm across. As the coal is heated, computer-reconstructed images obtained from the CT scanner, complemented by computer graphics, indicate changes in the electron density of various internal parts of the lump of coal, and hence in its internal chemistry. The images show water oozing out of the inside of the lump in the early stages of heating, followed by the more volatile organic components of the coal.

Beyond learning the basic details of how coal burns, the technique should provide a means of fitting the type of coal to the procedure used to burn it. Among specific investigations suggested are the study of pyrites in coal and examination of the way in which limestone added to coal picks up sulfur inside the lumps, thereby cleaning up the coal. Other studies now in sight include injecting various gases, such as xenon, into lumps of coal and then monitoring the lumps by the CT scanner technique to discover how porous they are.

During the process of coal burning, vaporization of elements takes place. The vaporization temperatures of the volatile elements vary according to whether the element is in an elemental or an oxidized state. Therefore, temperature and oxidation potential of the furnace are important factors in determining trace-element enrichment or depletion in the various types of residues produced. The relative order of volatility for oxides, sulfates, carbonates, silicates, and phosphates is[4]

$$As \approx Hg > Cd > Pb \approx Bi \approx Tl > Ag \approx Zn > Cu \approx Ga > Sn > Li \approx Na \approx K \approx Rb \approx Cs$$

For elements in the elemental state this order is

$$Hg > As > Cd > Zn > Sb \geqslant Bi > Tl > Mn > Ag \approx Sn \approx Cu > Ga \approx Ge$$

For sulfides the relative order of volatility is

$$As \approx Hg > Sn \approx Ge \geqslant Cd > Sb \approx Pb \geqslant Bi > Zn \approx Tl > Cu > Fe \approx Co \approx Ni \approx Mn \approx Ag$$

The list of species boiling or subliming below or at 1550°C during coal combustion include As, As_2O_3, As_2S_3; Ba; Bi; Ca; Cd, CdO, CdS; $Cr(CO)_6$, $CrCl_3$, CrS; K; Mg; $Ni(CO)_4$; $PbCl_2$, PbO, Rb; Se, SeO_2, SeO_3; Sb, Sb_2S_3, Sb_2O_3; SnS; Sr; Tl, Tl_2O, Tl_2O_3; Zn, ZnS.

Of special interest is the distribution of elements between three major coal residues, i.e., bottom ash, fly ash, and flue gas. In Reference 4 this is presented for the case when bottom ash was 22.2% of total, fly ash 77.1% of total, and flue gas contained 0.7% fly ash of total ash. The resulting distribution, which is first given in reports by Schwitzgebel et al.[5] and U.S. Environmental Protection Agency,[6] is shown in Table 3.

A relation between particle size and the chemical composition of coal residues has been confirmed by a number of investigators. In all investigations, an inverse relationship is observed between the concentrations of certain elements and particle size. Davison et al.[7] demonstrated that lead, thallium, antimony, cadmium, selenium, arsenic, zinc, nickel, chromium, and sulfur were markedly increased with decreasing particle size of ashes derived from Indiana coal sources fractioned in the range of 0.65 to >74 μm. Similar observations on boron, cadmium, chromium, copper, manganese, nickel, lead, and uranium were mad25 μm. The concentrations of gallium, germanium, mercury, and lead showed approximately 3-fold increase between the coarse (>50 μm) and fine (<2 μm) fly ash of Australian coal sources.[8,9]

Table 4 shows the effect of particle size of fly ash on the concentration of 17 trace elements as reported by Ondov et al.[10,11] The range of concentration increase in the <3 μm fraction as compared with the >15 μm fraction was from 1.5-fold for manganese to 11.5-fold for cadmium.

The availability of elements in coal fly ash is not only a function of particle size and concentration but also of depth in individual coal fly-ash plerospheres. Linton et al.[12] have shown that beryllium, carbon, calcium, chromium, potassium, lithium, manganese, sodium, phosphorus, lead, sulfur, thallium, vanadium, and zinc were particularly concentrated on the particle surface as compared with their concentrations at a distance inside the particle 50 nm from the surface; however, the nature of the surface of particles is not known with certainty. Thus, the magnitude of the surface area of particles affects the amounts of the trace element retained by such mechanism. Swaine[8,9] estimated that a typical value for the surface area of fly ash is of the order of 1 m²/g. Surface areas of bottom ash, inlet precipitator ash (ash entering the precipitator in the flue-gas steam), and outlet precipitator ash (ash exiting the precipitator in the flue-gas stream) measured by Kaakinen et al.[13] were, respectively, 0.38, 3.06, and 4.76 m²/g.

Table 3
DISTRIBUTION OF ELEMENTS AMONG BOTTOM ASH, FLY ASH, AND FLUE GAS[5,6]

Element	Bottom ash (22.2%)	Fly ash (77.1%)	Flu gas (0.7%)
Aluminum	20.5	78.8	0.7
Antimony	2.7	93.4	3.9
Arsenic	0.8	99.1	0.05
Barium	16.0	83.9	0.09
Beryllium	16.9	81.0	2.0
Boron	12.1	83.2	4.7
Cadmium	15.7	80.5	3.8
Calcium	18.5	80.7	0.8
Chlorine	16.0	3.8	80.2
Chromium	13.9	73.7	12.4
Cobalt	15.6	82.9	1.5
Copper	12.7	86.5	0.8
Fluorine	1.1	91.3	7.6
Iron	27.9	71.3	0.8
Lead	10.3	82.2	7.5
Magnesium	17.2	82.0	0.8
Manganese	17.3	81.5	1.2
Mercury	2.1	0	97.9
Molybdenum	12.8	77.8	9.4
Nickel	13.6	68.2	18.2
Selenium	1.4	60.9	27.7
Silver	3.2	95.5	1.3
Sulfur	3.4	8.8	87.8
Titanium	21.1	78.3	0.6
Uranium	18.0	80.6	1.5
Vanadium	15.3	82.3	2.4
Zinc	29.4	68.0	2.6

Let us also mention tars and pitches as coal products. Tars and pitches are aromatic materials encountered in coking and in coal conversion processes. Because of the different conditions to which coals are exposed in these two types of processes, the tars and pitches produced by each may differ significantly in composition; however, some insight into their complex composition may be gained from studies of the materials produced by coking.

Pitches are obtained through distillation of tars and represent from 30 to 60% of the tar components. Pitches have been estimated to contain about 5000 compounds. During carbonization, coal passes through two stages of decomposition: At 400 to 500°C plasticity begins to develop, and at 650 to 750°C, more advanced decomposition occurs. During these changes, the volatile products of the coal are released, and as they pass through the hot coke, the volatile products become involved in a series of secondary reactions. They emerge from the retort and are separated by fractional condensation or absorption into tar, ammonia liquor, benzole, and illuminating or heating gas.[14] The major reactions involved in the conversion of primary carbonization products into tars are (1) the cracking of higher molecular-weight paraffins to gaseous paraffins and olefins; (2) dehydrogenation of cyclohexane derivatives to aromatic hydrocarbons and phenols; (3) dealkylation of alkyl aromatic hydrocarbons, alkyl pyridines, and alkyl phenols; (4) dehydroxylation of phenols; (5) synthesis of polynuclear aromatic hydrocarbons by condensation of simpler structures with olefins; and (6) disproportionation of aromatic hydrocarbons to both simpler and more complex structures.

Table 4
TRACE ELEMENT CONCENTRATIONS (PPM) IN FLY-ASH PARTICLES OF DIFFERENT SIZES[10,11]

Element	Size Range (μm)			
	>15	8—15	3—8	<3
As	13.7	56	87	132
Be	6.3	8.5	9.5	10.3
Cd	0.4	1.6	2.8	4.6
Co	8.9	16.3	9	21
Cr	28	49	59	63
Cu	56	89	107	137
Ga	43	116	140	178
Mn	207	231	261	317
Mo	9.1	28	40	50
Ni	25	37	44	40
Pb	73	169	226	278
Sb	2.6	8.3	13	20.6
Se	19	59	78	198
U	8.8	16	22	29
V	86	178	244	327
W	3.4	8.6	16	24
Zn	71	194	304	550

The extent of these reactions is affected by the temperature of carbonization and the time of contact with the hot coke bed and the heated walls of the retort. Thus, tars that have been exposed to different carbonization processes vary in chemical composition. Distillation of crude tar yields a series of primary oil fractions and a residue of refined tar pitch, depending on the extent of the distillation process. Distillate oils obtained by steam or vacuum distillation of pitch or pitch crystalloids or from the coking of pitch are the only fractions from which pure chemical compounds can normally be isolated.

Different aspects of problems related to coal combustion have been discussed by a number of authors. Some of the more recent literature can be found in the list of references.[15-50]

II. MATERIAL BALANCE IN POWER PLANT

Trace element flows through a coal burning power plant have been studied in few cases. In the literature available, mass balance measurements at the Valmont Power Station[51] and T.A. Allen Steam Plant[52-57] and Four Corners Power Generating Plant[58] in the U.S. were described. It is of interest to describe these two measurements in some detail.

Valmont Power Station has a pulverized coal-fired boiler, with a special particle pollution abatement system. The flue gases are passed through a "high-efficiency" mechanical dust collector. The gas stream is then split up. One part (about 40%) goes to an electrostatic precipitator and the rest goes to a wet scrubber (see Figure 1).

In the work by Kaakinen and Jorden,[51] samples were collected from the streams indicated in Figure 1 which include coal; bottom ash (BA); ash from the mechanical collector hopper (MA) and electrostatic precipitator hopper (PA); fly ash from flue gas at the scrubber inlet (SI), scrubber outlet (SO), and electrostatic precipitator outlet

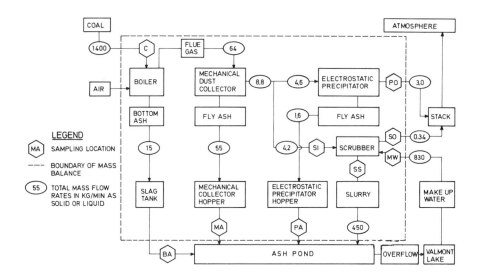

FIGURE 1. In the work by Kaakinen and Jorden,[51] samples were collected from the streams at locations indicated.

(PO); scrubber slurry (SS); and scrubber makeup water (MW). Elemental concentrations were determined by atomic absorption, wet chemistry, and x-ray fluorescence.

Mean concentrations for all samples are indicated in Table 5. From these measurements Kaakinen and Jorden[51] have calculated mass balances for 11 elements for which a full set of concentrations in Table 5 were available. Measurements of amounts of elements in the inlets and outlets of a pulverized-coal-fired power plant indicate that Al, Fe, Rb, Sr, Y, and Nb were at essentially constant concentrations in all outlet ashes, and that concentrations of Cu, Zn, As, Mo, Sb, Pb, ^{210}Po, and Se were generally lowest in bottom ash and increased progressively in fly ashes collected downstream toward the stack. A model was developed for the enrichment behavior in downstream ashes of this latter group of elements assuming that: these elements vaporize in the furnace and then condense or absorb into fly ash, smaller particles have larger specific surface areas and thus will have greater concentrations of these elements, and an increasing proportion of smaller particles in downstream fly ashes results in enrichment of these elements. Experimental enrichment values for Zn, As, Mo, and Sb agree with values predicted by this model.

The large imbalances found for Se and Hg support that major portions of the inputs of these elements left as vapors and/or very fine particles in flue gas. Naturally occurring mercury volatilizes when fossil fuels are burned in power generating plants. The vapor pressure of mercury at the temperature ranges encountered in the duct work and stacks of these power plants is high enough that condensation does not occur at the mercury concentrations encountered.

In a paper by Kalb,[59] a series of mercury mass balances was obtained at a coal-fired power plant by comparing the volatile and particulate mercury in the stack gas stream to the mercury initially in the coal, corrected for the mercury adsorbed and retained by the various ashes. These data were used to determine the fate of the mercury in the combustion process and to check the accuracy of the volatile mercury sampling procedure (gold amalgamation). The bottom ash had the lowest mercury concentration of the ash samples collected, and the mercury concentration increased as one proceeded through the ash collection system from the initial mechanical ash to the electrostatic ash. The mercury recovered in the various ashes represented about 10% of the total mercury introduced in the raw coal.

Table 5
TRACE ELEMENTS IN POWER PLANT SAMPLES AS REPORTED BY
KAAKINEN AND JORDEN[51]

Sample	Al (%)	Fe (%)	Cu (ppm)	Zn (ppm)	As (ppm)	Rb (ppm)	Sr (ppm)	Y (ppm)	Nb (ppm)
Whole coal	0.49	0.37	9.6	7.3		2.9	120	3.0	0.76
Bottom ash	8.8	6.6	82	58	15	48	1800	44	12
Mechanical collector Hopper ash	9.6	7.0	150	100	44	50	2400	61	16
Electrostatic-precipitator hopper ash	10.2	6.9	230	250	120	73	2500	68	19
Scrubber-inlet fly ash	9.0	7.4	280	360	130	51	2200	52	17
Electrostatic-precipitator outlet fly ash	9.2	7.4	320	370	150	56	2500	60	19
Scrubber-outlet fly ash	7.4	4.9	290	600	280	28	2500	31	18
Scrubber slurry	0.10	0.063	2.4	2.2	1.1	0.50	21	0.49	0.49

Sample	Concentration						Specific activity (Disintegrations per min/g)		
	Zr (ppm)	Mo (ppm)	Sb (ppm)	Pb (ppm)	Se (ppm)	Hg (ppm)	^{210}Pb	^{210}Po	^{226}Ra
Whole coal	13	0.99			1.9	0.070	0.61		0.031
Bottom ash	220	3.5	2.8	5	7.7	0.140	2.2	0.76	1.2
Mechanical collector Hopper ash	260	12	4.7	13	4.1	0.026	6.3	6.6	0.76
Electrostatic-precipitator hopper ash	210	41	14	66	27	0.310	23	28	1.8
Scrubber-inlet fly ash	160	54	14	110	73				
Electrostatic-precipitator outlet fly ash	190	60	18	130	62		38	42	0.89
Scrubber-outlet fly ash	80	110	22	340	440				
Scrubber slurry	1.8	0.53	0.10	0.91	0.33	0.014	0.29	0.30	0.0090

In a paper by Joensun,[60] 36 American coals were analyzed to determine their Hg content and thus the amount of Hg released to the atmosphere during coal combustion. In the testing procedure a sample is burned in O_2 and the fumes passed over hot (650° to 700°C) silver wire coils to complete oxidation. A heated (150°C) gold amalgamator then traps the mercury while most of the other fumes pass through. Any remaining organics are removed and the Hg amount is measured by a vapor detector. It was estimated, on the basis of 1971 coal consumption data, that the quantity of Hg emitted by the burning of coal in America is of the order of 3000 tons/year.

A study of trace element paths through a coal-fired power plant was conducted by Oak Ridge National Laboratory at the Tennessee Valley Authority 870-mW Thomas

A. Allen Steam Plant in Memphis, Tennessee. In-plant samples[53] were analyzed to determine emission characteristics and to obtain mass balances[52,54,56,61] for 37 elements. They concluded that Hg, Se, and probably Cl and Br were discharged to the atmosphere as gases. Some elements (As, Cd, Cu, Ga, Mo, Pb, Sb, Se, and Zn) were substantially more concentrated in fly ash than in slag. Other elements (Cr, Na, Ni, Sc, U, and V) were moderately more concentrated in fly ash than in slag, but several elements (Al, Ba, Ca, Ce, Co, Eu, Fe, Hg, K, La, Mg, Mn, Rb, Si, Sm, Sr, Ta, Th, and Ti) exhibited little preferential partitioning. The summary of measurements performed at Thomas A. Allen Steam Plant, and results obtained are presented in the book by Lyon.[62,63]

In the report by Bolton et al.,[52] mass balance measurements for 41 elements have been made around the Thomas A. Allen Steam Plant in Memphis, Tenn. For one of the three independent cyclone boilers at the plant, the concentration and flow rates of each element were determined for coal, slag tank effluent, fly ash in the precipitator inlet and outlet (collected isokinetically), and fly ash in the stack gases (collected isokinetically). Measurements by neutron activation analysis, spark source mass spectroscopy (with isotope dilution for some elements), and atomic adsorption spectroscopy yielded an approximate balance (closure to within 30% or less) for many elements. Exceptions were those elements such as mercury which form volatile compounds.

A mass balance for the various elements was calculated using the following equation:

$$Q_c(A) = C_c(A) \times (\text{g coal/min}) \tag{1}$$

$$Q_{p.I.}(A) = C_{p.I.}(A) \times (\text{g fly ash/min}) \tag{2}$$

$$Q_{S.T.}(A) = C_{S.T.}(A) \times (\text{g ash in coal/min} - \text{g fly ash to precipitator/min}) \tag{3}$$

for balance:

$$Q_c(A) = Q_{P.I.}(A) + Q_{S.T.}(A) \tag{4}$$

$$\text{percent imbalance} = \frac{Q_{P.I.} + Q_{S.T.} - Q_c}{Q_c} \times 100 \tag{5}$$

where $Q_c(A)$, $Q_{P.I.}(A)$, and $Q_{S.T.}(A)$ are the flow rates of element in g/min associated with the coal, precipitator inlet fly ash, and slag tank solids, respectively; and $C_c(A)$, $C_{P.I.}(A)$ and $C_{S.T.}(A)$ are the corresponding concentrations of element A in the coal, the fly ash collected in the precipitator inlet, and the slag tank solids. The flow of trace element into the plant with suspended particulates in inlet air was negligible. The concentrations of elements in coal, slag tank solids, precipitator inlet and precipitator outlet as reported by Klein et al.,[56] are shown in table 6. The authors have obtained the respectable balance for many elements. However, the results showed a consistent negative imbalance. This might be caused by the fact that soot blowing and air heater cleaning operations were not taken into account in the sampling. Because the flue gas sampling method was designed primarily to collect particulates efficiently, good balances were not obtained for elements forming volatile compounds. For example, more than 80% of the mercury entering with the coal is emitted with the flue gas as a vapor. The large imbalance for arsenic (−58%) indicates that a substantial portion of this element is also in the vapor phase of the flue gas.

Table 6
CONCENTRATIONS AND CONCENTRATION RATIOS BY KLEIN ET AL.[56]

	Element concentration (ppm)				Concentration ratios		
	Coal	Slag	Inlet fly ash	Outlet fly ash	Slag/ coal	Inlet fly ash/slag	Outlet fly ash/inlet fly ash
Al	10,440	102,300	90,900	76,000	9.8	0.9	0.8
As	4.45	18	110	440	3.6	6.1	4.0
Ba	65	500	465	750	7.7	0.9	1.6
Br	3.7	2	~4		0.5	2.0	
Ca	4,340	46,000	25,200	32,000	10.6	0.5	1.3
Cd	0.47	1.1	8.0	51	2.3	7.3	6.4
Ce	8.2	84	84	120	10.2	1.0	1.4
Cl	914	≤100	≤200		≤0.1	~1	
Co	2.9	20.8	39	65	7.2	1.9	1.7
Cr	18	152	300	900	8.4	2.0	3.0
Cs	1.1	7.7	13	27	7.0	1.7	2.1
Cu	8.3	20	140		2.4	7.0	
Eu	0.1	1.1	1.3	1.3	11.0	1.2	1.0
Fe	10,850	112,000	121,000	150,000	10.3	1.1	1.2
Ga	4.5	5	81		1.1	16.2	
Hf	0.4	4.6	4.1	5.0	11.5	0.9	1.2
Hg	0.122	0.028	0.050		0.2	1.8	
K	1,540	15,800	20,000	24,000	10.3	1.3	1.2
La	3.8	42	40	42	11.0	1.0	1.0
Mg	1,210	12,400	10,600		10.2	0.9	
Mn	33.8	295	298	430	8.7	1.0	1.4
Na	696	5,000	10,100	11,300	7.2	2.0	1.1
Ni	16	85	207		5.3	2.5	
Pb	4.9	6.2	80	650	1.3	12.9	8.1
Rb	15.5	102	155	190	6.6	1.5	1.2
Sb	0.5	0.64	12	55	1.3	18.8	4.6
Sc	2.2	20.8	26	36	9.5	1.2	1.4
Se	2.2	.080	25	88	0.0	310	3.5
Si	23,100	229,000	196,000		9.9	0.9	
Sm	1.0	8.2	10.5	9	8.2	1.3	0.9
Sr	23	170	250		7.4	1.5	
Th	2.1	15	20	26	7.1	1.3	1.3
Ta	0.11	0.95	1.4	1.8	8.6	1.5	1.3
Ti	506	4,100	5,980	10,000	8.1		1.7
U	2.18	14.9	30.1		6.8	2.0	
V	28.5	260	440	1,180	9.1	1.7	2.7
Zn	46	100	740	5,900	2.2	7.4	8.0

Ideally one should be able to calculate an exact balance of elements entering and leaving the steam plant since the weight of each constituent that enters in the coal and the weight of each constituent that leaves in the ash, slag, and gas can be determined. Failure to achieve perfect agreement (imbalance) can be a result of poor analytical techniques, improper sampling, loss of some waste product, or a combination of these. The smaller the imbalance, the greater the confidence in the measurement. An imbalance is calculated using the following formula:

$$\text{Imbalance} = (\text{Slag Flow} + \text{Fly Ash Flow} + \text{Gas Flow} - \text{Coal Flow}) / \text{Coal Flow} \tag{6}$$

In the studies of trace element balance at the Thomas A. Allen Steam Plant, closure for the balance study is very good for the majority of elements.[62] Serious negative imbalance does occur for Br and Cl. The impinger solutions were not analyzed for these elements. In addition, serious positive imbalances are present for Cr and V; perhaps some excess of these elements in the fly ash is due to corrosion of the boiler tubes. There were also difficulties with arsenic, probably due to sampling or analytical difficulties. From the data on concentration ratios three classes of partitioning behavior are observed:[62]

Class I: 25 elements — Al, Ba, Ca, Ce, Co, Cr, Cs, Eu, Fe, Hf, K, La, Mg, Mn, Na, Rb, Sc, Si, Sm, Sr, Ta, Th, Ti, U, and V — are readily incorporated into the slag. These elements are partitioned about equally between the inlet fly ash and slag. There is no apparent tendency to concentrate in the outlet fly ash.

Class II: 8 elements — As, Cd, Cu, Ga, Pb, Sb, Se, and Zn — are poorly incorporated into slag. These elements are concentrated in the inlet fly ash compared to the slag and in the outlet fly ash compared to the inlet fly ash.

Class III: Essentially Hg, Cl, and Br remain completely in the gas phase.

Two elements are unclassified: Mo should probably be included in Class II, and Ni probably in Class I.

Class I elements are not volatilized in the combustion zone, but instead form a melt of rather uniform composition which becomes both fly ash and slag. The slag is removed directly and quickly from the combustion zone, while the fly ash remains in contact with the cooling flue gas. The Class I elements remain in the condensed state, and hence show minimal partitioning between slag, inlet fly ash, and outlet fly ash.

Class II elements are volatilized on combustion. Since the slag is removed from the combustion zone, they have no opportunity to condense on the slag. They do, however, condense or become adsorbed on the fly ash as the flue gas cools. These elements are thus preferentially depleted from the slag (volatility effect), and preferentially concentrated on the outlet fly ash compared to the inlet fly ash (particle size effect).

Although the Oak Ridge-Allen Plant study is the most comprehensive, other studies of trace elements in coal-fired power plants have been reported. The U.S. Bureau of Mines[40] studied the fates of Hg, Cd, and Pb during coal combustion in a 100 g/hr combustor, in a 500 lb/hr furnace, and in commercial power plants. The Bureau of Mines also investigated the distribution of Cr, Cu, F, Hg, Mn, and Ni in coal with specific gravity separations to divide the coal into discrete fractions.

Roffman et al.[58] have studied a trace element balance on the units that employ scrubbers and the units that employ electrostatic precipitators. At the study by Roffman et al.,[58] concentrations of approximately 40 elements were determined in: coal, bottom ash, fly ash, scrubber effluents, liquid effluents, solid and gaseous stack emissions, ambient air soils, terrestrial flora and fauna, surface water in cooling pond and applicable streams, groundwater in the vicinity of cooling pond and waste ponds, bottom sediments, and aquatic flora and fauna. Data obtained through this extensive study were obtained after 10 years of power plant operation and were compared with available preoperational, baseline data. Mitigating effects of scrubbers and electrostatic precipitators on trace element emissions were determined.

By comparing the elemental inputs (coal concentrations converted to a coal ash basis) to the elemental outputs (bottom ash, fly ash and scrubber effluent), a mass balance can be calculated and some conclusions can be drawn regarding the volatilization of some elements.

Total output is the sum of discharge rates of bottom ash, precipitator fly ash, stack gases and stack particulates or bottom ash, scrubber slurry solids, scrubber liquid slurry, stack gases, and stack particulates. Ratios of output to input were generally between 0.5 and 2, but for most elements they were close to 1. The accuracy was greater for the unit using the scrubber than for the larger units using electrostatic precipitators. Considering that input and output values were based on estimated coal consumption values and by-product production rates, the degree of accuracy of these mass balances is quite good.

Table 7 lists calculated emission rates of trace elements. Examination of the data in this table indicates that selenium is a volatile element and that approximately 95% of the emitted mercury is in gaseous form, and scrubbers do not have any effect on these emissions. Scrubbers could reduce fluorine emissions by approximately 50%. Also, considering that the gaseous fluorine species include HF and SiF_4, the scrubbers probably remove, almost entirely, the acidic and potentially toxic HF while leaving the relatively untoxic species of SiF_4.

Yaverbaum[64] has reported studies of the distribution and potential emissions of biologically toxic trace elements during the fluidized bed combustion of coal (see also reports: PB 237 366, PB 237 754, and PB 246 116). The author has performed combustion experiments to study the distribution of trace elements in the combustion system. The approach is to make mass balances around the combustion system for as many of the trace elements as possible within the econimic limitations of the study. This involves sampling and analyzing all the solid materials charged to or recovered from the combustion system, including particulate matter entrained in the flue gas. The flue gas is also analyzed for the more volatile trace element species, Hg and F. Some of the results obtained for mercury and lead are shown in Table 8.

Fancher,[65] in his paper, discusses emission source from combustion of fossil fuels with special attention to electric generating stations. A mass balance study of Hg was undertaken at a power generating station in Hammond, Indiana; however, it was not totally accurate due to the low concentrations involved and the difficulty in determining the average Hg content in coal. The following conclusion resulted from the study: all Hg in the system results from the fuel itself; a majority (over 75%) of the Hg measured appeared in the flue gas while the remaining percentage appeared in the slag and ash.

The fate of mercury and some other elements during coal combustion is also discussed in the paper by Schultz et al.[40] Coal samples from Indiana, Iowa, Kentucky, Maryland, Missouri, Pennsylvania, and West Virginia were tested to determine the trace element distribution in various specific gravity fractions and to monitor the fate of various toxic trace elements present during combustion. Determinations for 8 elements (Hg, F, Cr, Cu, Ni, Mn, Cd, and Pb) were made for coal separated at the specific gravity fractions of 1.3 float, 1.3 to 1.4, 1.4 to 1.6, and 1.6 sink. Fly ash samples were tested only for Hg, Cd, and Pb. Atomic absorption analysis was used to determine the presence of the eight elements tested with the exception of F, which was determined by specific ion electrode. Specific gravity separations were performed using standard techniques. Combustion tests utilized a 500 lb/hr pulverized-fuel-fired furnace. Substantial data on trace element content is presented on coals from the Upper Freeport, Indiana V, Lower Kittanning, Stockton-Lewiston, Pittsburgh, and Hazard #4 seams. Trace element data on fly ash is presented from the Kentucky #6, Illinois #6, Pittsburgh, Tebo, and Weir seams.

Let us mention work by Lawasani[66] based on known concentrations of 15 elements determined in ash samples collected at several locations within a coal-fired power plant. A mathematical model was established based on a volatilization recondensation

Table 7
EMISSION RATES OF TRACE ELEMENTS FOR
UNITS OF FOUR CORNERS POWER
GENERATING PLANT (U.S.), AS REPORTED BY
ROFFMAN ET AL.[58]

Element	Units with scrubbers	%G	Units with electrostatic precipitators	%G
Ag	<8.6	—	<15	(63)
Al	>4,600	—	>600,000	(<1)
As	17	(32)	173	(8)
B	474	(3)	>2,800	—
Ba	464	(<1)	27,047	(<1)
Be	<1	—	28	(~5)
Bi	<8.6	—	<64	—
Ca	~4,622	(1)	>55,100	(<1)
Cd	~0.4	—	<20	(<1)
Co	<8.6	—	<65	(17)
Cr	<84	—	<109	(<9)
Cu	4.8	(1)	330	(<1)
F	480	(42)	28,700	(86)
Fe	680	(32)	>55,470	(<1)
Ga	<20	—	~210	(<5)
Ge	<8.6	—	~170	(6)
Hg	4.8	(96)	34	(95)
K	168	(52)	>55,100	(<1)
Li	<12	—	~1,009	(<1)
Mg	52	(25)	~55,031	(<1)
Mn	24	(37)	1,028	(3)
Mo	<8.6	—	<29	(<48)
Na	>820	—	>55,950	—
Nb	<8.6	—	<65	—
Ni	<8.6	—	~110	(<9)
Pb	1.2	—	270	(<1)
Sb	<1	—	<6.8	(<12)
Sc	<0.4	—	55	(<1)
Se	30	(80)	184	(24)
Si	548	(19)	>550,000	(<1)
Sn	128	(69)	<210	(90)
Sr	<17.6	—	1,604	(<1)
Ti	84	(4)	15	(63)
V	<26	—	~313	(<4)
Zn	464	(<1)	55	(<1)
Zr	<12	—	~1,610	(<1)

Note: Values in parentheses indicate the percentage behaving as gases or in very fine particulate form.

mechanism to provide a means for qualitative prediction of the behavior of other elements during coal combustion. Fly ash surface areas were determined experimentally by nitrogen adsorption, particle densities by helium displacement, and particle-size distributions by settling-velocity fraction. The enrichment of elements in specific ash streams was dependent upon the volatility of the elements at both combustion and lower temperatures, and volatility was in turn determined by the vapor-pressure of the chemical species in which the element existed at a specific temperature. Of the elements studied seven were geochemically classified as lithophiles and tended to occur as stable,

Table 8
MATERIAL BALANCES FOR SOME TRACE ELEMENTS IN
COMBUSTION EXPERIMENTS REPORTED BY YAVERBAUM[64]

Element Bed temperature	Hg 900°C		Pb 900°C	
Bed type	Alundum	Dolomite	Alundum	Dolomite
Input (concentrations)				
Coal (ppm)	0.07	0.07	1.6	1.6
Additive (ppm)	—	0.025	—	12
Initial bed	0.028	0.005	1.1	11
Total element in (μg)	6300	7600	160	620
Output (concentrations)				
Final bed (ppm)	—	—	60	11
Overflow	—	—	—	12
Prim. cyclone (ppm)	1280	990	130	16
Sec. cyclone (ppm)	130	420	4.3	27
Prim. filter (ppm)	150	110	7.4	96
Sec. filter (ppm)	10	—	0.6	—
Flue gas (ppb)	180	390	—	—
Total element out (μg)	1800	1900	200	590

nonvolatile mineral forms at combustion temperatures (e.g., Al and Fe). Concentrations of these elements were approximately equal in all ash streams and were therefore "non-enriched". The remaining eight elements were classified as chalcophiles and tended to occur as relatively volatile mineral forms at combustion temperatures (e.g., As and Zn). These elements increased in concentration in ashes downstream from the boiler and were thus "enriched". The developed vaporization-recondensation mechanism for volatile elements agreed well with experimental data; however, whether an element was partially or completely vaporized or partially or completely recondensed in the boiler was not predictable from the model.

Block et al.[67-70] have investigated the influence of the combustion parameters on the emission efficiencies. The fly ash samples were collected in two different combustion facilities, one typical for large installations and one for private home heating in Belgium. The fly ash samples were collected on filter paper and particle size information was obtained from cascade impactor sampling. The fly ash samples were analyzed using the same analytical procedure.

A comparative study of the composition of coal and its fly ash emissions during combustion proved that the elements are emitted in concentration ratios which are different from those in the original coal. Based on their mass-size distributions, the elements could be divided into three groups.

A first group of elements (Mg, Al, Si, Ca, Se, Ti, Cr, Fe, Co, Rb, Ba, Hf, the rare earths, Th: Group A) are associated with large particles and were found to be depleted in fly ash relative to their content in the coal. A second group of elements (As, Sb, Sc, In, Cu, Zn, Hg, Pb, Cl, Br, I: Group B) have a maximum concentration on small particles and are, compared with their concentrations in the coal, enriched in the fly ash by a factor of 2 to 10. For the third group of elements (Na, K, V, Mn, Ni, Ga, Mo, Ag, Cd, Sc, W, Au: Group C), the mass-size functions show an intermediate behavior. Their concentration ratios in fly ash and coal are near to unity.

This division into depleted and enriched elements appears to be roughly independent of the operating characteristics of the combustion facilities, but is mainly determined by the geochemical and physicochemical properties of the elements. The elements depleted in the fly ash (Group A) are primarily associated with the mineral fraction of coal (Groups I and III$_A$). They are lithophilic or siderophilic elements and most of them have oxide boiling points greater than the 1500°C estimated furnace temperature. The elements exhibiting enrichment in the fly ash relative to the coal are associated with the organic fraction of coal (Groups II and III$_B$). The oxides of these mainly chalcophilic elements volatilize completely at temperatures less than 1500°C.

III. ASH

Ash is one of the components determined in the process called the proximate analysis of coal, which also consists of the determination of the moisture and the volatile matter. The fourth component estimated in the proximate analysis is labeled fixed carbon. It is obtained from the sum of the other 3 components by subtraction from 100.

Ash is a product of the decomposition of silicate, sulfide, sulfate, carbonate, and oxide minerals which form the bulk of the inorganic matter in coal. This decomposition is accomplished by burning the coal powder in an oxygen atmosphere at 750 to 850°C. The ash is thus the noncombustible residue that forms when coal is ignited under controlled conditions. Fundamentally, the sources of ash can be subdivided into ash derived from adventitious mineral matter: the "inherent ash", formed through the decomposition of such minerals as kaolinite, montmorillonite, pyrite, gypsum, jarosite, calcite, and the "organic component ash" derived from the decomposition of plant material containing bound inorganic elements. The last type of ash is usually relatively insignificant. Once the ashing is completed at the high temperatures used, no distinction can be made as to the relative quantities of the inherent ash vs. the organic ash.[71-74]

The ash content of coals varies over a wide range. This variation occurs not only in coal from different parts of the world or from different seams in the same region but also in coal from different parts of the same mine. Before marketing, most commercial coals are cleaned or washed to remove a portion of what would be reported as ash in laboratory determinations. In any case, the ash determinations of significance to the electric utility are those made at the point of use.[75]

Coals may be classified into two groups based on the nature of their ash constituents. One is bituminous-type ash and the other is the lignite-type ash. The term "lignite-type" ash is defined as an ash having more calcium oxide, CaO, plus magnesium oxide, MgO, than ferric oxide, Fe$_2$O$_3$. By contrast, the "bituminous-type" ash will have more ferric oxide, Fe$_2$O$_3$, than calcium oxide, CaO, plus magnesium oxide, MgO.

According to Morrison,[75] the average constituents for power plant ash representative of the ash produced in the U.S. electric utility industry are the ones shown in Table 9.

During 1975, 75,000,000 tons of ash were produced in U.S. only. Table 9 indicates the tonnage of various ash components.

The ash content of U.S. coals has been determined by many authors.[77-129] As an illustration we show data by Gluskoter[76] for three regions of U.S. (see Table 10).

Considerable variations exist in geometric mean values for oxides of the major and minor elements in coal ash from the different areas. Table 11 shows data for the U.S. as compiled by Zubovic et al.[130] In general, the coal ash from the Interior province and the Appalachian region contains high Fe$_2$O$_3$ and low CaO, MgO, and Na$_2$O. Northern Great Plains and Rocky Mountain province coals are low in Fe$_2$O$_3$ and high in MgO and intermediate to high in Na$_2$O. The lowest geometric means for ash-fusion

Table 9
ASH COMPONENTS[75] AND THEIR ANNUAL PRODUCTION

Constituent	Symbol	Percent	Tonnage
Silicon dioxide	SiO_2	45.7	34,275,000
Aluminum oxide	Al_2O_3	26.0	19,500,000
Ferric oxide	Fe_2O_3	17.1	12,825,000
Calcium oxide	CaO	3.8	2,850,000
Sulfur trioxide	SO_3	2.6	1,950,000
Potassium oxide	K_2O	1.5	1,125,000
Titanium oxide	TiO_2	1.2	900,000
Magnesium oxide	MgO	1.2	900,000
Sodium oxide	Na_2O	0.6	450,000
Phosphorus pentoxide	P_2O_5	0.3	225,000

Table 10
ASH CONTENT OF U.S. COALS[77]

Area	Range (%)	Arithmetic mean (%)	Geometric mean (%)
Western U.S.	4.1 — 20	9.6	8.9
Eastern U.S.	6.1 — 25	12	12
Illinois basin	4.6 — 20	11	11

Table 11
GEOMETRIC MEAN VALUES FOR OXIDES IN COAL ASH FROM DIFFERENT REGIONS OF U.S.[130] (% by weight)

Compounds	Appalachian region	Interior province	No. Great Plains Province		Rocky Mountain Province		All U.S. coals
			Lignite	Subbituminous	Subbituminous	Bituminous	
SiO_2	42	29	19	26	44	47	35
Al_2O_3	24	13	9.1	13	18	18	17
CaO	1.4	3.6	16	14	5.5	5.2	3.8
MgO	.78	.71	5.8	3.7	1.5	1.3	1.3
Na_2O	.38	.29	1.7	1.4	.47	.75	.52
K_2O	1.7	1.3	.47	.43	.59	.53	.94
Fe_2O_3	12	24	5.8	5.9	4.8	4.7	9.1
MnO	.02	.05	.08	.07	.06	.04	.04
TiO_2	1.1	.64	.59	.76	.83	.86	.89
P_2O_5	.15	.18	.34	.62	.48	.39	.25
SO_3	1.9	3.6	14	12	5.3	3.5	4

temperatures are in ash with high mean concentrations for Fe_2O_3, CaO, MgO, and Na_2O. The highest fusion temperatures are in coal-ash samples where these components are in low concentrations.

Data on the chemical composition of coal ash and its relation to ash-fusion temperatures can be an important factor to be used when considering a coal reserve for a particular coal-conversion process. Engineers can ascertain well in advance of plant construction the need for additives to coal which will achieve a prescribed ash-fusion temperature.

The composition of coal ash varies widely; therefore we shall mention some of the reports on ash composition instead and present some general remarks.

Sondreal et al.[121] have studied composition of lignite ash. Lignite, like other coals, has been infused with a somewhat random mixture of mineral constituents derived from both the original plants and extraneous mineral inclusions added during the coalification process. Consequently, the composition of ash from lignite from different locations varies appreciably. Average characteristics for lignite ash weighed equally according to the nine locations investigated are shown in Table 12. Data shown on this table are compiled from the works of Selvig and Gibson,[119] Lowry,[131] and Henderson.[132] In Table 12 are included the observed high and low values of the property. The extreme values are generally characteristic of a particular sampling location. However, when data points are considered as a whole, properties are distributed more or less continuously over the ranges indicated.

Mukherjee and Chowdhury[133] have investigated ash of Indian coals. Their results are shown in Table 13. The major constituents of the ash are silica, alumina, and iron. The SiO_2 to Al_2O_3 ratios in the samples vary between 1.8 and 2.2. The principal clay mineral associated with Indian coals is kaolinite and the ratio of SiO_2 to Al_2O_3 falls close to that of kaolinite after correction for the free silica content. In the report by Mitchell and Gluskoter[104] ten samples of mineral-matter residue were obtained by the radio-frequency low-temperature ashing of subbituminous and bituminous coals. The low-temperature ash samples were then heated progressively from 400°C to 1400°C at 100°C intervals. Mineral phases present at each temperature interval were determined by X-ray diffraction analyses. The minerals originally present in the coals (quartz, kaolinite, illite, pyrite, calcite, gypsum, dolomite, and sphalerite) were all altered to higher temperature phases. Several of these phases, including kaolinite, metakaolinite, mullite, anhydrite, and anorthite, were found only in limited temperature ranges. Therefore the temperature of formation of the ashes in which they occur may be determined. Mineralogical differences were observed between coal samples from the Rocky Mountain province, the Illinois basin, and the Appalachians; and as a result of these mineralogical differences, different high-temperature phases resulted as the samples were heated.

Block et al.[70] have studied the concentrations of elements in coal as the function of the ash content of the coal. Some of their results are shown in Figure 2. Vanadium and silicon are increasing with the increased ash content of coal; this is not so for antimony and copper. Their concentration in coal starts increasing only after ash content reaches high value (20% for Cu, about 40% for Sb).

Abernethy et al.[77] have analyzed large numbers of coal samples from different parts of U.S. All laboratory samples of coal were prepared in accordance with the procedure described in Bureau of Mines Bulletin 638. The coal samples were ashed by a low-temperature procedure designed to minimize loss of volatile elements. A 10-g sample of pulverized coal was weighed into a flat-bottom dish 1.5 in. wide, 2 in. long, and 0.5 in. deep, fabricated from 36-gage aluminum sheet. The coal samples were ashed by heating slowly to 425°C in 4 hr and maintained at this temperature for 16 hr. The ash samples were transferred to small plastic bottles and analyzed by spectrochemical method according to Peterson and Zink.[135]

Spectrochemical analyses of ash from commercial coals showed that 22 trace elements occurred in most of the samples tested. Another seven elements — arsenic, bismuth, cerium, neodymium, niobium, rubidium, and thallium — were detected in many samples. For example, arsenic, with a limit of detection of 0.005% in ash, was found in 67% of the samples from the Eastern province, 41% of those from the Interior province, and 16% of the Western state ashes. Rubidium, with a limit of detection of

Table 12
APPROXIMATE LIMITS IN ASH COMPOSITION OF VARIOUS COALS[121]

Coal	Constituent (weight % of ash)									
	SiO_2	Al_2O_3	Fe_2O_3	TiO_2	P_2O_5	CaO	MgO	Na_2O	K_2O	SO_3
U.S. coals:										
Anthracite	47.7—67.7	24.7—43.5	2.1—10.2	1.1—1.8	0.08—3.7	0.2— 3.7	0.2— 1.2	—	—	0.1—1.1
Bituminous	7.1—68.5	4.1—38.9	1.8—43.6	0.5—3.7	0.05—3.1	0.7—36.4	0.1— 4.2	0.2—2.8	0.2—3.5	0.1—32.3
Subbituminous	16.7—58.2	4.1—35.0	2.7—18.9	0.6—2.3	0.02—3.1	2.2—45.1	0.5— 8.0	—	—	2.7—16.1
Lignite	6.3—45.2	6.3—22.5	0.9—17.8	0.1—0.8	0.0 —1.3	15.3—44.4	3.0—12.2	0.2—11.3	0.1—1.7	6.2—30.3
British coal: bituminous	25—50	20—40	0—30	0—3.0	—	1.10	0.5— 5.0	1—6	—	1—12
German coals:										
Bituminous	25—45	15—21	20—45	—	—	2—4	0.5— 1.0	—	—	4—10
Brown	7.0—46.3	6.0—29.4	16.6—26.0	—	—	4.1—43.0	0.9— 4.0	—	—	2.1—22.0
Australian coal: brown	0.1—41.6	0.4—36.9	2.1—29.8	0.1—0.3	—	tr.—36.4	0.7—19.5	0.5—6.2	0.1—1.1	8.0—33.1

Table 13
CONSTITUENTS (WT% OF TOTAL) OF INDIAN COAL ASH[133]

Sample	SiO_2	Al_2O_3	Fe_2O_3	TiO_2	Pb_2O_5	SO_3	CaO	MgO	Alkali and undetermined (by difference)
Overall	40.0	18.9	28.8	0.60	0.08	1.63	0.2	3.1	6.7
Gravity fractions									
1.30 (float)	33.6	18.3	23.6	0.32	0.19	6.19	2.0	4.4	11.4
1.30—1.50	45.4	20.2	27.6	0.40	0.23	0.27	Trace	0.9	5.0
1.50—1.80	48.6	21.5	22.6	0.62	0.08	0.40	Trace	0.7	5.5
1.80 (sink)	53.2	25.8	13.2	0.67	0.07	0.46	Trace	1.0	5.6

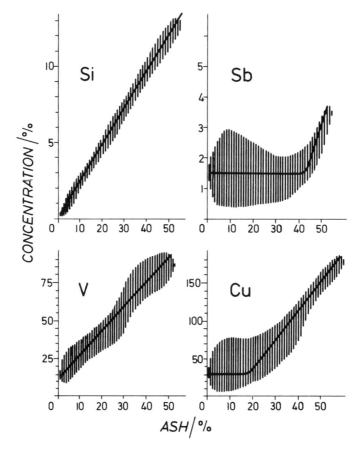

FIGURE 2. Concentrations of four representative elements (Si, Sb, V, and Cu) in coal as a function of ash content of the coal.[70]

0.001% was found in all samples from the Interior province, but in only 58% of the ashes from the western states. All state averages for manganese content in ash are less than the crustal abundance. The average content of chromium, copper, nickel, and rubidium in ash of the Western state coals usually is less than the crustal abundance for each element; however, most of the ashes from the Eastern and Interior provinces are enriched in these elements. Averages for the other elements generally show some enrichment in the ash.

The results obtained by Abernethy et al.[77] for the average trace element content in ash of coals from different U.S. regions are shown in Table 14.

A. Ash Fusion

The vast literature on relations between fusibility and chemical composition of coal has been reviewed by Watt.[136]

With the development of the pulverized coal furnace in the 1920s, followed by the development of the cyclone furnace in the 1940s, ash fusion temperature has become a major parameter in determining the applicability of a coal for use in a boiler unit. Ash fusion is also a factor to be considered in stoker furnace since a low fusion ash will clinker and cause loss of burning efficiency, but normally it is not a problem of major proportions.[137]

In the pulverized coal-fired boilers, ash fusion is a major consideration regardless

Table 14
AVERAGE TRACE ELEMENT CONTENT IN ASH (PPM OF ASH) OF COALS FROM DIFFERENT U.S. REGIONS[77]

Element	Ala.	E.Ky.	Md.	Ohio	Pa.	Tenn.	Va.	W.Va.	Ark.	Ill.	Ind.
Barium	1195	1077	450	438	703	1248	1273	910	1000	423	290
Beryllium	88	20	7	9	8	6	14	14	3	11	16
Boron	322	255	140	561	153	247	164	232	173	690	803
Chromium	207	260	140	235	244	200	253	222	300	252	182
Cobalt	198	212	150	144	175	136	182	202	550	131	226
Copper	150	156	75	80	125	116	171	132	55	71	91
Gallium	55	99	20	50	71	57	85	77	25	35	35
Germanium	46	64	7	59	49	35	41	46	10	116	139
Lanthanum	138	175	100	126	130	132	151	157	300	105	169
Lead	40	59	10	43	52	50	78	58	35	279	68
Lithium	812	1064	140	394	642	994	441	520	100	386	231
Manganese	208	361	30	207	205	234	540	249	150	621	245
Molybdenum	117	71	17	57	98	80	106	73	125	75	49
Nickel	186	217	125	203	195	168	281	212	325	211	308
Scandium	78	131	65	58	86	141	92	93	40	77	74
Strontium	1396	1538	900	511	943	1386	1240	1104	2500	697	660
Tin	24	63	5	13	11	19	30	17	12	22	7
Vanadium	338	400	225	236	330	354	417	348	350	297	327
Ytterbium	5	9	3	7	6	6	11	7	3	4	4
Yttrium	126	217	50	150	127	102	151	145	60	89	98
Zinc	243	203	200	284	222	242	291	201	190	1189	690
Zirconium	607	823	1100	805	680	460	559	738	600	755	945

Element	Iowa	Kan.	Mo.	W.Ky.	Ariz.	Colo.	Mont.	N.Mex.	N.Dak.	Utah	Wash.	Wyo.
Barium	300	150	183	468	400	795	3000	2250	2650	1122	1714	1967
Beryllium	10	5	10	15	10	6	12	8	2	3	4	28
Boron	833	250	667	752	500	494	475	361	337	861	314	411
Chromium	400	150	433	197	100	49	4	91	34	88	121	67
Cobalt	343	450	233	167	0	104	61	126	57	66	217	60
Copper	67	150	108	95	50	49	25	50	13	38	121	50
Gallium	70	20	65	40	50	32	39	34	20	30	59	17
Germanium	133	60	88	82	50	19	25	32	6	8	9	18
Lanthanum	133	150	100	115	0	129	97	150	96	131	133	50
Lead	200	100	267	69	40	31	38	40	22	24	25	7
Lithium	300	50	137	171	200	95	215	138	95	283	277	217
Manganese	433	300	350	201	100	216	456	165	300	157	121	160
Molybdenum	100	50	108	79	10	18	38	17	32	11	26	25
Nickel	567	550	767	170	50	53	26	69	14	51	114	47
Scandium	50	40	47	67	10	56	34	68	45	37	89	40
Strontium	667	900	417	578	1000	974	2612	800	2612	1457	3071	1167
Tin	9	10	16	27	10	23	9	16	13	13	139	12
Vanadium	300	150	375	341	100	125	97	213	94	117	429	167
Ytterbium	8	3	9	5	1	3	4	5	4	2	4	3
Yttrium	100	275	142	142	100	83	60	85	60	67	94	53
Zinc	1333	750	620	514	100	362	337	164	250	109	243	425
Zirconium	667	750	733	824	400	872	612	914	662	861	1286	450

of whether the furnace is a dry-bottom or slag-bottom furnace. In the dry-bottom pulverized-coal furnace, almost any coal can be used; however, a high-fusion coal is normally preferred to reduce the buildup of molten ash deposits on the superheater and reheater tubes. These deposits impair heat transfer and cause excessive pressure drops in the gas passages.[138] It has also been predicted that these ash deposits are a mechanism for boiler tube corrosion.[139]

The more recently developed cyclone furnace offers several advantages over the pulverized-coal-fired furnace, among them being a reduction of slag deposition on heat absorption surfaces, less fly ash production, reduced crushing of coal, efficient burning of low grade coals, and finally a potentially smaller furnace volume.[140]

The major limitation of the cyclone furnace is its unsuitability for using coals with fusion points above approximately 1310 to 1420°C. As these furnaces become more common, this factor will become increasingly important. Already there have been situations where utility customers with cyclone furnaces have ruled out otherwise suitable coal fuels because of higher than desirable fusion temperatures.[137]

However, there are steps which can be taken to lower the ash fusion temperature of a coal which may make the coal more suitable for use in a furnace. Several investigations have revealed that addition of such compounds as lime or ferric oxide to a coal will serve as a flux to reduce the fusion point of the ash. Also, alkali compounds such as Na_2O and K_2O have an even greater fluxing effect, giving reductions in the softening temperature of 24 to 28°C for each percent of alkali oxide present. However, the use of such alkali additives may be restricted because they may cause corrosion to boiler tubes. Boiler tube corrosion from ash fouling is accelerated when operating at temperatures above 620°C[138] which are commonly needed to obtain the highest thermal efficiencies. Thus, the benefits which can be derived from improved efficiencies resulting from the use of higher temperatures become less attractive when using higher alkali content coals both because of the accompanying increases in corrosion as well as from the decreased combustion efficiency which results from ash fouling.

While detailed quantitative knowledge of the constituents responsible for ash fouling is incomplete, it is known that bonded deposits have been attributed to, among other constituents, the quantities of chlorine and alkali elements in coal ash. Findings indicate that when using coals having an Na_2O content in excess of 0.5% a rapid build-up of boiler tube deposits result. Research by Reese et al.[139] more closely defines the origin of bonded deposits by postulating that the water-soluble portions of potassium and sodium constitute the principal deposit-forming constituents since they generally occur as simple compounds in coal ash while the water-insoluble portions of potassium and sodium occur in more complex compounds which are less likely to decompose and deposit on boiler tubes during combustion. For more discussion of this problem see work by Bucklen et al.[137]

Melting behavior of coal ash materials has also been recently studied by Vorres.[141] The melting behavior of coal ash materials is characterized by several temperatures relating to stages of deformation of cone-shaped ash samples on heating. Coal ash does not melt sharply like a pure compound, but rather softens over a temperature range as the temperature is increased. As it melts, the heated ash exhibits a plastic range between the solid and mobile liquid states. The temperature range corresponding to the plastic state, as well as viscosity and melting phenomena, depend on the composition of the ash and the gaseous environment. The purpose of this paper is to provide some framework for understanding the observed behavior, and to suggest use of a concept that should be helpful in further studies of that behavior. It is suggested that the ionic potential (valence divided by ionic radius) may be used as a measure of acids and bases in predicting coal ash melting and viscosity from the chemical composition. The role of an acid is that of a complex ion formed with anions provided by bases which in coal ash systems are oxide ion donors. In systems containing acids and limited oxide ion concentrations, polymers tend to form in the melts. Addition of bases reduces polymer size and would decrease viscosity.

Empirical correlation of coal ash viscosity with ash chemical composition is discussed by Barrick.[82] Stepwise regression analysis is used to correlate the coal ash viscosity vs. temperature relationship with the ash chemical composition. Empirical correlations are given for coal ash in an oxidizing atmosphere, coal ash in a reducing atmosphere, coal ash in an oxidizing atmosphere, and lignite ash in a reducing atmosphere. Each correlation consists of 8 equations that calculate 8 viscosity-temperature points ranging from 50 to 10,000 P. The "R^2" value exceeds 0.90 for all but 1 of the 32 equations. The correlations provide an improved method for predicting the ash viscosity vs. temperature relationship from the ash analysis.

Critical properties of lignite ash are summarized by Sondreal et al.[121] Here are some of their conclusions. Components which have been suggested to be important in determining the fouling tendency of higher rank coals include sodium oxide, chlorine, sulfur, and the total quantity of ash. The situations with lignites is different, because chlorine in lignite is present at or below 0.01 wt% of the coal, while the sodium oxide content ranged upward to about 0.6%. Based on the criteria used for other coals, the chlorine in lignite cannot be used as a measure of volatile alkali because of the low concentrations. Some experiences burning lignite have suggested that high sodium content can aggravate fouling.

Furthermore, a complicating factor with lignite is the relatively high calcium content. The calcium may play a supplemental role in the formation of alkali-bonded deposits. Interdependence in properties of mixtures of sodium and calcium sulfates is a subject which must be studied in the laboratory before any general theory on alkali-bonded deposits can be formulated for lignite ash.

The total percentage of sulfur and the forms of sulfur in coal can have an important influence on the occurrence and composition of ash deposits in boilers and on corrosion. Tests performed by Sondreal et al.[121] show that a significant portion of sulfur in lignite is retained in the ash. Using the standard laboratory ashing procedures, some 60 to 90% of the total sulfur in the lignite remains with the ash. In actual boiler operation, a smaller portion of the sulfur is retained. Only limited data are available, but these data indicate 10 to 30% of the sulfur remains in the fly ash from a pulverized-fuel-fired boiler, with somewhat higher percentages in the grate ash and the fly ash from a stoker-fired unit. Boiler deposits obtained after shutdown generally contain sulfate sulfur in excess of the sulfur-to-ash ratio of the lignite, indicating that the mineral content of lignite ash has an affinity for sulfur oxides. Data obtained for sulfur retention in ash of U.S. coals are shown in Table 15.

B. Ash Determination

The percent of ash in coal is usually determined by the standard method of the American Society for Testing and Materials (ASTM). This is a gravimetric procedure which involves programmed combustion of the coal and weighing the residual ash. It is time consuming, and inter-laboratory variations of 10% are acceptable for coals with more than 12% ash containing pyrite and carbonates. A number of techniques have been investigated for the continuous on-line measurement of the coal by nuclear techniques. The two most widely used are based on:

1. The backscattering of X- or γ-radiation, and
2. X-ray absorption

We shall discuss both approaches in some detail.

The backscattering of electromagnetic radiation has been discussed by many authors; e.g., Dijkstra and Sieswerda,[142] Hardt,[143] Cammack,[144] Vasil'ev et al.,[145] Clayton,[146] and Boyce et al.[85]

Table 15
AVERAGE RETENTION OF SULFUR IN
ASH OF U.S. COALS

	Retention (%)
Eastern area:	
Alabama, Illinois, Indiana, Kentucky, Maryland, Ohio, Pennsylvania, Tennessee, Virginia, West Virginia	4.6
Central area:	
Colorado, Kansas, Montana, New Mexico, North Dakota, Oklahoma, Wyoming	31.8
Western area:	
Oregon, Utah, Washington	22.8
Rank:	
Bituminous	9.6
Subbituminous	26.0
Lignite	77.6

Note: Results obtained by Sondreal et al.[121] by ashing in laboratory.

The most important considerations in using the backscattering of electromagnetic radiation in this application is the choice of incident radiation energy, and this is governed by variations in the composition and concentration of mineral matter in the coal. Iron is the dominant element, since the atomic number of iron ($Z_{Fe} = 26$) is much greater than the atomic numbers of other elements present at significant concentrations in the mineral content ($Z_{average} = 11$ to 12). For the technique to be successful, therefore, it is necessary to avoid the strong influence of iron on the intensity of the backscattered radiation.[85]

The initial approach[142] was to choose an incident radiation energy below 7.11 keV, the absorption edge of iron, since at such an energy the mass absorption coefficient of iron is close to that of other elements in the mineral matter. However, in practice, this entails crushing, drying, and grinding a continuous supply of coal to 0.2 mm, and this is a time-consuming and expensive operation. Although results of an acceptable accuracy can be obtained (except at high and variable concentrations of sulfur in coals of low ash content), equipment based on this technique does not generally meet the operational requirements to be found in a coal preparation plant.[85]

A method has been described which allows an incident radiation energy above 7.11 keV to be used, and includes compensation for any iron which is present.[116] By using a ^{3}H/Zr bremsstrahlung source, the demonstrable accuracy compares with that using an incident energy <7.11 keV, and gives a higher accuracy at the higher concentrations of sulfur. However, this technique also imposes similar requirements on the degree of coal preparation before measurement and therefore also involves long time delays (~ 15 to 20 min) before analyses are available. Such times are too long for most control applications.

The paper by Boyce et al.[85] discusses the problems of using a higher incident radiation energy, and shows that the optimum energy is about 15 keV and consistent with using plutonium-238 as the radiation source. At this energy a rapid crushing device can be used which is not highly sensitive to the moisture content of the coal. The need to dry and grind the coal to a maximum particle size (top size) of 0.2 mm is avoided and a top size of ~5 mm can be shown to be acceptable. Equipment based on this approach has now been in routine use for several years.

In practice, the penalty for selecting an exciting radiation of 15 keV results in a high sensitivity to variations in the concentration of iron in the ash. However, compensation for the effect of varying iron concentration can be achieved by arranging that the detector measures excited Fe K X-rays as well as the backscattered radiation. An aluminium filter is positioned over the detector window, and the filter thickness is adjusted so that the Fe K X-ray intensity just compensates for the decrease in backscattered intensity due to the presence of iron in the coal. Apart from iron, the principal elements in the mineral content which occur at concentrations which can significantly affect the validity of the two-component model are aluminium, silicon, chlorine, titanium, potassium, and calcium.

The accuracy of ash measurement by X-ray backscattering can be affected by the maximum particle size of the distribution, and also by variations in the particle size distribution itself.

Let us describe in some detail a method used by Brown and Jones[147] for calculating the ash content of coal. This paper describes a method for calculating the ash content of coal. In this method a curve is plotted which defines the relationship between the ash content and corrected backscatter, and is based upon the iron content of the sample. The standard deviation of the ash is calculated to be 1.17 for various particle sizes and moisture content of the coal.

The X-ray system used was a Columbia Scientific Industries Corporation Portable X-Ray Fluorescence Analyzer Minilab 700® with a 10 millicurie (mCi) curium-244 source. This source was chosen instead of the available 30 mCi ^{238}Pu or the 30 mCi ^{244}Cm source because it could be supplied under a general radioactivity materials license. The HRF analyzer could be powered by a variety of power supplies, either AC or DC.

Samples were obtained from the coal deliveries by an automatic sampler. The sampler took approximately 50 lb from each load of coal, ground the sample to a top size of 8 mesh and riffled it to provide the 8 mesh wet test sample. The drying of 8 mesh wet samples by ASTM methods gave the 8 mesh dry samples. The 60 mesh dry samples were prepared by grinding 8 mesh dry samples. The samples were then weighed to 20 ± 0.1 g and hand packed into a sample cell supplied with the instrument. The sample cell consisted of a plastic tube covered at one end by 0.001 in. polypropylene or Mylar®. To calculate the iron correction factor (CF), the theoretical backscatter was first calculated using the relation: Theoretical Backscatter = B/(Ash-A); where A and B were the constants of the theoretical curve (See Figure 2 from Brown and Jones,[147] in which A = 50.70955 and B = −0.000161).

The relationship between CF and the iron content was found to be hyperbolic. The CF values were plotted vs. the different coal "states"; the graph for all samples showed that little error was introduced if the CF values were calculated independent of moisture content and particle size. When the CF was multiplied by the iron count and added to the backscatter, the corrected backscatter was obtained. The relationship between the ash and the corrected backscatter is shown in Figure 3.

Fookes et al.[148] have proposed a method for the determination of the ash content of coal based on the measurements of mass absorption coefficients of coal at two X-ray energies. These measurements can be made using X-ray transmission or scatter techniques. Calculations based on transmission of narrow beams of X-rays have shown that ash can be determined to about 1% in coal of widely varying ash content and composition. Experimentally, ash content was determined to 0.67% by transmission techniques and 1.0% by backscatter techniques in coal samples from the Bulli seam, New South Wales, Australia, having ash in the range 11 to 34%. For samples with a much wider range of coal composition (7 to 53% ash and 0 to 25% iron in the ash),

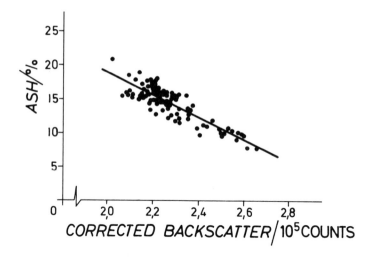

FIGURE 3. Relationship between the ash content and corrected backscatter as reported by Brown and Jones.[147] Solid curve is calculated assuming the relationship y = A + Bx.

ash content was determined by backscatter measurements to 1.62%. The method produced ash determinations at least as accurate as those produced by the established technique which compensates for variation in iron content of the ash by X-ray fluorescence analysis for iron. Compared with the established technique, it has the advantage of averaging analysis over much larger volumes of coal, but the disadvantage that much more precise measurements of X-ray intensities are required.[148] The possibility that the mass absorption coefficient of coal is an estimator of its ash content has been explored by a number of workers, particularly with regard to the application of radioisotopes to on-line analysis (see Renault[149]). One can assume that coal is a two-component mixture consisting of a fuel fraction and a noncombustible ash fraction.

The fuel fraction is composed of carbon and volatiles, and the ash fraction of mineral matter. At a wave-length of 0.7 angstroms, the mean mass absorption coefficient is 0.6 for the whole fuel fraction of a variety of New Mexico subbituminous coals. For the ash fraction the mean is 13.[149]

The mass absorption coefficient of the fuel fraction in the various coals is quite uniform; however, the ash is not, and this prevents the mass absorption coefficient from being used alone as a good estimator of ash concentration.

The intensity of Comptom scattering, Fe, and Ca characteristic radiation can be used to estimate the amount of ash in coal by X-ray fluorescence spectroscopy. In the work by Renault,[149] Mo, W, and Cr radiation were used to study several New Mexico coals, and the best results were obtained with Mo and W X-ray tubes. If the actual concentrations of Fe_2O_3 and CaO and the mass absorption coefficient at the molybdenum Compton wavelength can be determined, the regression equation:

$$\% \text{ Ash} = 24.2\,\mu^x - 6.28(Fe_2O_3) - 1.96(CaO) - 3.4 \qquad (7)$$

estimates the ash content with an average error of 0.5% ash at 0.71 angstroms.

Figure 4 from the paper by Renault[149] shows calculated ash vs. reported ash on 13 New Mexico coals from a variety of environments.

The correlation between calculated ash content on the basis of above assumption and measured ash content is very good. According to Renault,[149] using only a Cr tube

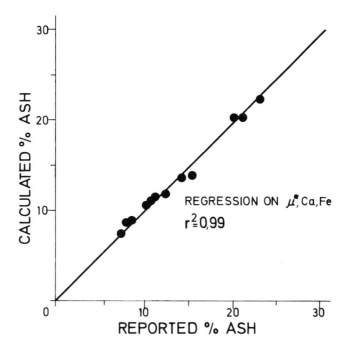

FIGURE 4. Reported ash vs. ash calculated from the multiple linear regression on mass absorption coefficient and concentration of iron and calcium.[149]

and scintillation counter and an air path and making only two intensity measurements — one on the Cr Kα Compton peak and the other on the Ca Kα line — and using carefully selected standards, one can estimate the ash in a heterogeneous group of coals with a correlation coefficient of 0.92. For better results, a W or Mo tube should be used; the measurements should be made in a vacuum, and background should be subtracted from the Ca peak. Using calculated mass absorption coefficients of the coals and their analyzed Fe_2O_3 and CaO concentrations, a much better fit can be obtained.

C. Bottom Ash

The inorganic constituents in coal and associated mineral matter are redistributed among the bottom ash, fly ash, combustion gases, and the FGD sludge, as a result of coal combustion. The primary product of breakdown of clay is mullite ($Al_6Si_2O_{13}$); of pyrite it is ferric oxide, and of calcite it is calcium oxides, while quartz and some silicates remain unchanged.

Liberation of ash from coal is accomplished in modern power plants by burning pulverized coal in steam-boiler furnaces. In a steam-boiler furnace, four major operations must be done:

1. Preparing coal and air
2. Converting coal into the elementary combustible components
3. Bringing the coal and air together in correct proportion at ignition temperature
4. Causing combustion to release heat energy for transfer to the water and steam in the boiler with enough remaining to maintain ignition of incoming fresh coal

All of these steps go on simultaneously in any furnace. Every coal particle goes

Table 16
CONSTITUENTS OF COAL, BOTTOM, AND FLY ASH[58]

	Major and minor constituents (%)		
Constituent	Coal	Fly ash	Bottom ash
Al_2O_3	3.0	27.4	26.2
SiO_2	5.7	57.2	59.1
MgO	0.08	0.68	0.6
Fe_2O_3	0.55	3.8	3.9
Na_2O	0.22	1.4	1.3
K_2O	0.15	0.9	0.93
CaO	0.72	4.5	5.0
TiO_2	0.19	1.5	1.4
Ignition loss		3.3	1.2
Ash	22.2		

through each of the steps in order. The raw coal must first be ground to a fine powder in a mill or pulverizer. This produces an enormous increase of surface area for a given weight of fuel.

When pulverized coal is burned in a dry-bottom furnace, about 80% of the ash originally in the coal leaves the furnace entrained in the flue gas. The remaining 20% of the ash leaves the bottom of the furnace and is known as dry bottom ash, which has a granular characteristic. On the other hand, with pulverized coal burned in a slag-tap furnace, as much as 50% of the ash leaves the furnace as fly ash while the other 50% falls to the bottom of the furnace where it is quenched with water and is shattered into a slag which is granular, black, and glassy. With the cyclone furnace, 70 to 80% of the total ash is removed from the bottom of the furnace as slag; and the remaining 20 to 30% leaves the furnace as fly ash in the flue gas.

Analysis of various ashes shows that the distribution of major elements is approximately the same in the bottom ash and fly ash fractions. However, for certain of the trace components, there is a very definite partitioning between the bottom ash and fly ash. For some elements there can be differences of an order of magnitude in the concentrations of trace elements between these two fractions; for example, Se exhibits this tendency.[150]

For the illustration let us mention results from the work of Roffman et al.[58] They have studied ash from Four Corners Power Generating Plant in the U.S. and their results are shown in Table 16. Trace elements in coal and ashes of the same power plant as reported by the same authors are shown in Table 17. Bottom ash and fly ash from unit equipped with electrostatic precipitators are shown.

In the recent report[4] a chemical composition of bottom ash and FGD sludge and element enrichment ratios in a 330-MW subbituminous coal-fired power plant are presented. Their results are shown in Table 18.

Concentrations of antimony, boron, cadmium, chlorine, erbium, copper, fluorine, iodine, selenium, silver, tin, and tungsten in bottom ash were either similar to, or less than, those found in the source coal. Manganese was somewhat unique, showing a concentration in bottom ash 41 times greater than in coal. Concentrations of zinc, lithium, and hafnium were from 11.5 to 18.4 times more concentrated in bottom ash than coal; those of strontium, beryllium, boron, cesium, gadolinium, samarium, scandium, terbium, thorium, thulium, and uranium from 5 to 10 times; and the remaining elements from 1 to 5 times more concentrated in bottom ash than in coal. In general,

Table 17
CONCENTRATION OF TRACE
ELEMENTS (PPM) IN COAL AND
ASHES[58]

Element	Coal	Bottom ash	Fly ash from precipitator
As	1.1	1.1	11
Be	1.5	5	6
Cd	<1	<0.7	<1.6
Cr	4	20	60
Cu	14	53	80
F	44	17	100
Hg	0.08	0.06	0.13
Mr	40	200	300
Ni	4	20	30
Pb	6.3	26	62
Sb	0.13	0.08	0.4
Se	2.7	1.5	6.6
V	20	50	200
Zn	6	<10	100
Ag	<0.2	<1	<1
B	80	200	700
Bi	<1	<10	<10
Co	<2	<10	<10
Mo	0.8	3	10
Nb	<2	<10	<10
Sn	<0.6	<3	<3
Zr	40	200	300
Ga	8	30	400
Ge	<6	<30	<30
Li	50	200	200
Sr	40	300	500
Ba %	0.1	0.5	1.0

the concentrations of most elements in the solids of FGD sludges did not differ by more than a factor of 2 (more or less) from those of the bottom ash. Exceptions were arsenic, bismuth, copper, fluorine, lead, selenium, and zinc, which were concentrated by more than a factor of 2 in FGD sludge compared with bottom ash, and barium, hafnium, iodine, lithium, terbium, and uranium, which were concentrated by more than a factor of 2 in bottom ash than in FGD sludge.

D. Fly Ash

Fly ash, sometimes called pulverized fuel ash (PFA), is extracted from the flue gas by means of electrical precipitators, cyclones, or collector bags. The coal burns rapidly at a temperature of 1500 ± 200°C, and as a result many of the minerals originally present in the coal undergo chemical and physical changes. Clearly the composition of the resulting fly ash will vary with the assemblage of minerals present in the original coal and their physical relation to one another. The composition will also depend upon the combustion temperature and the residence time at high temperatures, and possibly in the oxidizing or reducing conditions in the immediate environment of each ash particle. Many of the physical and chemical properties of size-classified fly ash depend on the relative size distribution of each fraction.[151]

In this section we shall mention some of the numerous studies of fly ash properties. Bibby[152] has analyzed 20 daily samples of fly ash, obtained from the combustion of

Table 18
CHEMICAL COMPOSITION OF BOTTOM ASH AND FGD SLUDGE (PPM) AND ELEMENT ENRICHMENT RATIOS IN A 330-MW SUBBITUMINOUS COAL-FIRED POWER PLANT[4]

			Enrichment ratio	
Element	Bottom ash	FGD sludge (solids)	Bottom ash coal	FGD sludge coal
Mn	3300	2200	41	28
P	1800	1700	2.8	2.7
Sr	970	860	7	6
Sb	0.49	0.82	1.1	1.9
As	5.4	17	4.2	13.1
Ba	320	110	2.5	0.9
Be	11	7.5	5.9	3.9
B	800	530	7.3	4.8
Br	1.5	0.98	0.9	0.6
Cd	0.67	0.89	1.1	1.4
Ce	200	100	3.6	1.8
Cs	4.3	2.8	6.5	4.2
Cl	26	17	0.3	0.2
Cr	110	180	3.8	6.2
Co	26	17	4.4	2.9
Cu	120	340	4.8	13.6
Dy	4.2	2.8	2.6	1.8
Er	0.48	0.48	0.9	0.9
Eu	1	0.69	1.2	0.8
F	91	300	0.3	0.9
Gd	0.69	1.1	5.3	8.5
Ga	27	48	7.3	12.9
Ge	2.8	4.0	4.9	7.0
Hf	5.7	1.8	18.4	5.8
Ho	0.31	0.20	2.1	1.3
I	0.6	0.2	0.5	0.2
La	31	20	3.4	2.2
Pb	8	27	2.8	9.3
Li	860	2	12.8	0.03
Lu	0.15	0.10	1.9	1.3
Mo	35	23	2.4	1.6
Nd	130	84	1.7	1.1
Ni	28	19	2.2	1.5
Nb	18	24	3.3	4.4
Pr	11	7.6	1.6	1.1
Rb	23	32	2.6	3.6
Sm	1.9	1.3	5.8	3.9
Sc	80	53	6.2	4.1
Se	0.87	5.8	0.5	3.6
Ag	0.14	0.19	0.6	0.8
Te	0.16	0.09	1.6	0.9
Tb	0.46	0.15	5.1	1.7
Th	13	18	5.0	6.9
Tm	0.26	0.12	6.5	3.0
Sn	2.0	3.1	1.1	1.7
W	3.7	5.0	1.1	1.5
U	23	7.2	9.2	2.9
V	180	280	2.5	3.8

Table 18 (continued)
CHEMICAL COMPOSITION OF BOTTOM ASH AND
FGD SLUDGE (PPM) AND ELEMENT ENRICHMENT
RATIOS IN A 330-MW SUBBITUMINOUS COAL-FIRED
POWER PLANT[4]

Element	Bottom ash	FGD sludge (solids)	Enrichment ratio Bottom ash coal	Enrichment ratio FGD sludge coal
Yb	1	1.4	3.7	5.2
Y	75	50	4.4	2.9
Zn	10	66	11.5	76
Zr	62	76	2	2.5

subbituminous Waikato coals. The contents of MgO, SO_3, B_2O_3, and total and uncombined CaO in the New Zealand material were high when compared with most fly ash produced in Europe and the U.S. The fly ash was separated into three major components, namely a magnetic fraction, and nonmagnetic acid-insoluble and acid-soluble fractions. These three fractions were related to three major mineral forms found in the coal: respectively, hydrated iron carbonates and oxides, quartz and aluminosilicates, and calcium minerals. The variable composition of the fly ash arises from differences in the proportions of these three mineral suites in the original coal. The composition was also found to vary with particle size. It is suggested that this was due to differences in the particle sizes of minerals in the coal and to increased thermal decomposition of minerals in the smaller particles of fly ash. A low-density fraction was separated and found to be similar in composition to cenospheres found in British and American fly ash.

Results obtained by Bibby[152] for some of the samples analyzed are shown in Table 19 and Table 20. Table 19 gives the analysis of the acid-insoluble fractions from 5 samples and the analysis of the combined sample. Table 20 gives the analysis of the acid-soluble fractions from 6 samples. The magnetic fraction is included in Table 20 since it is also acid-soluble.

It can be inferred from these results that three groups of minerals were deposited in the coal by different mechanisms, and that these groups of minerals are related to the magnetic, acid-soluble, and acid-insoluble fractions found to be present in the PFA. The evidence also suggests that relatively little chemical interaction between the different minerals took place during the combustion process.

The major components of acid-insoluble fractions were SiO_2 and Al_2O_3, together with minor amounts of TiO_2, Fe_2O_3, CaO, MgO, Na_2O, and K_2O. X-ray diffraction showed the presence of quartz and mullite ($3Al_2O_3 \cdot 2SiO_2$) in varying amounts. From the chemical analysis it would appear that the insoluble fraction is predominantly quartz and mullite, most probably derived from quartz and kaolinite in the coal.

The main components of the acid-soluble fraction were CaO, Fe_2O_3, and Al_2O_3, together with significant amounts of MgO, SO_3, B_2O_3, and uncombined CaO.

The amount of magnetic material separated from the fly ash samples is included in Table 20, and ranged from 0.45 to 18.1%. X-ray diffraction showed the major iron compounds to be hematite and magnetite. The ratio of the two oxides was found to be variable, and may reflect the residence time in the flame or the combustion temperature, since magnetite is the stable oxide at high temperatures.

Table 19
COMPOSITION OF ACID-INSOLUBLE FRACTION OF FLY ASH[152]

	Sample					
	1	5	9	12	17	Combined
SiO_2	38.38	53.48	45.49	49.10	33.07	45.14
Al_2O_3	10.13	21.80	14.09	15.99	5.60	13.16
Fe_2O_3	0.88	1.09	0.94	0.93	0.55	0.66
TiO_2	0.96	1.81	1.27	1.43	0.47	1.23
CaO	1.16	1.38	0.99	1.34	0.84	0.39
MgO	0.36	0.41	0.29	0.42	0.20	0.06
Na_2O	0.63	0.42	0.43	0.55	0.84	0.73
K_2O	0.31	0.61	0.40	0.50	0.37	0.44
Total (%)	52.81	81.00	63.90	70.26	41.94	61.71

Table 20
COMPOSITION OF ACID-SOLUBLE FLY ASH[152]

	Sample					
	1	2	3	4	5	6
SiO_2	0.88	0.82	0.56	0.54	0.66	0.70
Al_2O_3	6.32	4.48	4.39	4.24	2.61	6.04
Fe_2O_3	5.03	4.60	5.53	4.16	2.74	5.10
TiO_2	0.40	0.38	0.42	0.36	0.38	0.56
CaO	21.65	16.84	16.92	12.31	9.42	18.90
MgO	2.97	2.37	2.31	2.05	1.44	2.52
SO_3	2.30	1.23	2.20	1.11	0.45	1.32
Na_2O	0.27	0.25	0.24	0.16	0.08	0.35
K_2O	0.04	0.04	0.04	0.03	0.02	0.07
Mn_2O_3	0.05	0.05	0.04	0.02	0.03	0.06
P_2O_5	0.07	0.05	0.06	0.04	0.03	0.04
B_2O_3	1.52	1.27	1.45	0.97	0.75	1.72
Magnetic fraction	5.5	6.8	4.6	3.5	0.45	4.2
Total (%)	47.00	39.18	33.76	29.49	19.06	41.58
Uncombined CaO	3.80	3.80	3.31	2.14	1.30	4.80
Loss on ignition	1.2	1.3	3.2	1.8	3.5	2.2

Many other research groups have performed the analysis of fly ash. Table 21 shows data reported by Smith and Larew[153] and Thornton et al.[154] In the report by Smith and Larew,[153] five sources of fly ash were included in the analysis. Four of these were bituminous fly ash while the last was a lignite fly ash. They all had a low carbon content (<6%) and varied physical or chemical properties (first five samples shown in Table 21). The last sample shown in Table 21 was collected by an electrostatic precipitator. The coal was a low sulfur coal from Campbell County, Wyoming, and was pulverized before injection into the burner.

A compilation of values for chemical composition of fly ash published in literature has been presented in a recent report (see Reference 4). In this report the range of fly ash composition represents literature values taken mainly from Block and Dams,[69] Bern,[155] Swaine,[156,157] Martens,[158] Mulford and Martens,[159] Martens et al.,[160] Doran and Martens,[161] and Page et al.[162] The average composition of fly ash was taken from

Table 21
FLY ASH ANALYSIS[153,154]

Component (%)	Coal Origin					
	Ft. Martin	Black Dog	Amax	Albright	Harfield's Py.	Campbell, Wyoming
Al_2O_3	23.8	21.1	23.6	27.4	24.6	13.0
SiO_2	47.8	39.5	45.6	53.6	48.5	34.0
Fe_2O_3	17.2	7.4	14.3	10.4	14.3	6.0
CaO	3.8	21.9	6.0	1.3	4.1	20.0
MgO	1.2	5.9	1.2	0.7	1.1	6.0
K_2O	2.3	0.5	1.8	2.1	1.7	0.8
Na_2O	0.8	0.6	1.0	0.3	0.7	2.8
SO_3	0.1	1.4	0.2	<0.1	<0.1	13.7
TiO_2	—	—	—	—	—	1.0
Loss on ignition 950°C	1.2	1.7	5.7	2.0	3.1	—
Specific gravity	2.39	2.28	2.43	2.12	2.34	—
Blaine fineness, cm^2/gm	2404	4210	4339	1980	2456	—

data of Swanson,[163] Schwitzgebel et al.,[5] and Page et al.[162] for ashes derived from western U.S. coal sources. No values for thorium and uranium are presented in this table, probably because of lack of published data. The results of this compilation are shown in Table 22. The major components of fly ash — silicon, aluminum, calcium, magnesium, iron, sodium, potassium, and titanium — make up from 35 to 70% of the total elemental composition of fly ash. Fisher et al.[151] indicated that the composition of fly ash from western U.S. coals could be approximated by the formula,

$$Si_{1.00}Al_{0.45}Ca_{0.051}Na_{0.047}Fe_{0.39}Mg_{0.020}K_{0.017}Ti_{0.011}$$

Additional papers about coal ash and its composition and properties can be found in the list of references at the end of this chapter (see References 164-288).

E. Ash Utilization

The literature on ash utilization is rather voluminous. Symposia and seminars are often held on the subject of ash utilization, and the proceedings are published afterwards. Here is some of the more interesting literature for the reader in this subject: Davis,[198] Dent,[199] Rose,[289] Bland,[290,291] Bishop et al.,[292] and Faber et al.[293]

There are no precise data on how much of ash is being produced in the world. If the experience of the U.S. may be used as an example, even with its bias towards higher quality coals, a conservative estimate of world production of ashes would amount to about 425 million tons, comprised of 345 million tons of coal ashes, and 80 million tons of lignite ashes. Knowing of the widespread availability and use of lower quality coals for energy purposes throughout the world, one might reasonably assume a theoretical world production of ashes to total about 500 million tons.[294] Figure 5, a chart, shows a distribution of world production of ashes by major producing countries in 1974.[294]

Estimates of ash production and actual reported data, where such data are available, amount to 222,633,000 metric tons or about half of the conservatively estimated total world production in 1974. It must be emphasized that these data represent only a partial estimate of the total world output and that not all the data were reported officially

Table 22
CHEMICAL COMPOSITION OF FLY ASH[4]

Element (%)	Range	Average
Al	0.1—17.3	11.7
Si	19.1—28.6	26.6
Ca	0.11—12.6	4.4
Mg	0.04—6.02	0.9
Na	0.01—0.66	0.6
K	0.19—3.0	0.8
Fe	1—26	4.1
Mn	0.01—0.3	0.02
S	0.1—1.5	0.2
P	0.04—0.8	0.13
Sr	0.03—0.3	0.09
Ba	0.011—0.5	0.3
Ti	0.16—0.7	0.5

Element (ppm)	Range	Average
Ag	0.04—5	0.3
As	2.8—6300	9
Be	3—7	5.5
B	48—618	390
Br	0.7—5.3	1.8
Cd	0.7—130	1.0
Ce	22—133	150
Cs	—	5
Cl	—	47
Cr	10—690	54
Co	7—49	12.6
Cu	14—1000	63
F	100—610	201
Gd	—	1.4
Ga	15—93	33
Ge	—	3.3
La	12—72	70
Pb	7—279	48
Li	50—1064	70
Mo	7—117	10
Ni	10—4300	34
Nb	—	20
Rb	—	43
Sb	1.6—202	1.6
Sc	3.7—141	15.6
Se	0.2—134	7.6
Sn	3.3—63	3.3
Te	0.02—0.2	0.2
Tl	0.8—1.7	1.2
V	50—1000	119
Zn	36—1333	72
Zr	50—1286	200

by the countries listed. The disposal of ash resulting from the use of coal has become a major problem to many utilities.

In the coming discussion we shall limit ourselves to fly ash. Efforts to alleviate disposal problems have led to the development of several applications for fly ash, but these use only limited amounts of fly ash at present. Some of the established fly ash outlets are as additives in ready-mixed concrete, building blocks, precast concrete

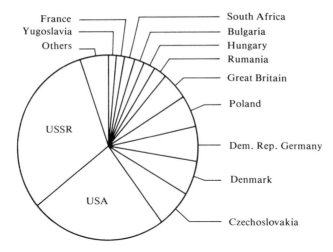

FIGURE 5. Distribution of the world production of ashes by major producing countries.[294]

products, slag and clay bricks, bituminous mix filters, road base choking materials, and lightweight aggregates. Fly ash is also used in limited quantities as metal polishing agents, mild abrasives, grouting components, and sand substitutes in sand blasting.

As an example, Table 23 shows ash collection and utilization in the U.S. as presented in the paper by Markon.[294]

These figures have not changed significantly since then. These figures are encouraging, but still fall far short of the acceptance rates being achieved in the U.K. and in Europe. The 1974 results in these countries — the most current statistics available — show Finland utilized 84% of its ash, 72.5% in France, 50% in England, and 38.2% in Poland.[293]

The use of lignite fly ash has been discussed in several papers by Manz,[239] the Economic Commission for Europe Ash Symposium,[239] and at 1967 and 1970 Pittsburgh ash symposia.[293] Table 24 shows production and utilization of ash from lignite in 1974.[239]

Fly ash for use as an admixture in concrete has received more attention than other current applications. The use of fly ash as a concrete admixture adds desirable properties to the concrete if the fly ash has pozzolanic properties. A pozzolan is a siliceous material which has no actual cementing properties. However, in the presence of water, the pozzolan will react chemically with calcium hydroxide to form insoluble compounds possessing cement properties. For more discussion of this specific aspect of fly ash utilization, see papers by Bucklen et al.[137] and Faber et al.[293]

Probably the largest potential volume outlet at present for fly ash is in the lightweight aggregate field in which the fly ash is pelletized and bloated to provide a good quality product. This potential is illustrated by the fact that each cubic yard of lightweight concrete requires approximately 1 ton of lightweight aggregate. Fly ash can also serve as an acceptable mineral filter in other types of mixes, and the failure to adopt it on a more wide-scale basis is probably due primarily to a lack of promotion.

Utilization of fly ash is the subject of considerable research by utilities, independent research organizations, and educational institutions. It is conceivable that in the near future the decision of whether to use coal or an alternate fuel may be made on the basis of a profitable utilization of fly ash.[137]

Analyses by many research groups have shown that fly ash contains most of the

Table 23
ASH COLLECTION AND UTILIZATION IN U.S.[294]

	Fly ash (10^9 kg)	Bottom ash (10^9 kg)	Boiler slag (10^9 kg)	Total ash (10^9 kg)
1. Total ash collected	36.6	13.0	4.4	54.0
2. Ash utilization				
a. Used in Type 1-P cement — ASTM 595-71 or mixed with raw material before forming cement clinker	0.4	<0.1	—	0.4
b. Partial replacement of cement in concrete or concrete products	0.5	—	—	0.5
c. Lightweight aggregate	0.1	0.1	—	0.2
d. Stabilization and roads	0.3	0.5	1.1	1.9
e. Filler in asphalt mix	0.1	<0.1	<0.1	0.1
f. Miscellaneous	0.5	0.8	0.9	2.2
Subtotal	1.9	1.6	2.1	5.3
3. Ash removed from plant sites at no cost to utility	0.4	0.5	0.1	1.0
4. Ash utilization from storage	0.9	0.6	0.1	1.6
5. Total utilization (tons)	3.2	2.7	2.3	7.9
6. Total utilization (%)	8.7	20.7	52.2	14.6

Table 24
PRODUCTION AND UTILIZATION OF ASH FROM LIGNITE IN 1974, 10^6 kg [239]

		Commercial utilization							
Country[9]	Prod.	Concrete	Cement kilns	Roads	Lgt. wt. Agg.	Bricks	Structural fill	Misc.	Total	1974 prod. %
Bulgaria	2500			5	100		50	10	165	6.6
Canada	304	12					20	4	36	
GDR[a]	13115	547		39.5	201			787	2542	19.4
FRG[b]	7600	70	60	150			100	100	480	6.3
France	433	43.2	106.3	29.8			57.1	0.1	186.9	43.2
Hungary	4195		82		118				200	4.8
Italy	360	27					9		36	10
Poland	4939		56			7	130	34	227	4.6
Rumania	3806	140				11		16	167	4.4
Turkey	417	1.04							1.04	0.24
USA	1252	2.1		32.7			67.3	4.5	106.6	8.52
Yugoslavia	3354					190	10		200	5.96

[a] German Democratic Republic
[b] Federal Republic of Germany

essential nutrients required for successful plant growth and that it is an effective amendment for both agricultural and surface-mined soils.[158,165,184,259,295] The neutralizing effect of alkaline fly ash on acidic soils has been known for some time.[213,255]

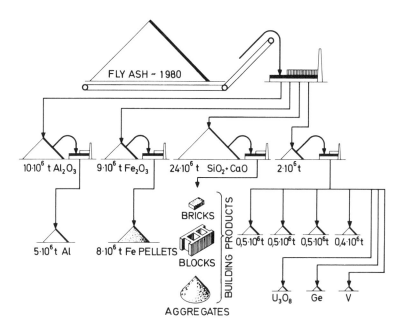

FIGURE 6. Joseph Pursglove vision of fly ash utilization.

In the experiment by Wochok et al.,[286] studies were performed in the greenhouse. The experiments demonstrated that each soil site has its own particular set of characteristics and ability to support plant growth. Abandoned strip mines possessed poor soil qualities and were acidic, ranging in pH from 2.5 to 5.0. These soils alone do not support vegetation very well, but the addition of fly ash results in good seed germination and subsequent plant growth.

Martens and Beahm,[237] have conducted micronutrient investigations in order to evaluate the plant availability of boron, molybdenum, and zinc in fly ash. The authors have found that fly ashes from burning of different coals vary widely in nutrient content and in chemical and physical properties. Weathering changes the nutrient content and chemical properties of fly ash. Data reported by Martens and Beahm[237] show that, compared with nonweathered fly ash, a much higher level of a weathered fly ash could be incorporated into soil without adversely affecting plant growth. These data suggest that a higher level of weathered fly ash could be applied to soil for correction of nutrient deficiencies, for modification of physical soil properties, and for partial substitution of soil in fill areas where plants are to be grown. Data reported also show that certain fly ash samples could be used as a soil additive for correction of molybdenum deficiency of plants. The authors caution the use of fly ash as an amendment to agricultural soils until research determines whether the material supplies elements that would be entering the food chain.

The idea of metal recovery from ash has received very much attention. This subject will be discussed in detail in Part V of this chapter. We shall finish this chapter with the vision of Pursglove[296] presented in Figure 6. This represents the ultimate goal in ash utilization.

IV. DESULFURIZATION

In general, the fate of sulfur dioxide emissions involves photo-oxidation in the atmosphere to form sulfur trioxide which, under humidifying conditions, becomes sulfuric acid or sulfate aerosol. Residual sulfur dioxide and sulfuric acid or sulfates are scavenged from the atmosphere by vegetation, eventually being discharged into the sea by the rivers of the world, along with sulfur accumulated from weathering rocks, and sulfur applied as fertilizer. Some of the sulfate is directly deposited into the ocean via rain or dust. When an excess occurs, the atmosphere-to-land portion of the sulfur cycle is unsound.[297]

Sulfur content of utility coal is controlled in some countries by government regulations. For example, the U.S. "Clean Air Act" of 1963 greatly broadens the activities of the Federal government in the air pollution field and specifically points out the need for development of techniques to extract sulfur from fuels. The removal of sulfur compounds from the stack gases of coal-burning power plants has been a national issue for the past decade. A great deal of progress has been made in developing systems for commercial applications. However, some fundamental issues remain, and a number of complex problems continue to defy solution.

Ecological consequences of the disposal of sludges resulting from flue-gas desulfurization remain unknown. Two types of pollutants associated with the desulfurization of flue gases have been identified: localized acid mist and scrubber sludge. No data has been presented that show that scrubbers normally emit large quantities of acid mist. However there is sample evidence that sludge disposal may present formidable problems.

In view of the increasing seriousness of air pollution problems, it is probable that utilities in the future will impose more stringent sulfur specifications upon coal suppliers. A preferable alternative to this would be the development of low-cost methods to remove the sulfur at the power plants before firing or development of means to extract the sulfur dioxide gases from the flue gases. In line with the latter, a pilot plant at the Pennsylvania Electric Company Seward Generating Station demonstrated the technical feasibility of extracting sulfur dioxide gas from the flue gases and converting it to usable sulfuric acid. Such an arrangement would recover 90% of the sulfur in the coal. Also, the resulting sulfuric acid produced from this process might find use as a leaching agent to beneficiate coal ash as part of a coal-associated mineral processing complex at power plants for the production of titanium, silica, iron, and alumina. The beneficiation of power plant derived coal — associated minerally by use of sulfuric acid produced during coal desulfurization — should receive careful consideration concerning technical and economic feasibility.

However, when sulfuric acid is produced on a continuous basis, it is important that the market be capable of continuously absorbing the supply. Sulfuric acid is very difficult to stockpile, particularly the large volumes that could be produced at a power plant. Storage costs are high compared to recovered sulfur due to the larger space requirements and the need for special handling. If it is not possible to sell the recovered product, the acid must be neutralized prior to disposal. Another problem in marketing of sulfuric acid arises from the quality of the recovered product. For some FGD processes, the resulting sulfuric acid, is not of a commercial grade.

The ease of storage of sulfur enhances its potential as a marketable product. Large quantities of sulfur can be easily transported and stockpiled if necessary. The cost of processing recovered SO_2 gas to produce sulfur can be expensive. Since hydrogen sulfide may be used in this procedure, sulfur production is most suitable where hydrogen sulfide is also being recovered as a by-product. The conversion to sulfur, therefore, may not be attractive to a utility.

The references on recent work in the area of coal desulfurization can be found in papers by Condry.[298,299] The chemistry of coal desulfurization is discussed in the book by Meyers.[297] According to Meyers, one method for chemical desulfurization of coal which has been extensively tested is ferric sulfate leaching. In the book this method is described in detail, starting with the basic chemistry and proceeding through process development and engineering design and cost estimation results. The entire spectrum of chemical desulfurization technology is presented. As mentioned earlier, sulfur occurs in two major forms (pyrite and organic) in coal which are separately located in the inorganic and organic portions of coal, respectively. This division of the coal matrix into inorganic and organic portions plays a significant role in the chemistry of coal and is particularly significant in the design of methodology for the penetration, contact, and chemical reaction of the various sulfur functional groups. Conceivably, a chemical reagent could be dissolved in a hydrophobic solvent such that the reagent comes into intimate contact only with the organic carbon matrix of the coal, while another reagent, dissolved in water, would come into more intimate contact with the inorganic portion of the coal matrix. This could then be a means of obtaining part of the selectivity required for an effective economic desulfurization process. However, because of the wide reactivity of a given chemical reagent, it is not possible to obtain a high degree of chemical selectivity between the organic and inorganic portions of the coal matrix in many cases.

The chemical reactions which can effect desulfurization of inorganic or organic functional groups present in coal are numerous.

The reactions of iron pyrite which have been observed are theoretically capable of proceeding at temperatures below the 400°C decomposition point of coal. A solvent which will dissolve significant amounts of iron pyrite without decomposition is yet to be discovered. Therefore, removal of pyritic sulfur from coal requires a chemical transformation such as:

(1) A displacement reaction

$$:\ddot{N}: \quad Fe-S_2 \rightarrow :\ddot{N}: \quad Fe + S_2^{-2} \qquad (8)$$

(2) An acid base neutralization, which may produce hydrogen sulfide, an iron salt,

$$4H^+ + FeS_2 \rightarrow 2H_2S + Fe^{+2} \qquad (9)$$

(3) An oxidation reaction which converts the persulfide ion to a soluble, volatile or otherwise removable form

$$[O] + FeS_2 \rightarrow Fe^{+2} + 2S + [O^{-2}] \qquad (10)$$

(4) A reduction reaction which would remove pyritic sulfur as hydrogen sulfide

$$FeS_2 + H_2 \rightarrow FeS + H_2S \qquad (11)$$

$$FeS + H_2 \rightarrow Fe + H_2S \qquad (12)$$

The reactions of iron pyrite reported in the literature are almost exclusively oxidation and reduction reactions. No displacement or neutralization are reported.

Organic sulfur is thought to occur in four major forms: mercaptans or thiols, sulfide, disulfide, and aromatic ring sulfur as exemplified by the thiophene system.

Potential desulfurization methods for coal may be classified into six groups: solvent partition, thermal decomposition, acid-base reaction, reduction, oxidation, and displacement.[297]

Solvent partition — The organic portion of the coal system is soluble in varying degrees in a vast number of organic solvents which can dissolve essentially all of the organic matrix, depending on the specific solvent used and the temperature of extraction. In order for "desulfurization" to take place as a result of the solvent extraction of coal, the solute must contain a higher ratio of organic sulfur to carbon than the residue. The higher this ratio, the higher the desulfurization selectivity. In order to enhance the solubility of the sulfur-containing coal molecules, it could be possible to add coordinating agents to the solvent which would form soluble salt which could be decomposed after extraction for recovery and recycling. The selectivity of solvent partition of organic sulfur would be enhanced if the coal macromolecule could be cleaved so that the organic residues associated with dissolved organic sulfur are relatively low in molecular weight.

Thermal decomposition — Since the coal begins to decompose rapidly at temperatures near 400°C, thermal decomposition of organic sulfur compounds would have to take place below 400°C, and the cleavage of sulfur would occur in preference to or predominate over the cleavage of other coal compounds. While organic sulfur compounds are subject to pyrolysis, so are the other components of the coal macromolecule such as amines, ethers, acids, etc., and C—O bonds are more easily broken than C—S bonds. Therefore, a catalytic approach to pyrolytic decomposition, in which a catalyst selectively promotes the decomposition of organic sulfur, is the only possibility likely to be successful.

Acid-base neutralization — Mercaptans or thiols, which are weakly acidic, are capable of being neutralized and brought into aqueous solution by strong bases. This is not true of sulfides, disulfides, or aromatic sulfur compounds. Thus, a caustic leaching process limited to acid-base neutralization would remove only thiols. As in solvent partition, a preliminary lowering of the molecular weight of the coal macromolecule would aid the selectivity of this type of treatment.

Reduction — Organic thiols, sulfides and disulfides, as well as aromatic organic sulfur, are susceptible to reduction by molecular or nascent hydrogen or hydrogen donor systems sulfides and thiols and would give hydrogen sulfide as a product, while under mild reduction conditions disulfides would give hydrogen polysulfides (under more vigorous-reducing conditions, the product would be H_2S). Hydrogen polysulfides are thermally unstable and would tend to dissociate into hydrogen sulfide, an elemental sulfur. Any elemental sulfur which survived the reducing treatment could then be removed from the reactants by careful vaporization or by extraction with a solvent selective to sulfur. The desulfurization of coal with hydrogen at temperatures above 400°C is well known (coal liquefaction) and gives rise to liquid and gaseous hydrocarbon products.

Oxidation — Organic thiols, sulfides, and disulfides may be oxidized to sulfones, then to sulfoxides, and finally to sulfonic acids (thiols are first oxidized to disulfides). The sulfonic acids can then be hydrolyzed in boiling water to eliminate sulfuric acid as the sulfur product which can be filtered away from the solid coal matrix. Thus, at every former sulfur position which has been oxidized and hydrolyzed, an alcohol functional group would remain. The key to a desulfurization process based on oxidation would be the minimization of the oxidation of the remainder of the coal matrix, which is a difficult task since amines, alcohols, esters, and olefinic groups are as easily oxidized as sulfur-containing units.

Nucleophilic displacement — Although many sulfur compounds are nucleophilic, a strong nucleophile can attack the sulfur atom of a sulfide or disulfide giving an alkyl or thioalkyl leaving group, or the nucleophile can attack a carbon atom bonded to a sulfide group. These reactions can be visualized as sulfur removal methods if the leaving group contains few carbon atoms relative to the number of cleaved sulfur atoms and is either soluble or volatile, so that the sulfur-containing moiety can be filtered or volatilized away from the coal matrix.[297]

It would be an impossible task to cover all the published reports on sulfur removal from the coal and related activities. We shall mention here a number of research projects which have been reported in scientific literature in this field.

Cole et al.[300] have studied the various forms of S present in coal samples from Marian Country, Iowa, in relation to: reducing S in Iowa coals by mechanical means; the possibility of recovering S from coal "brasses"; utilizing high-S coals by adapting methods used in the commercial production of S and S products from industrial gases; and examining the economic feasibility of recovering the concentration of iron sulfides found in associated rock formations. Included in this report are the drilling logs of samples taken during this study, which contain the following information: location of the drill hole, collar elevation, depth and description of material, and the percent S present. In conjunction with the drill hole logs are tables containing proximate and ultimate analyses, Btu, ash fusibility, and analyses of the various sulfur forms of the coal in each drill hole.

Kestner et al.[301] examined the reduction of sulfur forms in coal by magnetic separation. Samples of Upper Freeport, Redstone, Sewickley, and Pittsburgh coal seams were tested in a laboratory using a high-intensity magnetic separator. Variations of the feed size-range, separator size setting, and intensity of the magnetic field were tested to determine the conditions for optimum separation of the particles. In addition, tailings from the tests were subjected to microscopic examination of sulfur forms. This report also contains a table of mineral impurities in coal and their mass susceptibility to a magnetic gradient which was compiled from several other sources.

Leonard et al.[302] have reviewed desulfurization methods not based on specific gravity or froth flotation. This report also contains some information concerning the effects of mineral properties on coal cleaning processes. The S reduction methods covered in this survey are grouped into three major classes: electrical (which includes magnetic and electrostatic coal cleaning), thermal, and chemical (including some leaching techniques and bacterial oxidation). Research work reviewed in this report noted the following: reducing the size of coal to 16 mesh or less and then treating with a stream-air atmosphere appeared to increase the amount of pyrite separated from coal during magnetic separation; that the physical properties of mineral matter, particle size and shape, surface moisture, and the rank of coal appeared to have an effect on the amount of S removed by electrostatic means; and the amount of desulfurization achieved by thermal methods appears to be related to the form of S and the particle size.

Cavallaro and Van Eck[303] have discussed preparation characteristics and desulfurization potential of Iowa coals. Ten high-sulfur coal seams from south-central Iowa were sampled, crushed, and washed to determine if physical desulfurization could remove sufficient S to allow compliance with current utility fuel standards. Geologic sections of each channel sample were presented showing some pyrite and mineral source occurrences within the seams, i.e., shale, siltstone, and fireclay; detailed pyrite sulfur analyses of sized gravity fractions were tabulated. After channel samples were prepared by crushing and sizing to three size ranges, the coal was gravity cleaned in static float-sink booths, and the eight fractions were analyzed for proximate analysis, FSI, HGI, and ash fusability. Although pyrite and total sulfur were reduced by crushing, results were felt biased due to a decrease in yield. The washability tests showed that pyrite content could be reduced, but most coals were still high in sulfur and would require both physical desulfurization and SO_2 removal after combustion to be burned as utility coal.

Chung[304] has performed a series of experiments to recover Fe and S in the mineral residue from coal liquefaction, and to convert this mineral residue to mineral wool. The mineral residue composition of Kentucky, West Virginia, North Dakota, Washington, and Wyoming coals was tabulated for 24 elemental oxides as determined by both wet chemical and quantitative spectrochemical analyses. Iron and sulfur recovery tests were done using liquefaction mineral residue from Kentucky coal and two synthetic mineral residues. These were heated in electric furnaces using graphite reactors at temperatures ranging from 1100 to 1600°C. The mineral residues were reduced both with and without water. To produce mineral wool, the residues were mixed with fluxing agents (Ca or Na) and heated in a reducing atmosphere at temperatures ranging from 1200 to 1320°C. Fibers were then drawn from the molten glass. The results of the initial recovery tests indicated the yields of Fe approached 100% with proper supplies of water. Elemental sulfur was collected from the exhaust gases. Problems existed in the quality of mineral wool produced from the mineral residues due to the presence of carbonaceous material and FeS in the mineral residues.

Meyers[305,306] has discussed the chemical desulfurization of coal. If the desulfurization of coal through chemical leaching of the pyrite sulfur content can be accomplished without otherwise changing the chemical or physical structure of coal, this process could have an important impact on providing clean energy from coal. This paper describes the Meyers Process for chemical desulfurization of coal which utilizes aqueous ferric sulfate to selectively oxidize and leach the pyritic sulfur from coal. The technology is being developed under the sponsorship of the Environmental Protection Agency. Process conditions and experimental results are presented, and a test plant is described.

Leonard[307] has discussed the possibility of further lowering sulfur in intensively cleaned coal at mines. Considering the mineral matter of coal to be a function of the ash-to-sulfur ratio in the raw coal, the author believes that this ratio could be used as an index to predict the ash and sulfur composition of coal cleaned at different gravities. Plotting the content of raw ash against the sulfur content of the products from specific gravity fractions indicated the relationship between ash and sulfur. Graphs showing the ash-sulfur relationship, for different ash-sulfur ratios, are presented for channel samples taken from the Upper Freeport, Lower Kittanning, and Pittsburgh coal seams. On examination it was noted that a coal with a low ash-sulfur ratio could not be cleaned to the same extent as a coal with a high ash-sulfur ratio. The author concluded that additional work could be done to identify areas of low ash-sulfur ratios in present coal supplies and that the relationship found for samples of the entire coal seam may not hold true for bench samples of coal.

Stewart et al.[308] have presented a brief review of studies and investigative work directed to the beneficiation of Kentucky coals. The work included general coal property analysis, selected seam analysis, pyrite separation methods, combustion characteristics of selected coals, and coal blends and characteristics of resulting ash. The technical and economical aspects of high sulfur coal utilization through blending and through stack gas cleanup strategies were included in the work. Also presented are some initial results from recently acquired laboratory facilities for comprehensive testing of coal and related materials.

Attar et al.[309] have discussed the transformation of sulfur functional groups during pyrolysis of coal. In experiments, samples of bituminous coal containing variable amounts of added iron pyrites were pyrolysized at 1300°F in a stream of argon, and pyrolysis products analyzed by gas chromatography. Preliminary data on the kinetics of the transformations of sulfur functional groups in coal shows that: (1) hydrogen sulfide from loose FeS_2 crystals evolves at different rates than hydrogen sulfide from FeS_2 that was in the original samples; and (2) pyrolysis of coal in the presence of FeS_2 causes a decrease in the benzene-soluble, thiophenic organosulfur compounds in the oil. The difference in the kinetics of the release of H_2S is probably due to different mass-transfer limitations. The reduction in the benzene-soluble organosulfur compounds is probably due to catalysis of the hydrodesulfurization on the FeS surface. Since thiophene derivatives are found in the oil, even when FeS_2 is added to the coal, their source must be the organic matrix and not dehydrocyclation reactions on the sulfide surface.

Ganguli et al.[310] have described some of the experimental results showing the feasibility of removing sulfur, particularly organic sulfur, from high-sulfur coals by a simple method of low-temperature chlorinolysis followed by hydrolysis and dechlorination. At first the process concept of this chlorination method is described. Experiments, results, and discussion of this method of desulfurization are described for two bituminous coals. It is demonstrated that a simple method of coal desulfurization by chlorinolysis of powdered moist coal in methyl chloroform at 74°C and 1 atmosphere can remove up to 70% organic sulfur, 90% pyritic sulfur, 30% sulfate sulfur, and 70% total sulfur from a high-sulfur bituminous coal. After hydrolysis, the chlorinated coal is dechlorinated by heating at 300 to 350°C for 2 hr.

Ponder et al.[311] have discussed principal technologies for continuous SO_2 emission control: (1) use of coal that occurs naturally with an acceptably low-sulfur content; (2) physical and chemical coal cleaning; and (3) flue-gas desulfurization (FGD) processes. Also discussed are some additional technologies being developed that could make significant contributions to SO_2 control after 1980: coal gasification, coal liquefaction, and fluidized-bed combustion of coal.

Stambaugh et al.[312-315] have described coal desulfurization by the Battelle hydrothermal coal process. This process not only competes favorably on an economic scale with other desulfurization processes, but it also has significant technological advantages. The process should produce no significant amount of sludge for disposal. Its primary end products are clean solid fuel and elemental sulfur — which can be marketed or easily stored — and potentially recoverable metal values. In addition, the process, or modifications of the process, has the potential for producing improved feedstocks for gaseous and liquid fuels and for producing coal solutions which could be a source of coal chemicals.

Friedman and Warzinski[316] have done some work on chemical methods for desulfurizing coal. According to these authors, chemical cleaning removes more sulfur than physical separations, less than coal conversion, at costs and investment which intermediate. Fuel value losses are small for chemical methods, comparable with the best of the physical separation procedures. In addition, certain chemical treatment methods effect useful changes in coal composition and properties. The principal chemical desulfurization methods involve treatment, at elevated temperature, with either basic media or oxidizing systems. Aqueous sodium hydroxide, at 225°C, removes almost all of the pyrite from coal while subsequent washing with dilute acid removes most of the remaining mineral matter. At 300°C, similar treatment can remove, in addition, up to 40% of the organic sulfur from coals and also destroy coking properties. Oxidative desulfurization, using steam and air, also removes most of the pyritic sulfur and 20 to 40% of the organic sulfur, under milder conditions than those required for alkali treatment.

Lin et al.[226] have presented the results of pilot-scale studies of sulfur and ash removal from coal by high-gradient magnetic separation (HGMS). Work was done on both the liquefied coal and the raw pulverized coal in water slurries. The effects of residence time, field intensity, packing material and density, slurry concentration, and recycling on the trade and recovery of the wet separation of sulfur and ash from water slurries of Illinois #6 coal were quantitatively examined. The HGMS was effective in reducing the weight percent of total sulfur, ash, and inorganic sulfur by as high as 40.35 and 80%, respectively; while achieving a maximum recovery of about 95%. The results provide the first experimental verification of the applicability of Bean's magnetic filtration model in quantitatively correlating the data obtained from the pilot-scale beneficiation of coal slurries by the HGMS.

Cohen[317] has reviewed the basic requirements for high-gradient, high-intensity magnetic separation. The principal innovations of the superconducting magnet system design for this work include an annular quadrupole magnet and corresponding slurry channel operating in an open field without secondary poles. The transverse secondary circulation set up in the slurry through the primary orbital flow is used to transport all particles into the region of high field intensity. Two Pennsylvania coals with high pyrite contents were treated. Test conditions and results are described and discussed. One of the coals yielded 81% recovery of "clean" coal and over 83% rejection of pyrite. Possible further improvements are discussed.

Fine et al.[318] have discussed a process for the desulfurization of finally divided coal by flash roasting and magnetic separation. This process is expected to produce coal containing approximately one third of its initial sulfur content, coal chemicals, elemental sulfur, and a solid waste product that contains large amounts of sulfur, iron, and to a lesser extent, coal. A flow chart and energy balance for the possible process and preliminary experiment results from a laboratory scale model are presented.

Hendel and Winkler[319] have experimented with desulfurization of coal-oil mixtures by attrition grinding with activated iron powder. High bituminous coal mixed with fuel oil and iron powder and then ground in a laboratory ball mill at 250°C lost about 50% of sulfur in the first 5 min. When the iron powder was specially activated, the removal of sulfur from the same coal-oil mixture reached about 60% under the same grinding condition. The studies showed that oils rich in aromatic compounds increase the desulfurization. The studies also proved that grinding by attrition is much more beneficial than mixing alone. Economical large-scale operations are discussed by the authors in a conceptual design. The laboratory studies proved that the sulfurous fuel oil will also become partly desulfurized with iron. In the continuous ball mill (desulfurizer) there is a concurrent flow of coal, fuel oil, iron powder, and iron sulfide. The slurry of desulfurized coal and fuel oil can be distributed by tank cars, tankers, and pipelines. The slurry of fuel oil and coal powder can be used in fluidized-bed combustion for steam and power generation.

Condry[298] has investigated environmental effects of various desulfurization methods. Coal combustion, run-of-the-mine, alkali absorbent, metal oxide, absorption and adsorption, and alkaline earth desulfurization process and catalytic oxidation and reduction methods were evaluated for possibly dangerous environmental agents.

In the paper by Vasan and Willett,[320] a review of desulfurization technology is followed by a detailed description of the Holmes-Streetford Process, developed in the U.K., which has become the accepted cleanup technique for the coal-based gas industry. Operating costs are compared for various desulfurization systems. In conclusion, it is pointed out that the current efforts to reestablish the technology of coal-gasification for U.S. conditions and larger scale exploitation hinges upon proper gas-conditioning and gas-desulfurization to meet current EPA and OSHA standards. Such technology is available and is based on proven plant experience in several coal-based gasification units outside the U.S. The proven reliability and ease of operation of the

Holmes-Streetford Process is attested to by its adoption in nearly 36 plants around the world with 19 of these operating on coal-derived gases.

Ness et al.[247] have discussed flue gas desulfurization using alkaline fly ash from Western coals. The effectiveness of using the inherent alkalinity of fly ashes derived from Western coal for flue gas desulfurization was investigated in four separate pilot-scale studies and two commercial-scale installations; the properties of Western coals, their SO_2 emissions, and the design and performance of selected scrubbers are discussed. Although most utility scrubbers operating in the western U.S. have been installed for the purpose of particulate removal and operate without added reagents such as lime or limestone, it has been noted that appreciable amounts of SO_2 are removed from the flue gas. This SO_2 removal occurs as a result of the inherent alkalinity of the fly ashes produced from Western coals. Ash content of Western coals varies from 4 to 20%. Alkali content of this ash varies from under 10% to over 50%, and exhibits high contents of sodium oxide, magnesium oxide, and calcium oxide. This alkali content is highest in lignite and progressively less prevalent in subbituminous and bituminous coals. In this report a selected analysis of Western coals is provided, as well as discussion of SO_2 removal and scaling rate as a function of Na concentration. It is concluded that future utility scrubbers for Western coals will be designed for the primary purpose of SO_2 removal and not particulate removal and that scrubbers and electrostatic precipitators will be used in series for maximum effectiveness. The three most significant factors to be considered in designing scrubbers for Western coals are: (1) the low concentration of SO_2 to be removed, (2) the inherent alkalinity of Western coal fly ashes, and (3) the tendency to operate at a high state of oxidation, producing sulfate and not sulfite.

Murray[321] has described magnetic desulfurization of Illinois coals. High-extraction magnetic filtration (HEMF) was used successfully to process kaolin. This was the first successful commercial application of a new level of magnetic separation equipment and processing technology which resulted from the joining of four major concepts: discovery of the importance of retention time in mineral separation, development of very high gradient matrix collectors, high intensity fields in wet magnetic separators (up to 20 kilogauss), and modern design of large high-field magnets. His paper reviews works on the subject described in the literature along with information on data attained by several researchers related to the degree of sulfur removal from coal. Economic considerations are included. Inorganic sulfur (65 to 90%) can be removed from the coal by HEMF processing of coal slurry. The coal product from the HEMF process is relatively clean both from sulfur and ash content. However, fine pulverization to liberate the pyrite is necessary before magnetic filtration.

Boateng[322] has described desulfurization of coal by flotation of coal in a single-stage process in which coal was floated out and pyrite was depressed. Up to 90% of the pyritic sulfur content of bituminous coal could be removed at 75% coal recovery. The process was applied to three Canadian and two U.S. coals. Higher coal recoveries were obtained for low-sulfur coals; up to 94.4% coal recovery was possible with 18.2% pyritic sulfur removal. The effects of particle size, temperature, slurry density, and flotation time were studied. From a simplified rate equation, a nonintegral order was obtained for the flotation process.

Friedman et al.[323] have presented results of an experimental study in which desulfurization of coal has been performed by coal treatment with air and steam at elevated temperature. Experimental procedure is described, and measuring results are tabulated and evaluated. It is demonstrated that treatment of coal with compressed air and steam at 150° to 200°C represents a practical method to desulfurize to acceptable levels a sizable percentage of the available coal in the eastern U.S. at a cost in money and fuel value less than coal conversion and to an extent greater than can be achieved by physical depyriting methods. A scheme of an air-steam coal desulfurization plant is included.

Wilson,[324] in his report, describes the principle of Dry Table separation and discusses its application to pyrite removal from coal. Results of experimental rows of the system are presented which confirmed that the Dry Table can reduce the sulfur content in a coal through pyrite removal. The extent of the coal-pyrite separation is a function of pyrite liberation and the physical property differences between the free flowing coal and pyrite particles. However, there are cases where even coal containing unliberated pyrite can be separated into coal products of low and high pyritic sulfur contents. The Dry Table is best employed as a rougher; it has a separation performance similar to that of a Baum Jig. It can be used alone or in conjunction with existing coal cleaning equipment, and it is especially applicable where the use of water is restricted due to limited supply, freezing, or costly treatment prior to discharge or reuse.

Lin and Liu[325] have used the latest data from pilot-scale studies of sulfur and ash removal from both pulverized coals suspended in water slurries and liquefied SRC coal by the HGMS to design conceptual processes for the desulfurization. Estimates of magnetic desulfurization characteristics and conceptual process requirements, as well as installation and processing costs, have been determined. The results indicate that the magnetic desulfurization appears to be attractive when compared to other approaches for the desulfurization, in terms of costs and performance.

Sareen[326] discusses the sulfur removal from coals using an ammonia/oxygen system which removes almost all of the pyritic sulfur and up to 25% of the organic sulfur in about 2 hr. Because organic sulfur removal necessarily implies coal carbon losses, a balance must be otained between the amount of organic sulfur removed and the thermal losses that can be economically tolerated from the coals being cleaned. A process description is followed by a report on experimental investigation in which this process has been tested. It is demonstrated that increasing reaction time and ammonia concentration improved the extent of organic sulfur removal. Changing the NH_3 concentration had no apparent effect on the pyritic sulfur removal. Ammonia concentration has no apparent effect on oxygen uptake by the coal. Both the ammonia concentration and reaction time have an effect on the BTU and carbon value. As much as a 13% Btu and 10% carbon loss may be realized by using a 3 $M\ NH_3$ solution and reacting the coal for 2 hr. The large carbon losses are due to the formation of coal acids.

The paper by Hamersma et al.[327] is concerned with the Meyers Process, which utilizes a regenerable aqueous ferric sulfate leaching unit to chemically convert and remove the pyritic sulfur content of the coal as elemental sulfur and iron sulfate. Although only pyritic sulfur is removed (organic sulfur remains), the Meyers Process has wide applicability for converting U.S. coal reserves to a sulfur level consistent with present and proposed government sulfur oxide emission standards. In this process, aqueous ferric sulfate is used at 90 to 130°C to selectively oxidize the pyritic sulfur content of coal to yield iron sulfate and free elemental sulfur. The iron sulfate dissolves in solution while the free sulfur is removed from the coal matrix either by vaporization or solvent extraction. The leaching (oxidizing) agent is then regenerated as a similar product, iron sulfate, is removed by liming and/or crystallization. Experimental data are presented which were investigated in this study. Because of the relatively high pyritic sulfur and low organic sulfur contents, and high production (70% of current U.S. output) of Appalachian coals, the Meyers Process appears to have major impact in this area.

The paper by Fleming et al.[328] presents an experimental investigation in which effect of process variables on the results of hydrodesulfurization of coal by using Flash Desulfurization Process was studied. Results presented show that pretreatment of the coal enhances the removal of sulfur and produces a solid fuel that can be burned in conformance with the EPA limits of 1.2 lb $SO_2/10^6$ Btu.

Haldipur and Wheelock[329] have described the experimental work which was undertaken to determine the feasibility of desulfurizing a high sulfur bituminous coal from an Iowa mine by treatment at moderately elevated temperatures in a fluidized bed reactor with a small fluidized bed reactor to determine the extent of desulfurization and coal weight loss for different conditions of temperature and gas composition. Also the treatments were applied to both run-of-mine coal and beneficiated coal. In addition, the off-gas composition was measured during some experiments to determine the distribution of various sulfur and other compounds. Finally, consideration was given to the possibility of desulfurizing the off-gas and using it as a clean fuel to burn along with partially desulfurized coal char in the same plant in order to meet air pollution control regulations.

In the paper by Kor,[220] results of an experimental program on desulfurization of coal and coal char at various temperatures and pressures are presented. It was found that the desulfurization of coal char in mixtures of H_2 and CH_4 takes place in two distinct stages. In the first stage rapid desulfurization is accompanied by gasification. These two processes were shown to be interrelated, the relationship being dependent on the char preparation temperature only. The second stage desulfurization was found to proceed at a much slower rate and is being controlled by the slow reduction of pyrrhotite to iron. The authors studied the sulfidation of coal char and synthetic chars in order to obtain a better understanding of the interaction between chars and gas mixtures containing H_2S. It is shown that in the absence of impurities such as iron, the ability of chars or carbons to absorb significant amounts of sulfur in a sulfidizing gas such as an H_2-H_2S mixture depends on the state of crystallinity of the carbonaceous material. The sulfur absorption decreases with increasing crystallite size.

Huang and Pusifer[330] have discussed sulfur removal from coal during heat treatment in gaseous atmospheres of 3 different gases: nitrogen, hydrogen, and a 6% oxygen-94% nitrogen gas mixture. With each gas, both the temperature and holding time at temperature were varied. It is demonstrated that temperature is the most important factor influencing desulfurization in the three gases tested. The holding time at temperature is less important except in the case of hydrogen above 600°C where the sulfur content of the sample decreases with increased holding time. The experimental evidence indicated that this was due to the direct reaction of hydrogen with sulfur in the organic portion of the coal which takes place only at temperatures of 600°C or above. It was observed in the experiments that some inorganic sulfur was transformed into organic sulfur during the gaseous treatment. The fractions of inorganic and organic sulfur released as gases and the fraction of inorganic sulfur transformed into organic sulfur were estimated for each of the gas atmospheres and conditions of temperature and time used.

Stambaugh et al.[314] in their paper have described a coal desulfurization process which is based on heating a water-coal slurry and a leaching agent at moderate temperatures and pressures. Known as the Battelle Hydrothermal Coal Process (BHCP), it has successfully extracted up to 70% of the organic sulfur in both batch laboratory experiments and small-scale prepilot plant continuous operation to produce a clean, solid fuel for electric utilities, industrial boilers, schools, and any facility using coal as the source of energy. The process has both economic and significant technological advantages. Its primary end products are clean solid fuel and elemental sulfur — which can be marketed or easily stored — and potentially recoverable metal values. Sodium and calcium hydroxides are used as leaching agents.

Evans[331] has reviewed usable alternatives to stack-gas scrubbing which are now available for combating problems of air pollution. Coal beneficiation has experienced a number of evolutionary and innovative changes that have been brought into focus only recently. Evans lists a number of the processes proposed for reducing the sulfur content of coal. These processes range in coal-cleaning effectiveness from relatively simple improvements of conventional coal-cleaning techniques to processes that are, in sophistication and efficiency, just short of solvent-refined-coal technology. Particular attention is focused in this discussion on the magnetic separation and CFRI process (developed in India). Other processes reviewed are (1) the multi-steam coal cleaning system, (2) the Ilok process, (3) the SURC process, (4) the media-only process, (5) the Meyers process, and (6) the Ledgemont process.

Warnke and Deurbrouck[126] have described the basic research in the area of organic sulfur removal.

Kor[220] has described in his paper desulfurization tests on Illinois #6 coal and coal char at temperatures of 600 and 800°C and pressures of up to 10 atm performed in H_2, He, CH_4, and mixtures of H_2 and CH_4. Particular emphasis was placed on the effect of temperature, pressure, and methane content of the gas on the rate of S removal and the final S content. Optical micrographs and electron microprobe analysis plates are included for several samples. Electron probe analysis showed that the coal originally contained particles having a composition approaching FeS_2. The X-ray spectrum of the coal char used in the experiments showed that no FeS_2 was present; these particles had been completely reduced to pyrrhotite. The desulfurization of Illinois #6 coal char in mixtures of H_2 and CH_4 takes place in 2 distinct stages. The first stage proceeds rapidly and is dependent on the char preparation temperature only. The second stage proceeds much more slowly and is controlled by the slow reduction of pyrrhotite to Fe.

Guruz and Celebi[333] have investigated the extent of sulfur removal during the pyrolysis of Turkish lignites. They were not able to find a correlation between the extent of desulfurization by pyrolysis and the relative amounts of forms of sulfur in the parent lignite.

Ratcliffe and Pop[332] have described the chemical reduction of SO_2 with North Dakota lignite to be a facile reaction which occurs at a relatively low temperature of 600 to 650°C. Under optimum conditions, the reaction chemistry can be controlled to allow 85 to 90% conversion of SO_2 to free sulfur in a single-stage reaction. Major by-products of the reaction are CO_2, H_2O, and a free-flowing ash. The high sulfur yield from this reaction exceeds the calculated thermodynamic gas phase equilibrium value of 66 to 70%. The higher experimental yield was found to be due in part to a catalyzed re-equilibration of the gaseous products in the exit line. With lignite and low-rank coals, the mechanism of SO_2 reduction appears to involve reaction of hydrocarbons within the pores structure and thus allows complete conversion of the volatile matter with no tar formation. Volatilization and tar formation successfully compete with SO_2 reduction in bituminous coals under the same reaction conditions.

V. TRACE ELEMENT RECOVERY

Trace metals and minerals in coal and its ashes could supply much of the demands of the world if they were to be recovered from these materials. The problems encountered in utilization of these sources are not small, however. For example, during 1975, the electric utility industry in the U.S. burned approximately 410 million tons of coal for the production of electrical energy. The corresponding ash production was approximately 63 million tons (41 million tons was fly ash, and 22 million tons was dry bottom

Table 25
THE ANNUAL PRODUCTION OF
DIFFERENT ASH CONSTITUENTS FOR
2600 MEGAWATT POWER PLANT [334]

Constituent	Symbol	Percent	Tonnage
Silicon dioxide	SiO_2	45.7	502,700
Aluminum oxide	Al_2O_3	26.0	286,000
Ferric oxide	Fe_2O_3	17.1	188,000
Calcium oxide	CaO	3.8	41,800
Sulfur trioxide	SO_3	2.6	28,600
Potassium oxide	K_2O	1.5	16,500
Titanium oxide	TiO_2	1.2	13,200
Magnesium oxide	MgO	1.2	13,200
Sodium oxide	Na_2O	0.6	6,600
Phosphorus pentoxide	P_2O_5	0.3	3,300
Totals		100.0	1,000,000

ash and boiler slag). Ash production in the U.S. has increased to approximately 75 million tons in 1980. Power plant ash is now classified in terms of production as the sixth most abundant solid mineral, according to figures compiled by the U.S. Department of the Interior Minerals Yearbook.

Let us consider in some detail examples described by Morrison[334] in which he considers an ash recycling complex located adjacent to a 2600 mW power plant. The annual ash production of this size plant is about 1,100,000 tons. By applying the constituents of the representative ash produced in the U.S. and their percentages to the 1,100,000 tons, one obtains tonnage of different oxides as shown in Table 25.

Here is the dream, by Morrison,[334] of a power plant ash recycling complex to be built sometime in the future:

The complex will have one major plant whose function is to separate the ash into the various constituents just described. Within this separator plant, we will have many complex systems which will include metal oxide separators which use magnetism, air and mechanical-type separators. There will be crushing, sintering, blending and density control systems. Acid bath mixing and continuous centrifuges will help separate the constituents. Out of this plant will flow the various constituents to other plants which will treat the minerals by conventional systems. One conveyor will bring 502,700 tons of silicon dioxide, followed by 41,800 tons of calcium oxide. The two minerals will be stored in massive silos and will be fed to different sections of the plant where glass products, specialized super-lightweight aggregates, special filters for plastics and paint, non-shrinking casting sand, high-temperature refractories (both block and castable), fiberglass insulation, and many types of building blocks will be produced. The next stream will carry 286,000 tons of aluminum oxide to an aluminum reduction plant where aluminum ingots will be made; 286,000 tons of alumina will produce 143,000 tons of pure aluminum. Perhaps the aluminum could be the most rewarding part of the complex.

Another conveying system will carry 188,100 tons of extracted ferric oxide to another plant. Inside this plant, we have a series of pelletizers. The iron dust is pelletized into one-inch balls and sintered to make iron pellets. These pellets now have a weight of about 167,000 tons of iron, which could be turned into 95,000 tons of steel — enough steel to justify construction of a mini-steel mill. This mini-mill could specialize in production of steel products for use in making mine roof bolts, concrete reinforcing rods or any special steel products needed in the local area.

Thus far the complex has utilized 92.6 percent of the 1.1 million tons of ash. The remaining 7.4 percent consists of oxides of sulfur, potassium, titanium, magnesium, sodium and phosphorous. These remaining oxides could possibly be used by the local chemical industry.

On more realistic grounds, let us describe the progress made up to now. A good review article on this subject is written by Torrey.[150]

Sulfuric acid is the major chemical produced in the U.S. As early as 1944, Mitchell[335] attempted to recover pyrite from coal waste to be used for sulfuric acid production. The method used was mechanical separation with jig tables. Although borderline quality raw materials could be obtained from coal waste by this method, the economic picture at the time was not optimistic, particularly in view of the capital outlay required for concentrating the pyrite.

Stimulated by a shortage of sulfur and sulfuric acid, and an increased interest in environmental contaminants in the late 1960s, more workers turned their attention to pyrite recovery from coal waste. Sun and Savage[336] developed a comparatively successful flotation process, in which separation took place in stages. The process involved dual flotation, using collectors and frothers, first to separate the coal from mineral matter, then pyrite from the remaining clay minerals in the waste.

The chemistry of the formation of acids in coal-processing wastes is as follows: the initial step in acid production is the air oxidation of pyritic substances (pyrite, marcasite) in the wastes to form $FeSO_4$ and SO_2:

$$FeS_2 + 3O_2 \rightarrow FeSO_4 + SO_2 \qquad (13)$$

Further oxidation in the presence of water results in the formation of sulfuric acid:

$$2SO_2 + O_2 + 2H_2O \rightarrow 2H_2SO_4 \qquad (14)$$

Under some circumstances, the acid concentration in the drainage from coal mineral wastes can run as high as 5 wt%. The most facile formation of sulfuric acid occurs where the pyrites are very finely divided and where waste pile structure and climatic conditions are conducive to a good flow of air and water through the waste materials.

Overall, the stoichiometry of the formation of sulfuric acid from pyritic wastes can be written as follows:

$$2FeS_2 + 7O_2 + 2H_2O \rightarrow 2FeSO_4 + 2H_2SO_4 \qquad (15)$$

This situation is much more complex than it appears; the above equation is very simplified and does not reflect the mechanism of the reaction.

A magnetic pyrite separation process, aimed generally at pollution control via sulfur reduction of coals for electric utility use, was developed by Ergun and Bean.[337] This process was also thought useful for the recovery of iron or sulfur. An attempt was made to increase the magnetic susceptibility of pyrite above its normal value by using heat to rapidly convert a part of the iron into a ferromagnetic state. The most promising result was obtained using dielectric heating of the coal.

Iron recovery from coal is also described by Jensen.[338] The carbon in coal minerals can be gasified with steam in a fluid-bed reactor to produce carbon monoxide and hydrogen for fuel or synthetic gas. Elemental iron is produced within the coal mineral particles during the gasification process. Hydrogen made available by the steam-carbon reaction probably reduces the iron. The process is intermediate between diffusion controlled and rate controlled. Free iron can be produced from the mineral residues of coal mineral gasification by proper addition of a flux such as calcium oxide. With care, nearly quantitative recovery of the iron can be obtained. Many processes have been used for alumina extraction from coal wastes. Kelley et al.[339] present a detailed study of the thermodynamics of the conversion of various aluminum-containing compounds to alumina. These included a high alumina clay (20% available alumina as kaolinite) and alunite. Also coal waste has been burned in the presence of alkaline-earth compounds to solubilize the silicates in the ash, from which alumina could be extracted.

Less has been done on the technology of the recovery of specific trace elements from coal. In addition to uranium, some work was done on recovery of copper, zinc, lead, silica, phosporus, manganese, molybdenum, titanium, and germanium.

A. Aluminum Recovery

Possible methods of extracting alumina from coal-associated refuse developed until 1976 as listed in bibliography by Condry.[298] A list of 24 publications is presented with their contents abstracted.

Alumina extraction from clay was the objective of the concerted research effort by the U.S. Department of Commerce during World War II, when it appeared that foreign supplies of bauxite might be endangered. These efforts culminated in the construction and operation of a pilot plant, but when the possibility of disruption of foreign sources ceased, plans for a semi-commercial plant were terminated.[137]

More recently, a privately-owned coal company constructed and operated a large pilot plant in which coal-mine waste was used as the alumina source. Here again the economics of the extraction process did not prove out, and the pilot plant was taken out of operation in early 1963; further field testing was suspended.[614]

At present, the Anaconda Company is operating a small pilot plant in the western U.S., which extracts 5 tons of alumina a day from clays containing 26 to 27% aluminum oxide. Though Anaconda states that further work and tests are needed, they feel that their particular adaptation of the acid extraction process using a caustic or sinter purification step will be competitive with alumina produced from imported bauxite using the conventional Bayer process.

The situation that precipitated the "crash program" for alumina extraction during World War II, i.e., the possibility of a disrupted bauxite supply, is to a certain extent still present today. In 1960, 81.8% of the U.S. consumption of bauxite was imported. Therefore, the need for an economic process for alumina extraction from high alumina clays and shales is still a matter of justifiable concern.

The successful development of an economic alumina extraction process from clay could have far-reaching effects upon a significant portion of the coal industry. One of the major constituents of materials rejected in coal beneficiation processes is alumina, frequently averaging better than 30 to 35% Al_2O_3. Coal refuse piles which have been standing for many years could be recovered for their alumina contents and subsequently erase the problem of refuse disposal and unsightly gob piles.

Four proposed schemes for recovering aluminum values are:[186]

1. Extraction of alumina by an acid process
2. Lime sinter or lime-soda sinter processing followed by leaching to remove sodium aluminate from which alumina can be separated by crystallization
3. Hydrochemical treatment in a modified Bayer process with added lime to separate insoluble calcium silicate and a sodium aluminate solution from which alumina can be crystallized
4. Selective chlorination in several steps to remove aluminum chloride

Any of the above processes could then lead to recovery of aluminum metal.

1. Production of Alumina from Clay by Hydrochloric Acid Leach

According to Bucklen et al.[137] most of the hydrochloric acid processes patented or proposed have the following features in common: dehydration, leaching, filtration and concentration.

Dehydration — The crushed clay is calcined at a temperature of 650°C for 1½ hr.

This conditioning step destroys any organic matter, increases the solubility of alumina, and decreases the tendency of silica to go into solution in later stages of treatment. Calcining at about 650°C is optimum since clays calcined at lower temperatures, i.e., below 540°C, yield reduced alumina recovery, while higher calcining temperatures, i.e., 815°C or higher cause alumina and silica to fuse to form a compound which is insoluble in the 20% hydrochloric acid solution used for leaching.

Leaching — The calcined clay, ground to approximately 65 mesh, is leached with 20% hydrochloric acid for 1 hr at a temperature of at least 70°C. At 20% concentration, the hydrochloric acid solution is not greatly changed, even when boiled.

Filtration — The slurry is filtered to separate pregnant liquor rich in aluminum, iron, and trace chlorides from the solid residue containing silica, titanium, etc. The filter washings are returned to an HCl absorption tower for recycle.

Concentration — The clarified pregnant liquor is evaporated to produce a mixed crystal slurry and an aqueous condensate. The evaporated hydrochloric acid is recovered.

The mixed crystal slurry resulting from evaporation can be treated by one of the processes for iron removal to produce high grade alumina (isopropyl ether extraction, gas precipitation, gas precipitation-isopropyl ether extraction, caustic soda digestion process).

2. Production of Alumina from Clay by Sulfuric Acid Leach

Many processes have been proposed for extracting alumina from clays or shales by sulfuric acid leach. Most of these processes have some features in common such as clay preparation, roasting, acid leach, filtration, dehydration, decomposition, and acid or solvent recovery. A number of minor variations in operating conditions involving such process variables as temperature, time, and acid strength are to be expected. The major significant difference between processes is the method for iron removal from the aluminum sulfate solution; some processes using interrupted vacuum crystallization, others using electrolytic, chemical, and ethanol purification methods.

3. Production of Alumina from Clays by the Sulfurous Acid Process

The methods used in the sulfurous acid process are similar to those used in other acid processes for alumina production. The major difference is that hydrothermal treatment (autoclaving) is used instead of calcination as a means to decompose the intermediate sulfite compounds. The process consists of the following principal steps: clay preparation, leaching, filtration, autoclaving, thickening and autoclaving, thickening and filtration, caustic leaching, filtration, precipitation, thickening, washing and filtration, decomposition.

4. Lime-sinter Process

In order to release alumina from a refractory aluminosilicate compound, it is necessary to provide a reagent which has a stronger affinity for silica than does alumina. The chosen reagent is calcium oxide provided by limestone in the lime-sinter or Pederson process used on highly siliceous bauxites. Investigators have found that to release the alumina from an aluminosilicate compound, the ratio of calcium oxide to silica should be 2.0 for the silica, and also there must be sufficient calcium oxide to give a CaO to Al_2O_3 ratio of 5:3.

Extensive work has been done by Grim et al.[340] with the lime-sinter and lime-soda sinter processes. They studied the effects of CaO to Al_2O_3 ratios and of sintering temperatures on the recovery of alumina from various types of clays. Extraction of alumina from the sintered material was done with sodium carbonate solutions. In general, the highest extractions were obtained with a CaO to Al_2O_3 ratio of 5:3 in the sintered material, considering only that part of CaO beyond that required to react with silica to give dicalcium silicate, $2CaO \cdot SiO_2$. Generally the alumina yields were 80 to 95% when optimum sintering temperatures were used. For most of the clays, the optimum sintering temperature range was 1370 to 1390°C when the optimum calcium oxide to alumina ratio was used. On the basis of this and other work, similar treatment appeared promising for extraction of alumina from fly ash.

Cavin[341] performed some exploratory, sintering experiments with fly ash at 1320°C (maximum temperature on the available furnace). With CaO to Al_2O_3 ratios of 1:2, 1:4 and 1:6, the extraction of alumina was less than 0.3%. At a ratio of 1:8, the extraction was 2.1%. Undoubtedly the low sintering temperature was responsible for the low yields.

Chou et al.[186] proposed a process for recovery of alumina from fly ash by the lime sinter method which is outlined here:

1. Grinding of fly ash to separate magnetic adhering to non-magnetics
2. Magnetic separations send magnetic fraction to be used as from ore
3. Milling of fly ash with limestone
4. Sintering or limestone-fly ash mixture
5. Cooling and grinding of sintered material
6. Leaching of sintered material with sodium carbonate solution
7. Filtration of sodium aluminate extract from dicalcium silicate sludge and other impurities
8. Controlled precipitation of alumina trihydrate from the extract
9. Filtration of the alumina trihydrate from the extract
10. Calcination of the alumina trihydrate to the anhydrous form
11. Recycle of the sodium carbonate from "9" back to the leaching stage "6"

Using the modification of lime-soda-sinter process, Funnell and Curtice[342] found that they could extract 90% of the available alumina from the ash. Some of the characteristics of the process, as applied to the used ash, may be of interest to those who may be considering this approach. In order to achieve a friable product in the first roast, it was found that the ratio of lime, soda ash, and water had to be closely controlled. Roast time and temperature also show a close relationship to the lime, soda ash, and water ratio. The particle size of the aluminum trihydrate showed a dependence upon the rate of flow of the carbon dioxide and the degree or amount of agitation in the precipitation step. Silica carry-over was minimal, and the iron gave no problems. Over all, this approach appears to offer the best method of extracting alumina from the fly ash produced at the Four Corners Generating Station near Farmington, New Mexico. Lignite which has an ash content of 22% is used as fuel at this station. An average chemical analysis of this ash is shown in Table 26.

Seeley et al.[117] have also examined the recovery of aluminum and other metals from fly ash for two new methods combining the sintering with diluted acid leaching. In the first method, the fly ash is sintered with a $NaCl$-Na_2CO_3 mixture at 1000°C, followed by successive leachings with water and $1\ N\ HNO_3$ or H_2SO_4. In the second method, the fly ash is sintered with a $CaSO_4$-$CaCO_3$ mixture at 1400°C, after which the sintercake is pugged with $36\ N\ H_2SO_4$ and then diluted to $4\ N$ for a second leaching period. Al recovery is increased in each method as the sintering temperature is increased; recovery is >95% at the optimum sintering conditions, and a pure Al_2O_3 product can be obtained if a solvent extension or an ion exchange circuit is included for Fe and trace-metal removal. The product option flexibility is provided in that Fe, Ti, Mn, U, and Th can also be recovered.

Table 26
ANALYSIS OF FLY ASH FROM FOUR CORNERS
GENERATING STATION, FARMINGTON, NEW
MEXICO[342]

Component	% by weight
SiO_2	54.9
Al_2O_3	29.3
Fe_2O_3	3.6
CaO	3.7
MgO	0.9
TiO_2	1.6
Na_2O	1.4
K_2O	0.96
P_2O_5	0.2
SO_3	0.2
H_2O	0.18
C	0.81
SrO	0.10
B_2O_3	0.07
ZrO_2	0.03
MnO	0.10
Be_2O_3	0.009
V_2O_5	0.003
BaO	0.02
PbO	0.007
CuO	0.002
ZnO	0.08
Cr_2O_3	0.006
NiO	0.003
SnO_2	0.001

An advanced chemical process to recover aluminum and other valuable metals from the wastes of coal combustion has been developed by researchers at Oak Ridge National Laboratory (ORNL), Tennessee, as reported in January, 1980 issue of *Industrial Research and Development*.[617] The Calsinter process could extract enough aluminum from fly ash and other coal wastes to satisfy much of the current U.S. demand for the valuable metal and significantly decrease the need for imports of bauxite, the chief aluminum ore. In addition, the process could recover useful quantities of other metals including iron and titanium, while eliminating hazardous pollutants and reducing the volume of the waste for disposal.

In the process, fly ash is heated or sintered to a temperature of 1200°C with mixture of limestone and gypsum. A sulfuric acid leach then is used to remove about 98% of the aluminum and more than 90% of other metals. The process provides an advantage for plant operators who have installed flue-gas desulfurization (FGD) equipment on their stacks to meet Federal clean-air standards. The equipment produces large volumes of waste, known as FGD sludge. However, that can be used in place of gypsum and limestone in the Calsinter process and produces equally good metals recovery.

According to Kealy et al.,[343] alumina from coal wastes could be produced on a commercial scale cheaper than the delivered cost of alumina produced from high-grade bauxite from outside the U.S. It was also stated that current technology makes this recovery "matter of fact".

B. Germanium Recovery

Interest in commercial recovery of germanium (and gallium) from coal and coal products is best illustrated by the great number of papers and patents.

In the U.K. some production of germanium from flue dust was started in 1950 by Johnson, Matthey and Company. Flue dusts from boilers and gas producers in England were estimated to have a potential yield of 2000 tons of germanium and 1000 tons of gallium annually if these metals could be recovered. Work on the extraction of germanium and gallium was described by Powell and others,[344] Lever,[345] Crawley,[346] Reynolds,[347] and Fletcher.[348] Production of germanium from coal flue dusts in the U.K. was estimated by Beard[349] in 1955 to be comparatively light, 0.5 to 1 ton annually.

Additional literature on germanium extraction includes: Williams,[350] Chirnside and Cluley,[351] Steward,[123] Losev et al.,[352] Losev and Nikiforova,[353] Bylyna et al.,[354] Pogrebitskii,[355] and Bunina.[356] Russian authors have discussed the recovery of germanium from coal by halogenation and ultrasonics, and laboratory tests showed that germanium can be extracted from Moscow brown coal by irradiated carbon tetrachloride. The coal field of Ubagansk, east of the Urals, was reported to contain germanium in amounts that may make its recovery worthwhile.

Contrary to East European countries, in the U.S., only the U.S. Bureau of Mines appears to be pursuing an active program related to the concentration and extraction of germanium from coal and coal products.[183] The Pagle-Picher Company, an early leader in this field, has suspended its work in this area after arriving at the conclusion that such a process is not economically feasible at this time.

When viewed as a source of germanium alone, the processing of coal ash does not appear promising due to the low concentration and declining market for germanium. However, steps can be taken to modify the situation in such a manner as to make such processing more attractive. For example, due to its unique occurrence in the coal bed (major concentrations in the top and bottom few inches), selective mining techniques could be employed to recover these portions of the seam separately, resulting in a significantly upgraded ash product after combustion. Also, the possibility of a multi-product recovery process presents itself since germanium is normally associated with gallium in coal. Examination of increments of coal in West Virginia seams also shows frequent high concentrations of beryllium, cobalt, chromium, and nickel associated with blocks containing high germanium and gallium content. If new processes were developed or present processes modified to recover a multiplicity of products from coal ash, it might alter the economics of germanium extraction considerably.[137]

The extraction of germanium from coal would consist mainly of the following steps as indicated on flowsheet shown in Figure 7.[357-359]

1. Coal is turned to produce the highest possible GeO_2 enrichment in the coal ash. The temperature is raised slowly to 900°C, allowing the carbonaceous material to burn at much lower rates so that the loss due to the formation of GeO which is more volatile than GeO_2 will be limited.
2. The enriched ash is classified using cyclones to thus eliminate coarse particles which are lean in germanium. Since the concentration of germanium in the ash is by sorption, the greatest enrichment occurs in the high-surface-area-containing finer sizes. Thus, significant germanium concentration occurs in ash finer than 60 μm.
3. Sulfuric, nitric, and hydrochloric acids as well as soda lye at various concentrations have been used for the extraction of germanium from coal ash. However, dilute sulfuric acid of 0.05 N was found to be superior to other acids and alkalies in extraction from the standpoint of recovery. The more favorable results obtained using sulfuric acid may be due to the fact the GeO_2 in the ash forms more hydrolyzable (water-soluble) salts with sulfuric acid than with hydrochloric acid, nitric acid, or soda lye.

EXTRACTION OF GERMANIUM FROM COAL

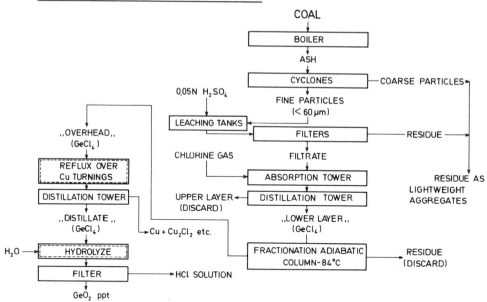

FIGURE 7. Schematic presentation of steps involved in the extraction of germanium from coal.[137]

4. Chlorine is admitted to the filtered germanium-bearing solution, which is distilled. The solution is then separated into two layers, the upper layer of which is discarded.
5. The lower layer containing crude $GeCl_4$ is fractioned in an adiabatic column at 84°C to separate the pure germanium tetrachloride which boils at this temperature. The residue which remains from this operation is discarded.
6. The overhead containing impure $GeCl_4$ is refluxed (impurities removed) over Cu turnings and again distilled.
7. The pure $GeCl_4$ in the distillate is hydrolyzed with water and filtered to give pure GeO_2.

In the operation of a process designed to recover germanium from coal, it is of critical importance to burn coal in such a manner as to retain the maximum amount of germanium in the coal ash.

Russian researchers investigated the effect of coal rank, heating rate, ash contents, sulfur contents, and ashing temperature on the loss of germanium during ashing.[360] The authors concluded that ashing temperature was insignificant; increased sulfur did not increase germanium loss (loss was insignificant in high sulfur sink fractions), germanium loss increased with heating rate increase and germanium loss varied from coal rank to coal rank.

C. Titanium Recovery

Steps included in titanium recovery process are shown in Figure 8 and they include:[137]

Magnetic separation — Fly ash is passed through a magnetic separator of suitable design, where magnetic iron oxide-bearing constituents are removed. This enriched iron-bearing constituent may find use with subsequent processing as a heavy media for mineral preparation plants or may be blended for blast furnace feed.

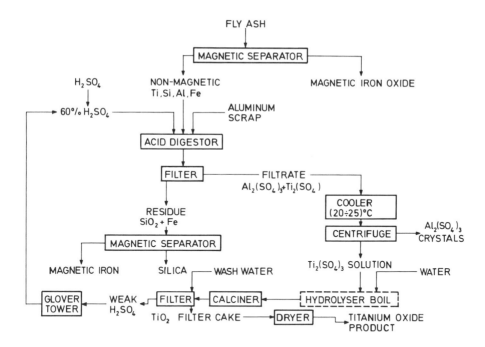

FIGURE 8. Titanium extraction from fly ash.[137]

Leaching — The nonmagnetic portion largely containing Ti, Al, Si, and Fe is then leached with 60% sulfuric acid using aluminum scrap as a reducing agent. The sulfuric acid used in this process conceivably can be recovered from scrubbed flue gases of the power plant furnishing the fly ash being treated. The foregoing reactions will proceed according to the general equations:

$$2TiO_2 + 3H_2SO_4 \rightarrow Ti_2O(SO_4)_3 + 3H_2O \qquad (16)$$

$$Al_2O_3 + 3H_2SO_4 \rightarrow Al_2(SO_4)_3 + 3H_2O \qquad (17)$$

The aluminum scrap serves a dual purpose: (a) it reduces titanium oxytrisulfate to titanous sulfate, and (b) it also reduces any iron oxide to a metallic state.

Filtration — The resulting solution is filtered through acid-resistant cloth to separate it from the silica residue, the already reduced metallic iron, and other impurities. The metallic iron in the residue can be substantially removed by magnetic separation. The iron recovered may be blended with the initially separated iron-bearing ash and pelletized as blast furnace feed or it may be disposed of as a point pigment. The silica can be digested with caustic soda or heated to fusion with soda ash for making sodium silicate for brick manufacture or it can be utilized for cement manufacture.

Cooling — The mother liquor is cooled to 20 to 25°C, at which temperature aluminum sulfate crystallizes out with water of crystallization leaving a concentrated solution of sulfuric acid containing titanous sulfate.

Centrifuging — The cooled liquid is centrifuged to separate crystallized $Al_2(SO_4)_3$ from $Ti_2(SO_4)_3$. The $Al_2(SO_4)_3$ can be purified and converted to alumina or may be marketed as aluminum sulfate after further treatment.

Hydrolysis — The recovered $Ti_2(SO_4)_3$ solution is diluted with water, boiled, and hydrolyzed to titanium hydroxide $Ti(OH)_3$.

Calcination — The Ti(OH)$_3$ is then calcined to produce TiO$_2$ which is allowed to settle in a vessel.

Filtering and Drying — The settled TiO$_2$ is then filtered and washed with water. The washed TiO$_2$ is dried as a final product. The weak sulfuric acid is concentrated and recycled to the process.

D. Uranium Recovery

Early investigations into the extraction of U from some of the first known deposits of lignites were conducted at the Battelle Memorial Institute[363] and at Oak Ridge National Laboratory. As prospecting and discovery work progressed, samples from larger deposits of much higher uranium grade became available. A testing program was initiated at the Raw Materials Development Laboratory, Winchester, Mass. in November 1954. Nearly 30 individual lignite samples were subjected to exhaustive laboratory tests including various methods of roasting, leaching, and uranium recovery in order to develop the most economical process for treating the ore. The results of this laboratory work were used in selecting a flowsheet for pilot plant studies of uraniferous lignites.

Relatively high-grade uranium-bearing lignite ores were discovered in southwestern North Dakota and adjacent areas of South Dakota in 1954.[361] Known reserves are estimated to contain about 3 million pounds of uranium oxide, or about 1% of the known reserves of uranium in the western U.S. Average uranium oxide content of the uraniferous lignites is approximately 6 lb/ton, compared with an average of 4 to 5 lb/ton for uranium ores that are being processed on a large scale in the 16 operating uranium mills in the U.S.[362]

At the request of the Raw Materials Division of U.S. Atomic Energy Commission (AEC), investigations on the production of a uranium concentrate from uraniferous lignites were conducted at the Grand Junction Pilot Plant during the period summary through August, 1957. These studies included process development work and amenability testing for 4 lignites: approximately 1750 tons of wet lignites were processed in the course of the test work.

The purposes of the pilot plant investigations were: (1) to project the proposed flowsheet of fluo solids roasting — two-stage acid leaching — CCD and/or filtration solvent extraction into an operable process, (2) to test the amenability of the various lignites mentioned above to that process, and (3) to obtain design data for possible commercial application. Existing pilot plant equipment was used wherever possible in completing the flowsheet; time and quantity of sample did not prevent complete investigations of all process variables. A portion of the program was devoted to applying standard mill equipment to the treatment of lignite in a manner to minimize operational difficulties. The results of this test work will serve as a guide in evaluating the economics of processing uraniferous lignites for their uranium value and in selecting a flowsheet and designing equipment for a process mill.

This work in 1954 to 1957 was largely summarized in two internal AEC reports, designated "WIN-54, Laboratory Investigation of Dakota Lignites", and "WIN-81, Pilot Plant Testing of Dakota Lignites".[364] The data presented cover laboratory metallurgical tests on more than 100 samples of uranium-bearing lignite, and the processing of 1750 tons of lignite in a pilot plant. Most of the work was done at the former AEC Raw Materials Development Laboratory located at Winchester, Mass., and at the former pilot plant at Grand Junction, Colo. Other studies were made at the Bureau of Mines Denver Coal Research Laboratory and its Salt Lake City Metallurgy Research Center.[362]

Breger and co-workers[365-369] have described geochemistry and mineralogy of uraniferous lignite as well as the extraction of uranium. In a paper by Breger et al.,[365,366] a series of analytical and chemical tests were described on a South Dakota lignite sample to determine the geochemistry and mineral association of U in lignite. Mineralogical and semiquantitative analyses were reported for the mineral gypsum, jarosite, quartz, kaolinite, calcite, and 28 trace elements. The lignite sample was sized at: -25×50 mesh, -50×100 mesh, -100×200 mesh, -200×325 mesh, and -325 mesh. The $+325$ mesh fraction of lignite was subjected to float-sink (specific gravity 1.7) to separate the mineral portion of the lignite. This sink fraction was used for mineral identification by polarizing microscopy, X-ray diffraction, and electron microscopy. Radioactivity measurements indicated that only 7% of the U in lignite was concentrated in the mineral portion of the sample. Spectrographic analysis of the sample indicated the presence and order of magnitude for concentrations of 28 elements. Samples of the lignite were also submitted for U extraction and leaching tests. It was found that as much as 98.6% of the U in the lignite sample could be extracted using hot, 6N HCl under conditions of continuous extraction for a period of several hours. Results of leaching tests utilizing H_2O, $1N$ HCl, $6N$ HCl, and $La(NO_3)_3$ indicated that the U held in lignite by ion exchange amounted to approximately 1.2% of the total U in the lignite, with the remaining 98.8% tied up in the organic portion of the coal.

Extraction of uranium from the Red Desert coal of Wyoming is described in the paper by Breger et al.[368] Samples of the Red Desert Seam in Wyoming were treated with $6N$ RCl to extract U from the coal. Semiquantitative analyses were tabulated for a -20 mesh size of the original coal, for 2 sequential acid extraction liquors, and for the coal remaining after extraction. The elements tested were: Al, Si, Ca, Mg, Fe, Na, B, Mn, Pb, Sr, Ti, Ni, Ce, Co, Ba, Nd, Cu, Mo, Cr, La, Ga, Zn, V, Y, Yb, Sc, Be, and Ag. For all four phases (raw coal, two extracts, and the coal after extraction), Al occurred in the highest concentration in the ashes. Almost 90% of the U contained in the Red Desert coal was removed by the acid extraction.

Effects of destructive distillation on the uranium associated with selected naturally occurring carbonaceous substances is described in another paper.[367] Tests were to determine if volatile U compounds could be formed during the course of the Fischer retort assay procedure. The materials tested: Chattanooga (carbonaceous) shale, from Tennessee; subbituminous coal, from Wyoming; coal from the Tennessee shale sample; and Swedish kolm were subjected to destructive distillation. The results were evaluated on the basis of percentage of U in the original sample, in the ash of the sample, in the distillation char, and in the ash of the char. The content of U in the test specimen was determined by the extraction-fluorimetric procedure. From the test results it was concluded that destructive distillation of carbonaceous substances results in no appreciable loss of U.

Details of the investigations performed for the recovery of uranium from lignites are described by Porter and Petrov.[370] The ash content of lignite has varied widely, the important samples containing 30 to 60% ash on a dry lignite basis. Because of the significant upgrading of uranium values obtained by ashing the lignite, and because recovery of uranium from raw lignite presented formidable processing problems, major emphasis has been placed on devising methods to recover uranium from lignite calcines. Proper roasting removes the problem of organic fouling of solutions, reduces the size of the leaching plant for an equivalent uranium throughout, and generally provides equal or better acid leaching efficiency with less acid than can be obtained from the raw lignite.

Restrictions have been necessary in roasting lignite in order to produce a calcine amenable to uranium recovery by acid leaching. It was noted that roasting temperatures higher than 600°C inhibited the solubility of uranium in acid solution. Roasting in a laboratory electric muffle furnace at a nominal temperature of 450°C until all

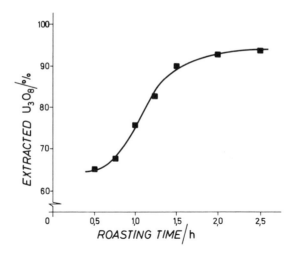

FIGURE 9. The effect of roasting time on the amount of uranium extracted, as described in the Report WIN-54 by Woody et al.[371]

carbon was burned provided calcine amenable to acid leaching procedures. The effect of roasting time on the amount of uranium extracted is shown in Figure 9.[371] When equipment of pilot plant scale was used (3-ft diameter multiple hearth or FluoSolids® roasters), there was a heat dissipation problem in controlling the combustion temperature of the roasting operation. Restriction of air may be only partially successful as a means of controlling the temperature since reduction of uranium, which is detrimental to acid leaching efficiency, may occur when the oxygen is too severely restricted. Moreover, it was found that almost complete removal of carbon is required for optimum leaching efficiency, and there must be enough oxygen to allow complete carbon combustion. Laboratory investigations have indicated that the carbon content of incompletely roasted lignite calcines can be controlled by flotation. Residual carbon can be floated out of the ash, either before or after acid leaching, and can be recycled to the roaster if practical operation of the roaster does not produce a carbon-free calcine. The grate roaster used for burning lignite ores is shown in Figure 10.[371]

Recovery of uranium from lignite ash by acid leaching required large quantities of sulfuric acid. For efficient recovery, the ash was contacted with enough sulfuric acid in an aqueous slurry of water and ore to maintain a leaching pH of about 0.5. Adequate mixing of pulp generally required a pulp density of about 40% solids, as severe thickening of the mixture occurred when the acid contacted the ash. High leaching temperatures (70° to 90°C) were beneficial to efficient uranium dissolution, but a good deal of the heat requirement was supplied by the heat of reaction of the acid and the ash. The sulfuric acid requirement in a conventional leaching circuit would be 500 to 1000 lb/ton of ash (150 to 600 lb/ton of raw dry lignite). However, as optimum leaching extraction required considerable free acid at the end of leaching, two-stage countercurrent leaching procedures were used with effect. Up to about half of the co-current leaching acid requirement was saved by this procedure, which utilizes the excess acid to neutralize the acid-consuming minerals in the ash. Conventional leaching of lignite ash, with its high acid requirement, also resulted in a solution containing relatively high concentrations of undesirable metal ions and salts. The high molybdenum, ferric iron, and sulfate concentrations precluded direct ion exchange treatment of acid liquors. When countercurrent leaching was carried out at high temperature (over 90°C),

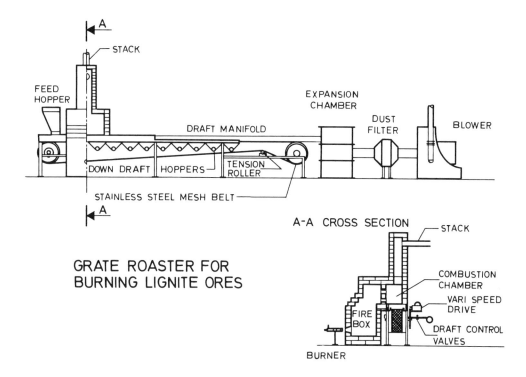

FIGURE 10. Roaster for burning lignite ores, after Woody et al.[371]

these concentrations were reduced as much as 90% by precipitation, and the resulting solution was rendered amenable to ion exchange after contact with activated charcoal to remove residual traces of molybdenum. All acid liquors produced from ashed lignite were amenable to solvent extraction procedures (see Figure 11).

Because of the high aqueous salt concentrations obtained by acid leaching of lignite ash, thickening and filtration were impractical unless flocculation aids were used. Flocculants produced satisfactory thickening rates when added as a dilute solution to the leach pulp. The amount of flocculant required was several times that for uranium ores treated on the Colorado plateau, owing to the viscous nature of lignite ash acid pulps. In most cases, filtration was practical only after a thickening stage. However, when the ash was properly handled, thickener and filter area requirements were lower.[370]

Thickener underflows and filter cakes produced from solid-liquid separations of lignite acid pulps retained a large amount of solution. The lignite ash, after acid leaching, could only rarely be thickened to 50% solids, and in some cases terminal thickening densities of 30 to 35% solids were encountered. Filter cakes generally retained about 50% moisture. This tendency of the ash to retain moisture resulted in a high soluble loss of uranium from conventional washing circuits. A combination of thickening and at least two stages of filtration appears necessary to avoid excessive loss of dissolved uranium in acid-leached residues.

Recovery of uranium by solvent extraction from sulfuric acid leach liquor by amines has been intensively studied.[372-375] There are commercially available, or potentially available, three long-chain amines that are excellent uranium extractants. These are triisooctylamine (TIOA), a tertiary amine manufactured by Carbide and Carbon Chemical Co.; Amine S-24, a secondary amine also available from Carbide and Carbon Chemical Co.; and Amine 9D-178, a secondary amine manufactured by Rohm and Haas.

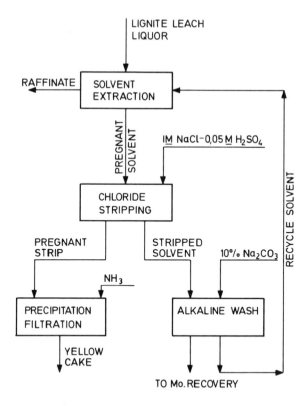

FIGURE 11. Flow sheet for the solvent extraction circuit.[371]

The principal advantage of the amines as uranium extractants is their selectivity for uranium. Of the ions commonly found in the lignite leach liquors, only molybdenum seriously interferes with uranium extraction. However, a very clean separation of uranium and molybdenum is easily achieved in the reextraction or stripping operation.

The solvent extraction is performed in four mixer-settler stages, which generally have been found adequate to give efficient uranium recovery. With a leach liquor of the composition shown in the foregoing table, an organic-to-aqueous flow ratio of 1:4 is enough to give nearly quantitative uranium recovery and 90 to 95% molybdenum extraction. This results in a pregnant solvent assaying 3.4 g U_3O_8 per liter and 1.2 g Mo per liter.

The uranium can be reextracted from the pregnant organic using $1 M$ NaCl to 0.05 M H_2SO_4 as the stripping solution. At an organic-to-stripping solution ratio of 6:1, 3 stages of conventional mixer-settlers are sufficient for 99.9% uranium stripping.

The uranium is easily recovered from the pregnant stripping solution, which will contain about 22 g U_3O_8 per liter, by precipitation with either ammonia or magnesium oxide. In either case, a final product assaying about 80% U_3O_8 and 0.2% Mo is obtained after drying.

Investigations by the U.S. Geological Survey indicate that lignites contain the highest concentrations of uranium followed by subbituminous B and C type coals. The higher bituminous and anthracite coals of the eastern and central U.S. contain negligible amounts of uranium.

While individual uranium minerals such as autunite, carnotite, etc. have been identified in some coals, the coals with the higher concentrations normally do not show the individual uranium minerals. Investigations indicated that the uranium occurred as an organo-uranium complex soluble at a pH less than 2.18.

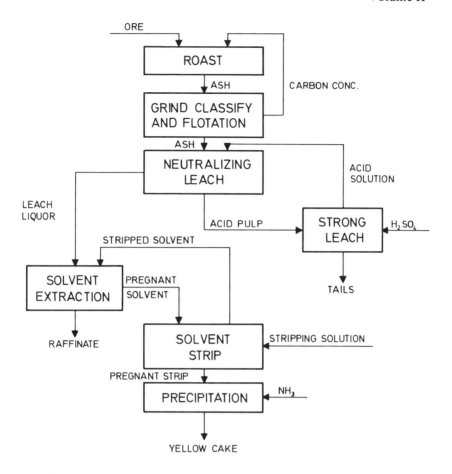

FIGURE 12. Lignite processing flowsheet of the Grand Junction pilot plant.[137]

When considering raw coal, the uranium concentrations generally are low, but by burning the coal and treating the ash, the concentrations can be upgraded considerably. Burning the coal also led to increased uranium recovery when using a leaching process. Therefore, it would appear that recovery of the uranium after the coal has been consumed as a fuel is most desirable. The major detriment to such a process is that the coals containing the greater uranium concentrations also contain higher ash concentrations, thus making them less desirable as a fuel. This consideration led to the idea of coal-uranium breeder,[376] which will be discussed later.

A common characteristic of lignite is the alkalinity of the coal and this alkalinity may have considerable bearing on the selection of a method for extracting uranium. This alkalinity is believed to derive from calcium compounds present in the lignite. An alkaline leach would be desirable for extracting uranium from highly alkaline lignite ash; however, most of the alkaline leaches using sodium carbonate have yielded reduced uranium recovery.

In the acid processes, leaching with sulfuric acid is preferable because it is equally as effective for extracting uranium as the more costly hydrochloric and nitric acid extraction media.

In the application of the sulfuric acid leach-ion exchange process to uraniferous lignite, ten major steps would normally be involved and they are as follows (see Figure 12, from report by Bucklen et al.,[137]):

1. The lignite is burned in a boiler at approximately 480 to 590°C.
2. The ash is concentrated by froth flotation to remove undesirable gypsum and limonite, which are major acid-consuming components.
3. The concentrate is filtered and the filtrate is recycled to the flotation cell.
4. The residue is then leached with 20% H_2SO_4 with a 20% pulp density at 82 to 88°C for approximately 2 hr.
5. The leached pulp is subsequently thickened to reduce the volume to be processed in a uranium extracting ion-exchange column.
6. The thickened leached pulp is filtered and a portion of the pregnant, or uranium bearing, filtrate is recycled to the leach tank while a second portion is concentrated in the uranium extracting ion-exchange column.
7. The filtered solids from step 6 are repulped, washed, and again filtered resulting in a product which can be used in the manufacture of building stone.
8. The pregnant filtrate from step 6 is passed through the ion-exchange column until the column becomes saturated with uranium concentrate which is evidenced by an increase in the quantity of uranium in the overflow or barren effluent. When saturation occurs, the filtrate flow is redirected from the saturated to an unsaturated standby ion-exchange column.
9. The uranium-saturated column is stripped of uranium concentrate by very slow flushing with 10% hydrochloric acid solution.
10. The stripped solution is treated with ammonia or other alkaline reagents and the resulting eluate is filtered, dried, and calcined.
11. The filtrate can be passed on to a molybdenum recovery plant if worthwhile concentrations of molybdenum are in evidence.

The preceding acid process can be modified in several ways to take into account variations of ash compositions or the availability of process materials. One such modification, specifically designed to treat lignite, calls for the addition of a sulfating material such as sulfides of iron, copper, and zinc followed by a sulfating roast which converts the uranium to a water-soluble form. Uranium is then water-leached or acid-leached and recovered by conventional processes.

Also, in every uranium recovery operation, consideration should be given to recovery of other metals or elements which may accompany the uranium. In some cases, recovery of other metals can be quite profitable besides conserving natural resources.

Reasonably assured resources of uranium in high-grade U.S. lignites have been estimated,[615] as of January 1, 1977, to be about 1900 metric tons U available at $80/kg U at a grade of 0.10% U.

Uraniferous coals of Karoo age have recently been discovered in the Northern Transvaal, South Africa. Uraniferous lignites have also been recognized in Canada and Spain and are likely to occur in other regions of the world. The Spanish deposits, which occur in the Ebro Valley, are estimated[616] to contain about 80,000 metric tons of reasonably assured resources and 64,000 metric tons of estimated additional resources probably recoverable at costs well in excess of $130/kg U.[1] Production rates would undoubtedly be restricted because of metallurgical problems in recovery of uranium from lignites.

Nature and solubility of the uranium in the Spanish radioactive lignites are described by Josa et al.[377] The authors describe the features of some 70 samples of radioactive lignites from various places in Spain (Huesca, Lerida, Teruel, Galicia, and Murcia) with uranium contents varying between 20 and 1200 ppm. They carried out experiments on extraction of the uranium from these ores both by direct treatment and after roasting to eliminate organic matter and bring about concentration. The acid method was considered for leaching of the uranium from the substances in question using agitation and static-bed techniques. Investigations were also carried out on the effect of the variables represented by grain size, amount of acid, temperature time and oxidants, in addition to those involved in the roasting process.

Spanish researchers have concluded that when lignites are used in a thermo-plant, the uranium becomes incorporated in the structure of the ash from the lignites, and it becomes practically impossible to recover it from this ash. On the other hand, when they tried to recover the uranium from the lignites themselves before combustion, they had very erratic results; one sample gave 90% recovery and another only 50%. Thus they charged liquids with organic material which they then passed over the resins and through solvents, but they had difficulty in extraction from the solvents. On the other hand, the ash recovered from the lignites calcined at below 500°C gave recoveries of more than 90%. It is believed that low-temperature conditions could give high recoveries, but the experiment was on small scale. In Spain, the aim of this investigation is to recover the thermal potential of the lignites as well as the uranium.

Present efforts in Spain for the recovery of uranium from lignites are directed towards using fluidized-bed combustion for the lignite burning process.

One of the most interesting ideas for uranium recovery is so-called coal-uranium breeder.[376] This combined facility which generated electricity or gas from coal and processed the ash to produce uranium could make available more nuclear energy than it consumed as fossil energy, depending on the reactor type and other variables. A considerable lessening of environmental impact is possible through the use of combined rather than separate coal plants and uranium mills. Potentially, a large fraction of U.S. uranium needs could be supplied at a reasonable cost and thus eliminate the apparent shortfall of sandstone uranium ores and delay the decision about the breeder reactor. More effort is needed to develop extraction techniques and to determine the uranium content of coals.

Figure 13 is a schematic of a combined coal power plant and uranium mill as presented by Smith.[376] It is labeled a Coal-Uranium Breeder (CUB) because, in some circumstances, it would actually make available more fuel than it consumes. The use of a combined coal-gasification plant and uranium mill as a CUB is also feasible. The inputs to the CUB are coal and water for cooling and processing. The outputs are U_3O_8 and electricity or gas, as well as various waste products and environmental emission.

The uranium mill section of the CUB requires sulfuric acid, steam, hot water, and electricity to process uranium ore. All of these are available from the coal plant. In fact the ash, hot water, and sulfur are often considered as waste products and their disposal results in a net cost to the coal plant. In a CUB they would create income.

Table 27 illustrates a possible mass balance for a CUB. The average annual U_3O_8 requirement for a light-water reactor (LWR) is about 131 short tons. This includes the requirement of the initial core and the annual refueling and is averaged over the lifetime of the plant. Thus, one year of operation of this 1600-MWe CUB (or a 230 Mft/day gasification plant at 58% capacity and 66% thermal efficiency) would provide enough U_3O_8 to fuel a 1000-MWe LWR for 1 year. The annual fuel requirement for an HTGR is about 75 tons of U_3O_8. Thus, a 1600-MWe or 230 Mft/day CUB would provide enough uranium for a 1750-MWe HTGR. This CUB in combination with an HTGR could be thought of as having a breeding ratio of 1.09 since it makes available 1.09 times more fuel than it consumes each year.

The exact materials, balance, and breeding ratio would depend greatly on the characteristics of the coal, on its ash, uranium, and Btu contents. In addition, the recovery percentage depends on the uranium concentration in the final ash, which depends both

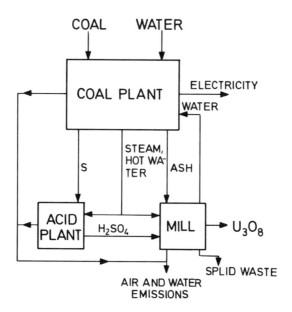

FIGURE 13. Coal-uranium breeder.[376]

Table 27
PARAMETERS OF A COAL-URANIUM BREEDER DESIGNED TO SUPPLY
THE URANIUM FOR A 1000-MWe LWR[376]

Power Plant	Coal	Annual flow (metric tons)
1600-MWe	3880 kcal/kg	Coal 4,800,000 (25 ppm U)
58% capacity factor	(lignite)	Collected ash 477,000 (250 ppm U)
37.5% thermal efficiency	10% ash	Sulfur captured 13,400
99.3% particulate capture	0.7% sulfur	H_2SO_4 produced 39,000 (95% conversion)
40% sulfur capture	25 ppm uranium	U_3O_8 produced 120 (85% recovery) (132 short tons)

on the original uranium concentration and the ash content. Recovery diminishes as the ash content increases because the uranium is less concentrated in the final ash. The breeding ratio could be much higher depending on the type of reactor, the type of coal, and the recovery technology.

VI. NEW TECHNOLOGIES OF COAL UTILIZATION

Efforts to develop new technologies of coal utilization are numerous in the U.S., Europe, and other countries. Here we shall mention some of the progress accomplished.

In 1974 a task force was formed in the U.S. (Montana) to initiate and pursue a broad-based program of research, development, and engineering in the field of open cycle *magneto-hydrodynamics* (MHD). This effort was based on the belief that MHD technology, although still early in its development and not fully proved as a practical generator of electric power, is the only nonnuclear, central-station electric power generation technology with significant potential for greatly improved fuel efficiency and direct operation with coal. MHD also offers lower environmental impact and a reduced pollution. Associated with this high efficiency is the reduced consumption of water, an important factor in the semiarid western regions. MHD generators burn coal at high temperatures to activate small amounts of potassium salt injected into the combustor and initiate ionization in the combustion gas, making it conductive. The gas then is expanded through a channel of electrodes enclosed in a magnet, where interaction with the magnetic field produces current in the gas that is drawn out directly through the electrodes.

In 1979 successful operation of a 20 MW thermal coal combustor at more than 2760°C and about 4 atm pressure was demonstrated at Avco Everett Research Laboratory Inc., Everett, Massachusetts. This combustor is one of three competitive designs funded in the U.S. with the hope of leading to eventual commercial use of MHD generators.

There are three ways of burning coal: (1) stoker grate firing, (2) blown pulverized firing, and (3) fluidized-bed combustion.

The fluidized-bed combustion is a new technology being under development since 1960. This method of coal burning is considered especially promising because it permits use of low grade, high-sulfur coal without the need for costly scrubbers and other environmental controls. The process uses a stream of air to suspend a bed of hot inert particles, generally limestone, which ignite coal fed into the bed. The heat generated by the burning coal is transferred to water tubes in the bed. Bed material and the coal fired through it are crushed to a size compatible with the chosen fluidizing velocity, which in turn is calculated from the quantity of air at bed temperature and pressure, and from the superficial heat of the bed.

The heat-absorption surface immersed in the bed is matched to the heat input, so that the bed temperature can be controlled between selected limits. Fluidized bed combustors operate in the range of 730 to 900°C, or about half that of conventional furnaces. Because combustion is effected at relatively low temperature, the coal-ash particles are kept below their softening temperature to prevent slagging and bonded-deposit formation. The same low temperatures reduce vapor pressures of alkali metal sulfates and chlorides, enabling most of the potentially polluting constituents to remain in the bed.

First-generation fluidized-bed combustion systems are presently considered to be capital-cost competitive with scrubber-equipped conventional coal-firing systems, with claimed higher efficiency. However, ongoing research is expected to eventually give the new technology a 15% capital cost advantage.[378]

During the fluidized-bed combustion of coal, the sulfur in the coal comes off as sulfur dioxide which is captured by the sorbent (dolomite or limestone) and converted into calcium sulfate. The lower temperatures decrease steam tube corrosion, allow the use of lower-grade coals (since these temperatures are lower than ash slagging temperatures), and also decrease NO_x emissions; they also prevent ash agglomeration: minerals exit overhead as fly ash that is collected in cyclones.[64]

Development of coal-fired fluidized-bed boilers is described in detail in Report PB 234 343.[64] The fluidized-bed boiler is a derivative of work by Fritz Winkler carried out before 1921, while the term "fluidized bed" was coined in the 1940s in the petroleum industry.

Figure 14 is a schematic drawing of a cross section of the proposed process used in a fluidized bed boiler. A water tube loop is the heat transfer surface, immersed in a fluidized bed, which has crushed coal and air fed into it from the bottom jets. The inert particles in the bed, preheated by oil burners, ignite the coal on contact. The term fluidization was invented to describe a certain mode of contacting granular solids with fluids and where the solids also behave with pseudofluidic characteristics. In contrast to the fixed bed, a fluidized bed is a bed in which the individual sand granules

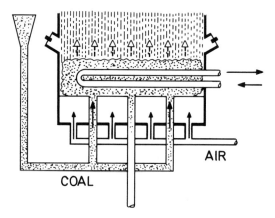

FIGURE 14. Schematic drawing of a cross section of a fluidized bed boiler.

are disengaged somewhat from each other and may be readily moved around with the expenditure of much less energy than would be required if the bed were not suspended in the air stream.

A fluidized-bed boiler is defined as a system which meets all of the following criteria:[64]

1. The primary function of the system is the generation of steam. Therefore, the materials of construction, the mode of operation, the arrangement, auxiliary power requirements, etc., are consistent with existing practices and economics in the conventional boiler field.
2. Fuel is added to and burned within a turbulent bed which has been termed a fluidized bed.
3. A significant fraction of the heat released by the fuel is extracted by heat transfer surfaces in direct contact with the burning-fuel bed.

All boilers can be viewed as chemical reactors in which, among others, the following reactions take place:

$$2H_2 + O_2 \rightleftarrows 2H_2O \qquad (18)$$

$$C + O_2 \rightleftarrows CO_2 \qquad (19)$$

$$S + O_2 \rightleftarrows SO_2 \qquad (20)$$

However, the fluidized-bed boiler more closely approximates what is normally considered a chemical reactor. In addition, operating conditions can be adjusted to promote reaction:

$$CaO + SO_2 + 1/2\ O_2 \rightleftarrows CaSO_4 \qquad (21)$$

This reaction may take place in any boiler but may be carried out more completely in a fluidized-bed boiler. The ability to adjust the combustion conditions so as to achieve a desired result is a feature which is unique in fluidized-bed boilers and constitutes one of its most significant advantages. Because of the relatively low temperatures at which the combustion takes place, the reaction

$$N_2 + O_2 \rightleftarrows 2NO \tag{22}$$

takes place rapidly and so less completely than in a conventional boiler.

The sulfur removal system in fluidized-bed combustion is based on the principle that a solid sorbent can trap the fuel sulfur in solid form as the coal is burned and prevent its release to the environment as gaseous sulfur dioxide (SO_2). Thermodynamic analysis shows which solids will react with sulfur dioxide under process conditions and, therefore, defines those sorbents which must be considered for use.

The calcium-based sulfur removal process has been developed more extensively than have those using alternative sorbents. Experimental work thus far has used limestone or dolomite sorbents as sources of calcium carbonate ($CaCO_3$) or calcium oxide (CaO) in the sulfur removal processes. More than 90% of sulfur dioxide emissions can be prevented and the fuel sulfur captured in a dry solid using limestone or dolomite as sorbent.

The fluidized-bed combustion of coal is now a well-established technology with large programs under way both in Europe and the U.S. The method is mainly applied for the burning of low-grade coals. Jung and Sanmore[379] have recently examined the behavior of very wet Victorian brown coal in a bed of sand fluidized, at temperatures around 1000 K, with either air or nitrogen. Small batches of coal with a narrow particle size range were added to the 76-mm diameter bed, and the times required for devolatilization and total combustion were recorded. Changes in particle water content, volatiles level, and particle size distribution were also measured. All the particles tested, up to 8.4 mm in diameter, dried rapidly and remained substantially intact throughout carbonization and combustion. Devolatilization was complete after about 60 seconds but extensive freeboard combustion of volatiles was evident. The water content of the coal had very little influence on burnout time. Char combustion dominated the overall combustion process and took place under kinetic control with significant pore burning.

A. Coal Conversion

During the process of coal conversion, the starting material is converted to gaseous, liquid, or solid hydrocarbon fuels. A major advantage of this technology is its use of coal which is in abundant supply in many industrial countries. The concept of coal conversion is deceptively simple. It involves primarily two basic steps: the cracking of heavy hydrocarbons into lighter ones and the enrichment of the resultant molecules with hydrogen.[380] Contrary to its concept, the application of coal conversion is not simple. It involves the handling of enormous amounts of a highly variable material often at high temperatures and pressures; it requires containment and control of both highly corrosive process materials and those that pose a possible health threat; and it calls for treatment and disposal of a voluminous solid waste and a possibly hazardous liquid or gaseous waste. Currently, coal conversion is practiced in several countries (including the U.S.) on a small scale at pilot and demonstration plants.

Considering bituminous coal as $(CH)_n$, a number of simple reactions could produce both liquid and gaseous products from coal including:[381]

$$2CH \rightarrow C_2H_2 \tag{23}$$

$$6CH \rightarrow C_6H_6 \tag{24}$$

$$4CH \rightarrow CH_4 + 3C \tag{25}$$

$$CH + 3/2\ H_2 \rightarrow CH_4 \tag{26}$$

$$2CH + H_2O \rightarrow CH_4 + CO \qquad (27)$$

$$CH + H_2O \rightarrow CO + 3/2\ H_2 \qquad (28)$$

$$2CH + O_2 \rightarrow 2CO + H_2 \qquad (29)$$

In the first reaction coal is passed through an electric arc. The large energy input essentially destroys the molecular structure of the coal. Only very stable fragments such as acetylene survive. Up to 35% of the carbon is converted to acetylene.

The second reaction can proceed only to the extent that benzene-ring structures exist in the original molecule. In addition, selective rupturing of bonds can yield a high percentage of benzene. All direct coal conversion processes that produce some liquid will yield significant amounts of benzene.

The third reaction occurs to some extent during coal heating, yielding a gas fairly rich in methane plus a high-carbon char or coke residue.

These three reactions may be considered as examples of thermal degradation or coal pyrolysis. Heat or other energy must be added to the coal in the absence of oxygen to raise the temperature to the point at which bonds can be ruptured.

The fourth reaction represents hydrogasification. Placing coal into contact with hydrogen at high partial pressures and elevated temperatures yields significant amounts of methane in the gases.

The fifth and sixth reactions illustrate the use of water as a possible source of hydrogen for gasification reactions. Some current gasification processes already have achieved the fifth reaction to some extent. But the sixth reaction generally is dominant.

The seventh reaction obviously is a partial oxidation. It is the primary reaction in many gasification processes.

In any real coal conversion reactor, several of the reactions occur as either overall or intermediate reactions. The final product mix will depend on the conditions maintained and the resulting equilibria and kinetics.[381]

Coal can be converted to either a gas (gasification) or a liquid (liquefaction), either underground *in situ* gasification or by mining the coal and transferring it to a conversion plant. However, because coal formations vary in size, depth, and composition, not all coals are equally suitable for a given conversion technique. Coal can also be processed thermally to produce a solid, high-carbon coke used commercially to reduce iron ore to metallic iron. The coking operation is basically similar to pyrolysis processes for synfuels production.

There is little information available on the fate of trace elements in such processes. When coals are heated in the virtual absence of free oxygen to about 400°C and above, they pyrolyze or decompose. Lower-molecular-weight substances evolve as volatile matter, leaving a residual char or coke. The carbon-rich char or coke may be used as a fuel for combustion or as a carbon source for steam gasification ($C + H_2O$ + heat $\rightarrow CO + H_2$) or as a reducing agent for the production of iron from its ore. A portion of the volatile matter can be condensed and recovered as liquid tars or oils. If the coal is slurried in a hydrocarbon oil and then heated under pressure, the same basic chemical reactions take place. However, in this process, referred to as liquefaction, substantially more of the coal is converted to liquids that may then be separated from any unliquefied coal.

Commercial slot-type coke ovens emit volatile pyrolysis products (tars, gases) during charging, through leaks in the oven, and during coke discharge and quenching operations. The fate of trace elements during these phases of operation is unknown. In contrast, synthetic-fuel processing occurs in closed and sealed systems, as in petroleum refinery operations. However, as with any process, there are specific effluent streams.

The effluents and products in which trace elements could be emitted to the environment are: dust, leachate, water vapor, glue gas, carbon dioxide, product gas, used catalyst, foil gas, sulfur, cooling-tower discharge water, ammonia, and sludge. Whether the trace-element content of any particular stream represents a potential hazard to the environment depends on three factors: the trace-element content of the prepared feed coal, the specific processing conditions, such as temperature and pressure (affecting the conversion of trace elements to labile forms), and the control measures exerted to reduce emissions. According to the report "Trace Element Geochemistry of Coal Resources Development Related to Environmental Quality and Health",[4] effluent streams that have the potential for being especially troublesome and that are unique to synfuel processes are the following:

1. The quenching and disposal of gasifier ash
2. Gas scrubber water and condensate
3. Acid-gas removal system
4. Tail gas from sulfur plants
5. Catalyst disposal
6. Product liquid utilization

B. Coal Liquefaction

When German planes roared aloft late in World War II, they burned gasoline made from coal. Even German "butter" was a synthetic made from coal. Behind this technology lay nearly a century of intensive research in Europe and America. In the 1800s it was discovered that the complex molecules of carbon, hydrogen, and oxygen in coal contained the building blocks capable of being rearranged into thousands of useful products.

German scientists observed that coal differs from oil primarily in that it contains less than half as much hydrogen. Working with catalysts at high temperatures and pressures, chemist Friedrich Bergius succeeded in adding hydrogen to coal until it liquefied. He won the 1931 Nobel Prize in chemistry. Two other Germans, Franz Fischer and Hans Tropsch, discovered a catalytic process for converting familiar coal gas into liquid fuel. A dozen Bergius plants and a smaller number using the Fischer Tropsch technology provided the German smorgasbord of coal products in World War II.

Coal is liquefied primarily by either pyrolysis or dissolution. In pyrolysis, coal is heated in the absence of air to drive off volatile matter so that, in addition to a liquid product, by-product gases and char are produced. Thus, the yield of liquid fuel is less than that from dissolution processes.

In dissolution, ground coal is dissolved in a solvent (usually a recycled solvent), the ash is filtered out, and the oil is heated by hydrocracking. As with upgrading of gas from coal, coal liquids can be catalytically upgraded with hydrogen to produce a synthetic crude oil. Hydrogenation is needed because natural petroleum contains 11 to 15% hydrogen, 83 to 87% carbon, and up to 4% oxygen, nitrogen, and/or sulfur; whereas, as mentioned earlier, coal contains only 5% hydrogen and 75% carbon. Dissolution processes can be distinguished as (1) those which use neither hydrogen nor a catalyst, (2) those which use hydrogen but no catalyst, and (3) those which use both hydrogen and a catalyst.

Diagram on Figure 15 presents a coal liquefaction generalized flow diagram. Major projects in the U.S. include SRC-I (Solvent Refined Coal), SRC-II, H-coal, and EDS (Exxon Donor Solvent). Characteristics of these key liquefaction processes are shown in Table 28.

Solvent refined coal (SRC) process produces a low-sulfur, low-ash, low-melting solid

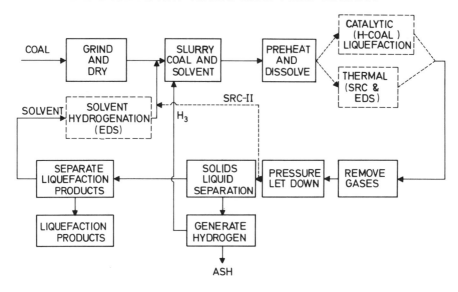

FIGURE 15. Coal liquefaction generalized flow diagram.

Table 28
CHARACTERISTICS OF KEY LIQUEFACTION PROCESSES

	SRC I	SRC II	H-coal	EDS
Coal types	Eastern	Limited	All	All
Product fuel	Solid	Liquid	Liquid	Liquid
Fuel quality		Intermediate	Highest-lowest	Intermediate
Hydrogen consumption	Lowest	High	High to intermediate	Intermediate
Scaleup proposed	Demo	Demo	Commercial	Commercial

fuel. Pulverized coal is slurried with a process-derived solvent, preheated, and sent to a reactor at 800 to 900°F and 1000 to 2000 psi where 90% of the coal organics are rapidly dissolved. The slurry is filtered to remove inorganic ash and the small amount of undissolved carbon. The filtrate is flash-distilled to recover the solvent and to isolate the products — a light liquid by-product and a heavy residual oil called solvent-refined coal, which cools to a low-sulfur, almost ash-free pitchlike material with a melting point of 300 to 400°F and a heating value of about 16,000 Btu/lb.

The Gulf Catalytic Coal Liquids (CCL), H-Coal, and Synthoil processes all convert coal to a liquid fuel using a catalyst and hydrogen gas in contact with pulverized coal slurried in a recycle solvent. The reaction conditions, catalysts, and procedures differ in each process, but because these methods are still being developed, the parameters will be varied to optimize the conversion.

Liquid fuel can be produced by catalytic liquefaction of synthetic gas. The SASOL plant in South Africa combines the Fischer-Tropsch method for catalytically converting synthetic gas to hydrocarbons with Lurgi high-pressure steam-oxygen gasifiers, which produce a gas consisting essentially of carbon monoxide and hydrogen. Two types of reactors are used: a fixed bed of iron catalyst in vertical tubes and a fluidized bed of powdered iron catalyst. By controlling the input to the reactors, the fixed-bed reactor produces a mixture of straight-chain, high-boiling hydrocarbons with some medium-boiling oils and diesel oil; whereas products of the fluidized-bed synthesis are mainly low-boiling hydrocarbons (C_1 to C_4) and gasoline with some oxygenated products and aromatics.

The literature on coal liquefaction is numerous and it is difficult to cover all reports published on this subject. We shall mention some here in order to indicate problems studied in laboratories and pilot plants.

Glenn et al.[382] have subjected three coal samples with different S content from the Pittsburgh coal seam to hydrogenolysis to determine if the amount of organic S present in the coal had an effect on the amount of benzene-soluble product which was produced. The samples of coal, all from the same mine, contained S values ranging from 0.83 to 4.37%. Samples were ground to -200 mesh and mixed with a catalyst (Adkins) and hydrogenated in a modified autoclave at 350 and 375°C for 72 hr. Analysis of the composition of the oils from both temperatures was performed by a chromatography procedure. Results from the hydrogenation of coals with different S contents indicated that, at 350°C, the greater the amount of S present the smaller the amount of conversion product produced and the amount of S in the coal hydrogenated at 375°C had no relation to the amount of conversion product produced. Analysis of the hydrogenated oils from the high and medium S coals, at both temperatures, indicated that they contained 90% of the original coal N and about 33% of the original coal S; however, the oil from the low S coal hydrogenated at 375°C contained about 50% of the original coal S. Generally, it was concluded that the role of S in the hydrogenation process is complicated, and additional testing will be necessary. A few generalized correlations were found between the S content of the coal and the following: the percentage of total conversion to benzene-soluble products; the percentage of original C in all bands; and the percentage of original S and N in the benzene-soluble residue.

Given et al.[383,384] have determined contents of 14 minor and trace elements in oils produced from 3 coals by the catalytic hydrogenation process of Gulf Research and Development Co. by emission spectroscopy. The results were compared with corresponding data for the original coals and the solid residues from the process. The contents of ash, sulfur, vanadium, lead, and copper are near or below the limits specified for an oil to be fired directly in a gas turbine while sodium and probably calcium are too high. Titanium appears to be somewhat enriched in the oils analyzed relative to other elements, suggesting its presence in organometallic complexes.

Concentrations of some elements in feed coal (B Seam, King Mine, Utah), solid residue, and oil are shown in Table 29.

The catalytic effect of the mineral matter constituents as estimated in the ash samples prepared from the various specific gravity fractions of a North Assam coal sample on hydrogenation has been studied in the paper by Mukherjee and Chowdhury.[133] Both iron and titanium present in coal have a catalytic effect on hydrogenation. Pyritic as well as organic sulfur present in the coal appear to be responsible for formation of the catalytically active form of iron sulfide from the iron present in the coal mineral matter. Of the aluminosilicates, kaolinite, which is the major component of coal mineral matter, was found to influence the conversion of coal to gas and benzene-solution products under the conditions of hydrogenation. Figure 16 shows the relation between mineral matter content and conversion. The plots for Fe and Ti resemble closely that relating conversion to the total mineral matter.

Filby et al.[385-387] have presented results on a study of the distribution and fate of 34 trace elements in the Solvent Refined Coal Process at the pilot plant located at Fort Lewis, Washington, and operated by the Pittsburg & Midway Coal Mining Co. under contract with the U.S. Department of Energy. Neutron activation analysis was used to determine Ti, V, Ca, Mg, Al, Cl, Mn, As, Sb, Se, Hg, Br, Co, Ni, Cr, Fe, Na, Rb, Cs, K, Sc, Tb, Eu, Sm, Ce, La, Sr, Ba, Th, Hf, Ta, Ga, Zr, and Cu in feed coals,

Table 29
CONCENTRATIONS OF SOME ELEMENTS IN FEED COAL, SOLID RESIDUE, AND OIL[383]

Element	Feed coal (ppm)	Solid residue (ppm)	Oil (ppm)
B	12	25	0.15
Co	25	12	0.12
V	12	25	0.06
Ni	12	5	0.15
Ti	75	150	5
Mo	—	25	0.08
Na	5000	1000	5
Pb	—	—	0.02
Zn	25	50	0.12
Mn	12	25	0.05
Cr	25	25	0.05
Cu	12	25	0.02
Cd	0.2	0.5	—
Be	0.4	0.8	0.0004

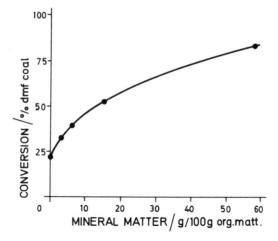

FIGURE 16. Relation between the mineral matter content and coal conversion.[133]

process solvent, Solvent Refined Coal (SRC), mineral residues, wet filter cake, by-product solvents, process and effluent waters, and by-product sulfur. The sample points were chosen such that the major process streams were adequately described and that the major input and output materials were included. Atomic absorption spectrophotometry was used to measure Pb, Cd, and Be in plant-derived solvents, effluent water and Hamer Marsh water. Specific methods were developed for analysis of a wide range of material compositions. The neutron activation analysis procedures were divided into short and long irradiation procedures for elements with short half-lives (less than 3 hr) and intermediate to long half-lives (8 hr to 5.2 years). Data are presented for preliminary SRC I process materials and also for a set of materials taken during operation of the pilot plant but not under equilibrium conditions.

Emissions of trace elements is a real environmental hazard associated with coal conversion processes and plants. An important objective of liquefaction processes is to remove much of the sulfur and the ash content of coal so that the resulting fuel can be burned without expensive stack gas scrubbers. Removal of a large fraction of the ash content will result in removal of a fraction of the trace element content of the coal. It is thus important that the fate and distribution of trace elements in the SRC I process be determined to assess the pollution potential of the product fuel (SRC), the environmental effects of emissions, and the effects of waste disposal. The distribution of trace elements during the liquefaction process is also important in determining the trace element material balances of the process and in evaluating the effects of such variables as coal type, autocatalytic effects, temperature, pressure, solvent composition, and degree of hydrogenation on the material balance.

In the work by Filby et al.,[388] two separate sets of samples were taken when the pilot plant had operated continuously for 7 days and composite samples were collected for each process fraction over a 24-hr period. These are designated Equilibrium Sets 1 and 2. A material balance was calculated for each element from the concentration data and the yields of each process fraction for each equilibrium set in the SRC process.

Approximate daily rates of production of trace elements in the 50 ton/day pilot plant are shown in Table 30. The amounts of each trace element produced daily have been calculated assuming 50 tons coal converted per day and concentrations of each trace element taken as those values for coals used in Equilibrium Sets 1 and 2.

Table 31 compares trace element levels in the feed coal and the corresponding solvent refined coal (SRC) to indicate the decrease in trace element levels in SRC, resulting primarily from the removal of mineral matter from the coal during the liquefaction process. Data for both equilibrium sets are shown. Table 32 shows the ratio of each trace element in SRC relative to feed coal (SRC/coal) and the percentage reduction in SRC compared to the coal. All elements, except Br, show a reduction when SRC is compared to coal. In both equilibrium sets there is an increase in Br relative to the feed coal; the reason for the increase is not clear.

While some metals may play a positive role in coal conversion, it is conceivable that some of the same metals in a different bonding state, or other metals, may deactivate the catalyst and impede the conversion process.[389]

In the report by Wakeley et al.,[390] solids accumulated in the reactor of a solvent-refined coal (SRC-I) pilot plant during processing of three coals were studied using optical microscopy and X-ray diffraction. A classification system was devised for each of the two groups of components: organic materials and mineral matter. The various organic components were classified by the extent of change from the original coal macerals, and by optical properties of different semi-cokes and other organic phases. Minerals were divided into four groups: those unchanged from the original coal, minerals which were physically degraded, minerals which were chemically or crystallographically transformed, and those minerals formed during processing of a subbituminous coal. Gold-tube carbonization experiments were performed on SRC to determine the conditions under which retrogressive reactions occur to form mesophases semi-coke. Autoclave experiments were designed to investigate the recrystallization of pyrite as pyrrhotites, and to determine the causes of carbonate-mineral formation in the reactor. Calcium carbonate was found to crystallize from the interaction of ion-exchangeable calcium and carbon dioxide, which are available when low-rank coals are processed.

According to Saxby,[391] the amount of oil that can be generated from a kerogen under natural conditions is an important parameter in source rock assessment. Similarly, the quantity of oil able to be formed from a coal or kerogen during much more rapid pyrolysis is a crucial factor in oil shale and coal conversion studies. Saxby[391] has

Table 30
PRODUCTION OF TRACE ELEMENTS IN KG/DAY FOR 50 TON/DAY PILOT PLANT: EQUILIBRIUM CONDITIONS[388]

Elements	Concentration in coal (Eq. Set 1)	Kg/day	Concentration in coal (Eq. Set 2)	Kg/day
Fe[a]	2.40	1200	2.90	1500
Al[a]	1.07	540	1.24	630
Ti	547	28	660	34
V	25.2	1.3	29.2	1.5
Ca	633	32	658	33
Mg	858	44	—	—
Cl	286	15	250	13
Mn	32.8	1.7	34.8	1.8
As	11.6	0.6	21.2	1.1
Sb	0.98	0.1	1.13	<0.1
Se	2.18	0.1	2.24	0.11
Hg	113.0	5.7×10^{-3}	114	5.7×10^{-3}
Br	5.79	0.3	3.24	0.2
Ni	18.0	0.9	12.4	0.6
Co	5.25	0.3	6.0	0.3
Cr	10.4	0.5	14.9	0.8
Na	124	6.3	144	7.3
Rb	3.57	0.2	11.9	0.6
Cs	0.75	<0.1	1.03	<0.1
K	1260	64	1840	88
Sc	2.10	0.1	2.71	0.1
Tb	0.43	<0.1	0.29	<0.1
Eu	0.26	<0.1	0.24	<0.1
Sm	2.59	0.1	2.29	0.1
Ce	24.2	1.2	30.5	1.6
La	7.52	0.4	8.0	0.4
Sr	97.2	4.9	134	6.8
Ba	31.0	1.6	53.2	2.7
Th	1.51	<0.1	2.16	0.1
Hf	0.34	<0.1	0.57	<.01
Ta	0.10	<0.1	0.10	<0.1
Ga	2.69	0.1	2.27	0.1
Zr	66.5	3.4	57.3	2.9
Cu	22.4	1.1	13.2	0.8

[a] % Concentration.

derived a relation for slow geological heating which is also valuable for comparing samples under other conditions. The percentage of oil, on a weight basis, that can be generated is given by: Oil = 66.7% H/C − 57.00/C −33.3. Typical results for 12 Australian coals and 12 world oil shales are given as examples (see Table 33). Many other reports can be found in the literature; for the partial list, see References 392-498.

C. Coal Gasification

Literature on coal gasification is voluminous; some references to the work performed are listed at the end of this chapter (References 499 to 605). Converting coal into an environmentally acceptable gaseous fuel requires nine basic and seven auxiliary steps; these steps include coal preparation and pretreatment, gasification, shift conversion, catalytic methanation, various cleanup and disposal procedures, and final

Table 31
COMPARISON OF TRACE ELEMENTS IN
INPUT COAL AND SRCI[388]

	Equilibrium Set 1		Equilibrium Set 2	
	Coal	SRC	Coal	SRC
Ti	530	465	660	490
V	30.1	4.63	29.2	13.7
Ca	633	72.8	660	123
Mg	1160	89	ND	96.0
Al[a]	1.18	0.02	1.58	0.042
Cl	260	160	290	99
Mn	34.0	20.3	35.7	14.2
As	12.5	2.00	20.1	2.27
Sb	0.76	0.060	1.43	0.06
Se	2.00	0.12	3.03	0.08
Hg	113	39.6	114	46.7
Br	4.56	7.74	3.72	4.93
Ni	14.9	3.0	12.4	2.0
Co	5.88	0.22	5.20	0.25
Cr	13.7	1.64	14.8	5.50
Fe[a]	2.11	0.030	2.38	0.026
Na	137	4.23	173	6.30
Rb	3.57	0.5	11.8	0.21
Cs	0.75	0.02	0.91	0.2
K	1550	4.72	2030	2.27
Sc	2.59	0.57	2.48	0.36
Tb	0.39	0.045	0.32	0.03
Eu	0.26	0.055	0.20	0.027
Sm	2.62	0.29	1.65	0.117
Ce	20.9	0.45	20.9	0.34
La	7.55	0.13	6.56	0.09
Sr	88.6	<5	158	<6
Ba	53.0	5.75	62.5	<2
Th	2.00	0.22	1.90	0.19
Hf	0.51	0.085	0.59	0.069
Ta	0.14	0.046	0.17	0.05
Ga	3.56	1.79	3.26	0.61
Zr	62.9	16.0	79.0	6.4
Cu	19.9	2.07	14.3	1.15

[a] % trace elements.

compression and dehydration of the product into pipeline gas.[403] The large number of named processes arise in an attempt to combine the various alternatives available in each step so that the entire procedure may be optimized. For example, a major research effort is directed toward optimizing the two key steps, gasification and methanation. The final products of the coal gasification processes are shown in Figure 17.

In principle, only three ingredients are needed to synthesize gas from coal — carbon, hydrogen, and oxygen. Coal provides the carbon, steam most often provides the hydrogen, and air provides the oxygen; however, the relative amounts of these elements are crucial.

Natural gas, which is primarily methane (CH_4), has a high hydrogen-to-carbon ratio (25% hydrogen, 75% carbon) and consequently has a high heating value. In contrast, a typical bituminous coal contains only 5% hydrogen, 75% carbon, and 20% other constitutents (sulfur and ash). Thus, when the object of coal gasification is to produce

Table 32
TRACE ELEMENT REDUCTION IN SRC COMPARED TO COAL[388]

Equilibrium Set No. 1			Equilibrium Set No. 2		
Element	SRC/coal	% Reduction	Element	SRC/coal	% Reduction
Ti	0.88	12	Ti	0.74	16
V	0.15	85	V	0.47	53
Ca	0.22	78	Ca	0.22	78
Mg	0.08	92	Mg		
Al	0.02	98	Mg		
Al	0.02	98	Al	0.03	97
Cl	0.61	39	Cl	0.34	66
Mn	0.60	40	Mn	0.40	60
As	0.16	84	As	0.07	93
Sb	0.08	92	Sb	0.04	96
Se	0.06	94	Se	0.03	97
Hg	0.35	65	Hg	0.41	59
Br	1.70	+70	Br	1.33	+33
Ba	0.11	89	Ba		
Th	0.11	89	Th	0.10	90
Hf	0.16	84	Hf	0.12	88
Ta	0.39	61	Ta	0.29	71
Ga	0.50	50	Ga	0.19	81
Zr	0.25	75	Zr	0.08	92
Cu	0.10	90	Cu	0.08	92
Na	0.03	97	Na	0.04	96
Rb			Rb	0.02	98
Cs	0.03	97	Cs		
K	0.003	100	K	0.001	99.9
Co	0.04	96	Co	0.05	95
Cr	0.12	88	Cr	0.37	63
Fe	0.01	99	Fe	0.01	99
Sc	0.22	78	Sc	0.15	85
Tb	0.12	88	Tb	0.09	91
Eu	0.21	79	Eu	0.14	86
Sm	0.11	89	Sm	0.07	93
Ce	0.02	98	Ce	0.02	98
La	0.02	98	La	0.01	99

a substitute natural gas, hydrogen must be added, usually by catalytic methanation. Without added hydrogen the gas will have a low- or intermediate-heating-value gas. Low- and intermediate-Btu gases, although less expensive and simpler to produce, are considered inadequate for intermingling with natural gas for pipeline transportation and would be used at or near the production site. Catalytic methanation converts a low-grade gasified product to a pipeline or high-Btu gas, and low- or intermediate-Btu process becomes a high-Btu process by adding methanation. Thus, because methanation is essential to high-Btu coal gasification processes, it is the subject of considerable investigation.[403]

Catalyst materials have been extensively studied. Of the metals identified as potential methanation catalysts, nickel surpasses all others as the catalyst of choice for commercial application; it is inexpensive, very active, and highly selective to methane. However, nickel can be deactivated or poisoned. Sulfur compounds are the primary poison, but nickel is also deactivated by high temperatures (above 450°C), carbon deposition, nickel carbonyl formation (below 260°C), or by the deposition of iron from iron carbonyl formed in the reactor vessel. Thus, gasification processes employing catalytic

Table 33
OIL YIELDS FROM AUSTRALIAN COALS[391]

Basin (Age)	Seam or district	H/C	O/C	(wt %)	Oil from 600°C pyrolysis[a]	Oil calculated[a]
Sydney (Permian)	Greta	0.93	0.065	7.9	21.6	23.0
Sydney (Permian)	Greta (torbanite)	1.28	0.037	5.9	49.8	47.0
Sydney (Permian)	Borehole	0.81	0.070	6.2	14.5	15.7
Sydney (Permian)	Bulli	0.64	0.039	10.0	6.6	6.4
Sydney (Permian)	Wongawilli	0.75	0.053	22.8	8.6	10.6
Sydney (Permian)	Lithgow	0.82	0.079	13.7	13.6	14.6
Gunnedah (Permian)	Hoskissons	0.72	0.105	14.9	8.0	7.4
Clarence-Moreton (Jurassic)	Rosewood (Walloon)	0.96	0.119	17.0	17.6	19.9
Bowen (Permian)	Bowen	0.63	0.030	11.3	6.1	6.2
Bowen (Permian)	Bluff	0.59	0.050	14.1	3.0	2.7
Bowen (Permian)	Moura	0.70	0.055	12.4	8.6	9.0
Callide (Triassic)	Callide	0.66	0.141	11.0	4.3	2.4

[a] Weight percent, dry basis.

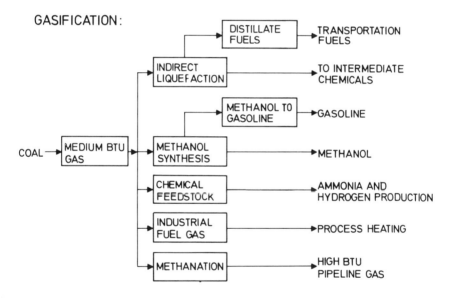

FIGURE 17. Final products of coal gasification processes.

methanation require rigorous cleanup of the product gas before it is allowed to contact the catalyst.

Many equipment designs have been proposed for coal conversion; however, most viable gasification processes can be identified by an operating temperature, the use or absence of pressure, and the type of reactor vessel. Gasification processes use one of three basic reactor bed types — fixed, fluidized, or entrained bed.

The Lurgi process, a fully commercial gasification method, employs a fixed-bed reactor operating at 800 to 1000°C. It uses either air and steam or oxygen and steam as the reactants, producing low- or intermediate-Btu gas. With methanation, the product is upgraded to a high-Btu pipeline gas.

The Winkler gasifier uses a fluidized bed maintained at a sufficiently high temperature (820 to 980°C) to crack all the tars and heavy hydrocarbons, leaving only heavy ash particles to be removed from the bottom of the gasifier. Winkler plants, producing a low- or intermediate-Btu gas, have operated since 1926.

Commercial gasifiers use either two or four opposing burners: the four-burner reactor can handle up to 850 tons of coal per day. The major advantage of an entrained-bed reactor, as used, for example, in the Koppers-Totzek process, is its ability to gasify any type or rank of coal, including the fines. Additionally, combustion takes place at about 1650°F, destroying tars, phenols, and light oils.

Many other techniques for gasification of coal are being explored; most employ the well-known methods with variations, but some use novel procedures (see references and the review work by Ensminger[403]). For example, the Bi-Gas process uses a two-stage entrained-bed gasifier that operates at system pressures exceeding 1000 psi. When followed by a fluidized-bed methanation step, the final gas has a heating value of pipeline gas — 900 Btu/scf.

The Synthane process also involves a two-stage pressurized gasifier in which the caking properties of the coal are destroyed in a fluidized-bed pretreatment before carbonization.

Table 34 shows some of better-known coal gasification processes for syngas production. They are at different stages of development and it would take too much space to describe work done in this area.

By definition, coal conversion is mandated to produce clean fuel from coal. As the result, much effort has been invested into sulfur removal from coal. Because organic sulfur is difficult to remove during coal pretreatment and because about 90% of the organic sulfur is converted to hydrogen sulfide in both gasification and liquefaction, this form of sulfur is estimated to be the major sulfur contaminant in gas streams. For example, of the total sulfur in a Lurgi product gas, hydrogen sulfide accounted for 95%; the remaining contaminants were carbonyl sulfide, 2.4%; carbon disulfide, 0.3%; mercaptans, 2.0% and thiophenes, 0.3%. Hydrogen sulfide is readily removed from gas streams, as is sulfur dioxide, but carbonyl sulfide and carbon disulfide are more resistant to acid-gas scavenging and can pass through sulfur traps, ending up in gases emitted to the atmosphere.[403]

Most of the sulfur in coal is expected to be recovered from conversion streams as elemental sulfur although the sulfur that reaches the stack gas as sulfur oxides and is collected by limestone scrubbing will end up as sulfate. Neither form of sulfur is particularly obnoxious environmentally, although the quantity to be disposed of is considerable: a 25,000-ton/day plant using a coal containing 4% sulfur will produce 1000 tons of sulfur daily. As opposed to throwaway disposal, new uses for elemental sulfur are being considered. Production of sulfur-based paving materials and plastics or use of sulfur in thermal and acoustical insulation are considered alternatives to landfill or minefill disposal.

A large number of components act as catalysts in the reactions of coal with hydrogen, oxygen, carbon dioxide, and water vapor, including many of the minerals in coal which act as internal catalysts.

Many studies on the chemical form of catalysts, the number of catalysts, and their particle size have been done. For example, Otto et al.[591] have evaluated gravimetrically the effects of alkaline earths on the steam gasification of two coal chars, one of low and one of high intrinsic reactivity. The chars were derived from coal powders which had been impregnated with Ca, Sr, or Ba, pelletized, and pyrolyzed in nitrogen. The additives increased the gasification rates in the order Ca < Sr < Ba. It follows from reaction kinetics that catalysis is caused by a large increase in the density of reaction sites, not by a lowering of the true activation energy. As shown by electron micro-

Table 34
GASIFICATION PROCESSES FOR SYNGAS PRODUCTION

Licensor	Type	Pressure (psi)	Status
Lurgi	Dry bottom, moving bed	400	Commercial for non-caking sized coal
Koppers-Totzek	Entrained	atm	Commercial for powdered coal
Winkler	Fluidized	atm	Commercial
Texaco	Entrained water slurry	350	150 T/D Operational at Oberhausen, W. Germany
		350	150 T/D Under construction at Muscle Shoals, Ala.
		350	150 T/D Operational at Plaquemine, La. (Lignite)
Shell-Koppers	Entrained	350	150 T/D Operational at Harburg, W. Germany
British Slagger	Slagging bottom, moving bed	400	300 T/D at Westfield, Scotland

graphs and elemental maps obtained by X-ray analysis, the strong catalytic effect is closely associated with the ability of the alkaline-earth species to spread over the char and to preserve contact with a freshly formed carbon surface as the gasification proceeds. The alkaline-earth catalysts are severely poisoned by hydrogen sulfide or sulfur dioxide from an external source.

In situ gasification of coal, which converts a coal seam into combustible gases, is capable also of producing a gas rich in carbon monoxide, carbon dioxide, hydrogen, hydrocarbon gases, and other gases. Depending on the gas injected (air, oxygen, or oxygen and steam), the heating value of the product gas can range from as low as 50 Btu/scf when air alone is injected to as high as 280 Btu/scf when an intermittent air-steam sequence is used.

The idea of underground gasification is very attractive. The method avoids mining and waste disposal and adds unaccessible, unworkable, or uneconomical coal deposits to the energy pool. Unfortunately, the gas has a low heating value and would either have to be used at the site or upgraded for transportation. Starting in the 1930s, different methods of underground gasification have been tested in the Soviet Union, at least one of which was tried on an industrial scale.

In situ gasification is not expected to replace mining, but it certainly can supplement external conversion. Two basic methods exist for underground gasification — shaft and shaftless systems. In the shaft method, large-diameter openings are cut into the seam (as in mining), whereas the shaftless method uses boreholes.

Early U.S. experiments with *in situ* gasification[606] showed the process to be technically feasible but incapable of competing economically with cheap, plentiful natural gas.

Recently (see *Industrial Research and Development,* January, 1980), a 21-day test burn was completed near Rawlins, Wyo. It was part of a program to demonstrate a process by which steeply-dipping coal seams — nonrecoverable using conventional mining technology — can be burned in place to produce gas. The recovered gas can then be converted into synthetic natural gas, methanol, ammonia, or liquid fuels. Both the test and evaluation of results comprised Phase III of a 5-year $13.5-million research and development contract awarded to Gulf Science & Technology Co., Pittsburgh, Penna., by the U.S. Department of Energy (DOE). Phases I and II involved planning, site selection, and a detailed geological evaluation of the site with environmental impact analysis. Phase IV would cover the preparation of preliminary design and cost estimates for a pilot-scale project including surface facilities to process the recovered gas for commercial applications.

The test burn took place in a 7 m thick coal seam about 122 m below the surface. Two wells were drilled, an injection well to convey combustion air, and a production well to recover the gas. Coal was ignited at the base of the injection well using an electric heater. Well linkage was accomplished by the backward burn technique which created a zone of relatively high permeability between the bottoms of the two wells.

Development of economically viable technology to utilize steeply dipping underground coal seams could add some 10^9 tons of coal to the U.S. energy resources. It could also provide an opportunity to convert high-sulfur coal into more environmentally-acceptable forms of energy. The project near Rawlins, Wyo. is one of six DOE efforts in underground coal gasification. Three other DOE funded contracts were handled by the Laramie Energy Technology Center, the Lawrence Livermore Laboratory, and the Morgantown Energy Technology Center.

For readers interested in underground coal gasification, one should mention useful citations from the NTIS data base: Cavagnaro,[607] Estep-Barnes and Kovach,[524] Fisher et al.,[608] Fleming,[609] Hommert and Beard,[610] Roehl et al.,[611] and Schrider et al.[612,613]

REFERENCES

1. Loevblad, G., Ed., Trace element concentrations in some coal samples and possible emissions from coal combustion in Sweden, Report IVL-B-358, 1977.
2. Bern, J., Residues from power generation: processing, recycling, and disposal, in *Land Application of Waste Materials*, Soil Conservation Society of America, Ankeny, Iowa, 1976, 226.
3. Maylotte, D. H., Reported at 1981 March Meeting of the American Physical Society, Phoenix, Arizona. See also *Ind. Res. Dev.*, May 1981, p. 64.
4. Trace Element Geochemistry of Coal Resources Development Related to Environmental Quality and Health, National Academy Press, Washington D.C., 1980.
5. Schwitzgebel, K., Meserole, F. B., Oldham, R. G., Magee, R. A., Mesich, F. G., and Thoem, T. L., Trace element discharge from coal-fired power plants, Int. Conf. Heavy Met. Environ., Symp. Proc., 1st, 2, 533-551, 1975.
6. U.S. Environmental Protection Agency, Coal Fired Power Plant Trace Element Study, A Three Station Comparison, Vols. I and II, Reports prepared by Radian Corporation, EPA Rocky Mountain Prairie Region VIII, 1975.
7. Davison, R. L., Natusch, D. F. S., Wallace, J. R., and Evans, C. A., Trace elements in fly ash-dependence of concentration on particle size, *Environ. Sci. Technol.*, 8, 1107, 1974.
8. Swaine, D. J., Trace elements in fly ash, *N. Z. Dep. Sci. Ind. Res. Bull.*, 218, 127, 1977.
9. Swaine, D. J., Trace elements in coal, *Trace Subst. Environ. Health*, 11, 107, 1977.
10. Ondov, J. M., Ragaini, R. C., and Biermann, A. H., Emissions and particle size distributions of minor and trace elements at two western coal-fired power plants equipped with cold size electrostatic precipitators, *Environ. Sci. Technol.*, 13, 946, 1979.
11. Ondov, J. M., Ragaini, R. C., and Biermann, A. H., Characterization of trace element emissions from coal-fired power plants, from Energy Res. Abstr. No. 29344, 1978, Report No. UCRL-80412, 1978, 21.
12. Linton, R. W., Loh, H., Natusch, D. F. S., Evans, C. A., Jr., and Williams, P., Surface predominance of trace elements in airborne particles, *Science*, 191, 852, 1976.
13. Kaakinen, J. W., Jorden, R. M., Lawasani, M. H., and West, R. E., Trace element behaviour in a coal-fired power plant, *Environ. Sci. Technol.*, 9, 862, 1975.
14. McNeil, D., The physical properties and chemical structure of coal tar pitch. In *Bituminous Materials: Asphalts, Tars, and Pitches*, Hoiberg, A.J., Ed., Interscience Publishers, New York, 1966, 139.
15. Alekhina, V. I. and Adamenko, I. A., Germanium compounds volatilizing during the pyrolysis and combustion of coal, Protsessy Term. Prevrashch. Kamennykh Uglei, 408, 1968.

16. Aleksandrov, I. V., Patrushev, S. G., Guseva, T. N., and Kamneva, A. I., Use of the perhydrol method to evaluate the tendency of coals toward spontaneous combustion, *Tr. Mosk. Khim.-Tekhnol. Inst.*, 80, 3, 1974.
17. Bertine, K. K. and Goldberg, E. D., Fossil fuel combustion and the major sedimentary cycle, *Science*, 173, 223, 1971.
18. Dik, E. P., Soboleva, A. N., and Surovitskii, V. P., Mechanism of formation of ash deposits on semiradiation heating surfaces during combustion of coals with high iron content, *Vliyanie Miner. Chasti Energ. Topl. Usloviya Rab. Parogeneratov, Mater. Vses. Konf.*, 2, 3, 1974.
19. Dik, E. P., Soboleva, A. N., and Suckhov, S. I., Sulfating of deposits during combustion of Berezovsk coal, *Vliyanie Miner. Chasti Energ. Topl. Usloviya Rab. Parogeneratov, Mater. Vses. Konf.*, 2, 63, 1974.
20. Edemskii, O. N. and Dering, I. S., Behaviour of the mineral part of Berezovskii and Irsha-Borodinskii coals during their combustion in a cyclone furnace, *Izv. Vyssh. Ucheb. Zaved., Energ.*, 17, 55, 1974.
21. Egorov, A. P., Laktinova, N. V., and Popinako, N. V., Study of trace element losses in the combustion of coal, *Khim. Tverd. Topl., Moscow*, 2, 30, 1979.
22. Egorov, A. P., Laktinova,N.V., Popinako, N.V., and Novoselova, I.V., Behaviour of some trace elements of coals during combustion at thermal electric power plants, *Teploenergetika, Moscow*, 2, 22, 1979.
23. Fejer, M. E., Ketels, P. A., Kouo, M.T., Nesbitt, J.D., and Schoffstall, D.R., Assessment application for direct coal combustion, from Gov. Rep. Announce. Index, U.S. National Technical Information Service, Springfield, Va., PB Rep. No. 263651, 1976, 268.
24. Friedrich, F. D., Lee, G. K., and Mitchell, E. R., Combustion and fouling characteristics of two Canadian lignite, Canadian Combustion Research Laboratory, Mines Branch Research Report R 208, 1969, 31.
25. Iapulacci, T. L., Demski, R. J., and Bienstock, D., Chlorine in coal combustion, *U.S. Bur. Mines Rep. Invest.*, 7260, 11, 1969.
26. Kapel'son, L. M., Shishkina, L. M., Kuznetsov, N. I., and Solomatina, T. V., Slagging of heating surfaces of boiler units during Donets coal combustion, from *Ref. Zh., Teploenerg.*, 2R111 (Abstr.), 45, 1971.
27. Kerley, R. V., Catalyzed combustion of coal with restricted air supply, French Patent 1547982, 1968.
28. Kerley, R. V., Composition for reducing smoke and sulfur trioxide production in coal combustion process. U.S. Patent 3927992, 1975.
29. Klein, D. H. and Andren, A. A., Trace element discharges from coal combustion for power production, *Water, Air, Soil Pollut.*, 5, 71, 1975.
30. Koma, T., Shimizu, H, Igarashi, T., Takahashi, H., and Ishizaki, T., Coal-dispered fuel combustion, Japan Kokai, Patent 78-47405, 1978.
31. Kunstmann, F. H., Harris, J. F., Bodenstein, L. B., and Van den Berg, A. M., The occurrence of boron and fluorine in South African coals and their behaviour during combustion, *Fuel Res. Inst. S. Afr., Bull.*, 63, 45, 1963.
32. Lebedov, I. K., Zavorin, A. C., and Sarapulov, G. A., Floating slugs in the combustion of Kansko-Achinsk coals, *Teploenergetika, Moscow*, 8, 21, 1978.
33. Lowblad, G. and Greenfelt, P., Heavy metals and other trace elements in black coal and their emission to the air at coal combustion. A literature Survey, Report IVL-B-345, 1977.
34. Mikhailova, A. I., Larina, V. A., and Vlasov, N. A., Combustion of coals during the determination of trace elements, *Khim. Tverd. Topl. Moscow*, 3, 15, 1978.
35. Natusch, D. F. S., Ed., Characterization of fly ash from coal combustion, Proc. Workshop on measurement technology and characterization of primary sulfur oxides emission from combustion sources, Southern Pines, N.C., 1978 (Report COO-4347-4).
36. Nelson, H. W., Mineral constituents in coal and their behaviour during combustion, in *Corrosion and Deposits in Coal and Oil Fired Boilers and Gas Turbines*, American Society of Mechanical Engineers, New York, 1959, 7.
37. Newall, H. F., The influence of inorganic compounds on the combustion of coal. III. The effect of water on constitution of added compounds, moisture, and mineral matter in coal, *Fuel*, 18, 13, 1939.
38. Page, A. L., Bingham, F. T., Lund, L. J., Bradford, G. R., and Elseewi, A. A., Consequence of trace element enrichment of soils and vegetation from the combustion of fuel used in power generation, Biannual Report, Southern California Edison Research and Development Series, 77-RD-29, Rosemead, Calif., 1977.
39. Schultz, H., Hattman, E. A., and Booher, W. B., Fate of some trace elements during coal pretreatment and combustion, *Am. Chem. Soc., Div. Fuel Chem. Prepr.*, 18, 108, 1973.
40. Schultz, H., Hattman, E. A., and Booher, W. B., The fate of some trace elements during coal pretreatment and combustion, in *Trace Elements in Fuel*, Babu, S.P., Ed., Am. Chem. Soc., Washington, D.C., Adv. in Chem. Series 141, 1973.

41. Seamans, R. C., Jr., ERDA's coal program for combustion, liquefaction, MHD, gasification, *Prof. Eng. (Washington, D.C.)*, 46, 21, 1976.
42. Smith, W. S. and Gruber, C. W., Atmospheric emission from coal combustion, an inventory guide, Public Health Service Publication No. 999-AP-24, 1966, 112.
43. Tager, S. A., Rabiner, Ya. P., Unge, M. K., Talumaa, R., Soone, L., and Ausmees, V., Behaviour of the mineral part of an acid composition during the high-temperature combustion of coal, *Teploenergetika*, 11, 42, 1975.
44. Strojan, C. L. and Turner F. B., Trace elements and sulfur in soil and plants near Mohave Generating Station in Southern Nevada, Conf. Proc. of Conf. Sens. Environ. Pollut. p. 537, Am. Chem. Soc., Washington, 1977.
45. Turner, F. B. and Strojan, C. L., Coal combustion, trace element emissions, and mineral cycles, DOE Symp. Ser., 45, 34-58, 1978.
46. Van Hook, R. I. and Shults, W. D., Eds., Effects of traces contaminants from coal combustion, Proceedings of a Workshop, Aug. 2-6, 1976, Knoxville, Tenn. ERDA 77-64, National Technical Information Service, 1976.
47. Volodarskii, I. Kh., Grekhov, I. T., and Shpirt, M. Ya., Behaviour of microelements of the mineral part of coals during combustion, *Khim. Tverd. Topl.*, 1, 157, 1970.
48. Weyesser, J. L. G., Coal analyses and their relationship to combustion characteristics, Illinois Engineering Experiment Station Circular Series No. 43, University of Illinois, 1942, 72.
49. Yeh, J. T., McCann, C. R., Demeter, J. J., and Bienstock, D., Removal of toxic trace elements from coal combustion effluent gas, PERC/RI-76/5, Pittsburgh Energy Research Center, Pittsburgh, Pennsylvania, 1976, 19.
50. Zahradnik, L., et al., Germanium distribution among combustion products, *Chemicky Prumysl.*, 9, 1959.
51. Kaakinen, J. W. and Jorden, R. M., Determination of trace element mass balance for a coal-fired power plant, Proc. 1st Annu. NSF Trace Contam. Conf., 1974, 165.
52. Bolton, N. E., Carter, J. A., Emery, J. F., Feldman, C., Fulkerson, W., Hulett, L. D., and Lyon, W. S., Trace element mass balance around a coal-fired steam plant, *Adv. Chem. Ser.*, 141, 175, 1975.
53. Bolton, N. E., Fulkerson, W., Van Hook, R. I., Lyon, W. S., Andren, A. W., Carter, J. A., Emery, J. F., Feldman, C., Hulett, L. D., Dunn, H. W., Sparks, C. J., Jr., Ogle, J. C., and Mills, M. T., Trace element measurements at the coal-fired Allen steam plant, Oak Ridge National Laboratories, Oak Ridge, Tenn. Progress report, February 1973 — July 1973.
54. Bolton, N. E., Van Hook, R. I., Fulkerson, W., Emery, J. F., Lyon, W. S., Andren, A. W., and Carter, J. A., Trace Element Measurements at the Coal-Fired Allen Steam Plant, Progress Report, June 1971-January 1973, ORNL/NSF/EP 43, Oak Ridge National Laboratory, Oak Ridge, Tennessee.
55. Klein, D. H., Andren, A. W., Bolton, N. E., Trace element discharge from coal combustion for power production, *Water, Air, Soil Pollut.*, 5, 71, 1975.
56. Klein, D. H., Andren, A. W., Carter, J. A., Emery, J. F., Feldman, C., Fulkerson, W., Lyon, W. S., Ogle, J. C., Talmi, Y., et al., Pathways of thirty-seven trace elements through coal-fired power plant, *Environ. Sci. Technol.*, 9, 973, 1975.
57. Klein, D. H. and Russell, P., Heavy metals, fallout around a power plant, *Environ. Sci. Technol.*, 7, 357, 1973.
58. Roffman, H. K., Kary, R. E., and Hidgins, T., Ecological distribution of trace elements emitted by coal-burning power generating units employing scrubbers and electrostatic precipitators, *Pap. Symp. Coal Util.*, 4, 192, 1977.
59. Kalb, G. W., Total mercury mass balance at a coal-fired power plant, Trace Elements in Fuel, Am. Chem. Soc., Washington, D.C., Advances in Chemistry, Ser. 141, Babu, S., Ed., 1975, 154.
60. Joensun, O. I., Fossil fuels as a source of mercury pollution, *Science*, 172, 1027, 1971.
61. Carter, J. A., Walker, R. L., and Sites, J. R., Trace impurities in fuels by isotope dilution mass spectrometry, in *Trace Elements in Fuel*, Babu, S.P., Ed., American Chemical Society, Washington, D.C., 1975.
62. Lyon, W. S., Ed., *Trace Element Measurements at the Coal-fired Steam Plant*, CRC Press, Boca Raton, Fla., 1977.
63. Lyon, W. S., *Trace Element Measurement at the Coal-fired Steam Plant*, Blackwell Scientific, Oxford, 1977, 144.
64. Yaverbaum. L., Fluidized bed combustion of coal and waste materials, Noyes Data Corp., Park Ridge, New Jersey, 1977.
65. Fancher, J. R., Trace element emission from the combustion of fossil fuels, Environmental Resources Conference on Cycling and Control of Metals, Beattelle, Columbus, Ohio, 1972, 109.
66. Lawasani, M. H., Model of fate of trace elements in coal-fired power plant, M.S. thesis, University of Colorado, 1974, 77.

67. Block, C. and Dams, R., Lead contents of coal, coal ash, and fly ash, *Water, Air, Soil Pollut.*, 5, 207, 1975.
68. Block, C. and Dams, R., Study of fly ash emission during combustion of coal, *Environ. Sci. Technol.*, 10, 1011, 1976.
69. Block, C. and Dams, R., Inorganic composition of belgian coals and coal ashes, *Environ. Sci. Technol.*, 9, 146, 1975.
70. Block, C., Dams, R., and Hoste, J., Chemical composition of coal and fly ash, *Meas. Detect. Control Environ. Pollut. Proc. Int. Symp.*, International Atomic Energy Agency, Vienna, 1976, 101.
71. Volborth, A., Miller, G. E., Garner, C. K., and Jarabek, P. A., Oxygen determination and stochiometry of some coals, Am. Chem. Soc. Div. Fuel Chem. Prepr., 22, 1977, Symp. on the New Tech. in Coal Anal., presented at 174th Am. Chem. Soc. Natl. Meet., Chicago, Ill., 1977, 9.
72. Volborth, A., Miller, G. E., Garner, C. K., and Jerabek, P. A., Material balance in coal. II. Oxygen determination and stoichiometry of 33 coals, *Fuel,* 57, 49, 1978.
73. Volborth, A., Miller, G. E., Garner, C. K., and Jerabek, P. A., Material balance in coal. I. Material balance and oxygen stoichiometry of six coals from Wyoming, *Fuel,* 56, 1976.
74. Volborth, A., Miller, G. E., Garner, C. K., and Jerabek, P. A., Oxygen in coal ash, a simplified approach to the analysis of ash and mineral matter in coal, *Fuel,* 56, 1976.
75. Morrison, R. B., Characteristics of Power Plant Ash, Construction Specification Institute, Huntington, West Virginia, Report June 1972.
76. Gluskoter, H. J., Illinois State Geological Survey, Circular 499, Trace elements in coal: occurrence and distribution, 1977, 154.
77. Abernethy, R. F., Peterson, M. J., and Gibson, F. H., Spectrochemical analysis of coal ash for trace elements, *U.S. Bur. Mines, Rep. Invest.,* 7281, 30, 1969.
78. Abernethy, R. F., Peterson, M. J., and Gibson, F. H., Major ash constituents in U.S. coals, *U.S. Bur. Mines, Rep. Invest.,* 9, 1969.
79. Anderson, C. H. and Beatty, C. D., Spectrographic determination of sodium and potassium in coal ashes, *Anal. Chem.,* 26, 1369, 1954.
80. Ball, C. G. and Cady, G. H., Evaluation of ash correction formulae based on petrographic analysis of mineral matter in coal, *Econ. Geol.,* 30, 72, 1935.
81. Barret, E. P., The fusion flow and clinkering of coal ash; a survey of the chemical background, *Chemistry of Coal Utilization,* John Wiley & Sons, New York, 1, 1945, 496.
82. Barrick, A. G., and Moore, G. F., Empirical correlation of coal ash viscosity with ash chemical compositions, American Society of Mechanical Engineers Paper 76-WA/Fu-3 for Meet Dec. 6, 1976.
83. Beckner, L. J., Trace element composition and disposal of gasifier ash, *Proc. Synth. Pipeline Gas. Symp.,* 7, 359, 1975.
84. Block, C. and Dams, R., Determination of trace elements in coal by instrumental neutron activation analysis, *Anal. Chim. Acta,* 68, 11, 1974.
85. Boyce, I. S., Clayton, C. G., and Page, D., Some considerations relating to the accuracy of measuring the ash content of coal by x-ray backscattering, International Atomic Energy Agency-SM-216/43, 135.
86. Bryers, R. W. and Taylor, T. E., An examination of the relationship between ash chemistry and ash fusion temperature in various coal size and gravity fractions using polynomial regression analysis, *J. Eng. Power, Transac. ASME,* 98, 528, 1976.
87. Cannon, R. M., Seeley, F. G., and Watson, J. S., Engineering analysis and comparison of new processes for the recovery of resource materials from coal ash, From Energy Res. Abstr. 1973, Abstr. No. 36936, Report No. CONF-790205-3, 18 p, 1979.
88. Cooley, S. A. and Ellman, R. C., Analysis of coal and ash from lignite and subbituminous coals of Eastern Montana, Energy Resources of Montana, 22nd Annual Publication of the Montana Geological Society, 1975, 143.
89. Cooper, J. A., Wheeler, B. D., Wolfe, G. J., Bartell, D. M., and Schlafke, D. B., Determination of sulfur, ash and trace element content of coal, coke and fly ash using multielement tube-excited x-ray fluorescence analysis, Proc. ERDA Symp. X-Gamma Ray Sources Appl., 1976, 169.
90. Cooper, J. A., Sheeler, B. D., Wolfe, G. J., Bartell, D. M., and Schlafke, D. B., Determination of sulfur, ash, and trace element content of coal, coke, and fly ash using multielement tube-excited x-ray fluorescence analysis, *Adv. X-Ray Anal.,* 20, 431, 1977.
91. Corey, R. C. and Meyers, J. W., Germanium in coal ash from power plants, *Coal Utilization,* 12, 33, 1958.
92. Curtis, C. W., Terror, A. R., and Guin, J. A., Particle size analysis in the SRC process by Coulter counter, Am. Chem. Soc. Div. Fuel Chem. Prepr., 22, 1977, paper presented at Am. Chem. Soc. Natl. Meet., 174th, Chicago, Ill., 1977, 125.
93. Duel, M. and Abbuell, C. S., The occurrence of minor elements in ash of low-rank coal from Texas, Colorado, North Dakota, and South Dakota, U.S. Geological Survey Bulletin 1036-H, 1956, 155.

94. Duel, M., Biochemical and geochemical origins of ash-forming ingredients in coal, Am. Chem. Soc., Div. Gas and Fuel Chemistry, Chicago, Ill., Meeting, September 7-12, 1958.
95. Dutcher, R. R., White, E. W., and Spackman, W., Elemental ash distribution in coal components — use of the electron probe, Proceedings 22nd Iron-Making Conf. Iron and Steel Division, Am. Inst. Mining Engineers, 22, 463, 1963.
96. Eddy, G. E., Sulfur and ash distribution of Pittsburgh coal, Mathies Mine, Pennsylvania: Evaluation of paleotopographic control, Ph.D. Dissertation, West Virginia University, Morgantown, W. Va., 1971, 69.
97. Ely, F. G. and Bernhart, D. H., Coal ash — its effect on boiler availability, in *Chemistry of Coal Utilization*, John Wiley & Sons, New York, 1962, 820.
98. Hall, R. H. and Lovell, H. L., Spectroscopic determination of arsenic in anthracite coal ashes, *Anal. Chem.*, 30, 1665, 1958.
99. Headlee, A. J. W. and Hunter, R. G., Elements in coal ash and their industrial significance, *Ind. Eng. Chem.*, 45, 548, 1953.
100. Hendrickson, T.A., Ash in coal, *Synthetic Fuels Data Handbook*, Cameron Engineers Inc., 1975, 147.
101. Jones, D. J. and Buller, E. L., Analysis and softening temperatures of coal ash from coals in the Northern Anthracite field, *Ind. Eng. Chem., Anal. Ed.*, 8, 25, 1936.
102. Levene, H. D. and Hand, J. W., Sulfur straws in the ash when lignite burns, *Coal Min. Process.*, 12, 46, 1975.
103. Masson, D. M., Characteristics of ash agglomerates from an ash-agglomerating gasifier, *Am. Chem. Soc., Div. Fuel Chem., Prepr.*, 22, 124, 1977.
104. Mitchell, R. S. and Gluskoter, H. J., Mineralogy of ash of some American Coals, Variations with temperature and source, *Fuel*, 55, 90, 1976.
105. Moody, A. H. and Langan, D. D., Fusion temperature of coal ash as related to composition, *Combustion*, 6, 13, 1935.
106. Moulton, L. K., Bottom ash and boiler slag, *U.S. Bur. Mines, Inf. Circ.*, 8640, 148, 1973.
107. Muter, R. B. and Cockrell, C. F., Analysis of sodium, potassium, calcium and magnesium in siliceous coal ash and related materials by atomic absorption spectroscopy, *Appl. Spectrosc.*, 23, 493, 1969.
108. Nicholls, P. and Selvig, W. A., Clinker formation as related to the fusibility of coal ash, *U.S. Bureau of Mines Bull.*, 364, 67, 1932.
109. Palmenberg, O. W., Relation of ash composition to fusing temperature in anthracite, *Ind. Eng. Chem.*, 31, 1058, 1939.
110. Paulson, L. E. and Fowks, W. W., Changes in ash composition of North Dakota lignite treated by ion exchange, *U.S. Bur. Mines, Rep. Invest.*, 7176, 18, 1968.
111. Peterson, R. E. and Prasky, C., Metalization of pelletized domestic iron oxide superconcentrates with lignite and coal in a rotary kiln., *U.S. Bur. Mines, Rep. Invest.*, RI-8179, 18, 1976.
112. Peterson, M. J. and Zuik, J. B., A semiquantitative spectrochemical method for analysis of coal ash, *U.S. Bur. Mines, Rep. Invest.*, 6496, 15, 1963.
113. Ray, S. S. and Parker, G. F., Characterization of ash from coal-fired power plants, Interagency Energy Environ. Research and Development Program Report, PB-265-374, EPA-600/7-77-010, Washington, D.C., 1977, 129.
114. Rees, O. W., Composition of the ash of Illinois coals, Ill. State Geol. Surv., Circ., 365, 20, 1964.
115. Rekus, A. F., The spectrographic determination of potassium in coal ash, *Appl. Spectrosc.*, 12, 141, 1958.
116. Rhodes, J. R., Daglish, J. C., and Clayton, C. G., A coal monitor with low dependence on ash composition, *Radioisotope Instruments in Industry and Geophysics, Proc. Symp. Warsaw, 1965*, International Atomic Energy Agency, Vienna, 1966, 447.
117. Seeley, F. G., Canon, R. M., and McDowell, W. J., Chemical development of new processes for the recovery of resource materials from coal ash, from Energy Res. Abstr. 1979, Abstr. No. 25257, Report, NO. CONF-790205-4, 1979, 35.
118. Seglin, L. and Lamb, R. H., Preliminary evaluation of a conceptual molten iron gasification process for the production of hydrogen, from Energy Res. Abstr. 1979, Abstr. No. 2669, Report, No. ORNL/Sub-78/13518/1, 1978, 62.
119. Selvig, W. A., and Gibson, F. H., Analysis of ash from United States coals, *U.S. Bur. Mines, Bull.*, 567, 33, 1956.
120. Somerville, M. H. and Elder, J. L., Comparison on trace element analysis of North Dakota lignite laboratory ash with Lurgi gasifier ash and their use in environmental analysis, in Environmental Aspects of Fuel Conversion Technology III, Ayer, G. A., Massoglia, M. F., Eds., Proc. Symp. Env. Aspects Fuel Conv. Tech. Hollywood, 1977 (Report EPA-600/7-78-063).
121. Sondreal, E. A., Elder, J. L., and Kube, W. R., Characteristics and variability of lignite ash from the Northern Great Plains Province, *U.S. Bur. Mines, Inf. Circ.*, 8304, 1966.

122. Sondreal, E. A., Kube, W. R., and Elder, J. L., Analysis of the Northern Great Plains province lignite and their ash: a study of variability, *U.S. Bur. Mines, Rep. Invest.,* 7158, 94, 1968.
123. Steward, K., Germanium in coal ash, *Gas J.,* 274, 279, 1953; **Gas World,** 137, 794, 1953.
124. Thiessen, G., Ball, G. C., and Grotts, P. E., Coal ash and coal mineral matter, *Ind. Eng. Chem.,* 28, 355, 1936.
125. Tufte, P. H., et al., Ash fouling potentials of western subbituminous coal as determined in a pilot plant test furnace, Proc. Am. Power Conf., CONF-760469-5, National Technical Information Service, Springfield, Va., 1976, 16.
126. Warnke, W. E. and Deurbrouck, A. W., Sulfur moisture, and ash reduction in coal preparation research, *Min. Congr. J.,* 63, 43, 1977.
127. Whitemore, D. O. and Switek, J., Geochemical controls on trace element concentrations in natural waters of a proposed coal ash landfill site, from *Gov. Rep. Announce Index,* 159, 86, 1977.
128. Wright, J. H. and Roffman, H. K., Coal ash — A potential mineral source, Inst. Environ. Sci. Annu. Tech. Meet., Mount Prospect., Ill., 1976, 163.
129. Yang, R. T., Shen, M. S., and Steinberg, M., Fluidized-bed combustion of coal with lime additives: catalytic sulfation of lime with iron compunds and coal ash, *Environ. Sci. Technol.,* 12, 915, 1978.
130. Zubovic, P., Hatch, J. R., and Medlin, J. H., Assessment of the chemical composition of coal resources, Proc. U.N. Symp. World Coal Prospects, Katowice, Poland, October 15-23, 1979, 68.
131. Lowry, H. H., Ed., *Chemistry of Coal Utilization,* Suppl. Vol., John Wiley & Sons, New York, 1963.
132. Henderson, P. L., *Brown Coal: Its Mining and Utilization,* Melbourne University Press, Melbourne, Australia, 1953, 53.
133. Mukherjee, D. K. and Chowdhury, P. B., Catalytic effect of mineral matter constituents in a North Assam coal on hydrogenation, *Fuel,* 55, 4, 1976.
134. Lee, R. E. and Von Lehmden, D. J., Trace metal pollution in the environment, *J. Pollution Control Assoc.,* 23, 853, 1973.
135. Peterson, M. J. and Zink, J. B., Semiquantitative spectrochemical method for analysis of coal ash, Bureau of Mines, College Park, Md., Report BM-RI-6496, 1964.
136. Watt, J. D., The Physical and Chemical Behaviour of the Mineral Matter in Coal under the Conditions Met in Combustion Plant, Part I. The Occurrence, Origin, Identity, Distribution and Estimation of the Mineral Species in British Coals, Literature Survey, British Coal Utilization Research Association, Leatherhead, Surrey, England, 1968, 121.
137. Bucklen, O. D., Cockrell, C. F., Donahue, B. A., Leonard, J. W., McPadden, C. R., Meikle, P. G., Mih, L. C., and Shafer, H. E., Coal associated minerals of the U.S., Part 7: Uses, specifications and processes related to coal-associated minerals, RDR-8(7), Coal Research Bureau, National Technical Information Service, Springfield, Va., PB-168, 1965, 116.
138. Lowry, H. H., *Chemistry of Coal Utilization,* 1945, Vol. I and II; and *Chemistry of Coal Utilization Supplemental Volume,* John Wiley & Sons, New York, 1963.
139. Reese, J. T., et al., How coal properties relate to corrosion of high temperature boiler surface, Paper read before the American Power Conference, March 1961.
140. Perry, J. F., Ed., Industrial furnaces, in *Chemical Engineer's Handbook,* McGraw-Hill, New York, 1962.
141. Vorres, K. S., Melting behaviour of coal ash materials from coal ash composition, Am. Chem. Soc. Div. Fuel Chem. Prepr., 22, 1977, *Symp. on the Prop. of Coal Ash, at Am. Chem. Soc./Chem. Inst. of Can. Joint Conf.,* Montreal, Quebec, American Chemical Society, Washington, D.C., 1977, 118.
142. Dijkstra, H. and Sieswerda, B. S., Apparatus for the automatic determination of the ash content of coal, 3rd Int. Coal Prep. Congr., *Ann. Mines Belg.,* 645, 1958.
143. Hardt, L., A rapid method for determining the ash content of coal by means of low-energy radiation, *4th Int. Coal Prep. Congr.,* National Coal Board, London, 1962, 101.
144. Cammack, P., On-stream ash in coal monitoring for profit, *Trans. Soc. Min. Eng. AIME,* 260, 361, 1976.
145. Vasil'ev, A. G., Klemoner, K. S., and Kramarev, O. B., Calculation of the errors in monitoring the ash content of a flow of coal, *Sov. Min. Sci.,* 12, 92, 1976.
146. Clayton, C. G., Applications of nuclear techniques in the coal industry, IAEA-SM-216/101, 85-118, Vienna, 1977.
147. Brown, H. R. and Jones, R. H., reported at 28th Annual Conf. on Appl. of X-Ray Analysis, Denver, Colo., August 2, 1979.
148. Fookes, R. A., Gravitis, V. L., and Watt, J. S., Determination of ash content of coal by mass absorption coefficient measurements at two x-ray energies, IAEA-SM-216/5, 167-182, Vienna, 1977.
149. Renault, J., Rapid determination of ash in coal by Compton scattering, Ca, and Fe x-ray fluorescence, presented at 28th Annual Conf. on Appl. of X-ray Analysis, Denver, Colo., August 2, 1979.

150. Torrey, S., Ed., *Trace Contaminants from Coal,* Noyes Data Corporation, Park Ridge, New York, 1978.
151. Fischer, G. L., et al., Size-dependence of the physical and chemical properties of coal fly ash, *Am. Chem. Soc., Div. Fuel Chem., Prepr.,* 22, 149, 1977.
152. Bibby, D. M., Composition and variation of pulverized fuel ash obtained from the combustion of sub-bituminous coals, New Zealand, *Fuel,* 56, 427, 1977.
153. Smith, L. M. and Larew, H. G., Technology for using waste sulfate in road construction, in CONF-760322, p. 114.
154. Thornton, S. I., Parker, D. G., and White, D. N., Soil stabilization using Western coal high calcium fly ash, in CONF-760322, p. 170.
155. Bern, J., Residues from power generation: processing, recycling, and disposal, in *Land Application of Waste Materials,* Soil Conservation Society of America, Ankeny, Iowa, 1976, 226.
156. Swaine, D. J., Trace elements in coals, *Ocherki Sovrem. Geokhim. Anal. Khim,* 482, 1972.
157. Swaine, D. J., Trace elements in Coal, in *Recent Contributions to Geochemistry and Analytical Chemistry,* Tugarinov, A. E., Ed., John Wiley & Sons, New York, 1976.
158. Martens, D. C., Availability of plant nutrients in fly ash, *Compost Sci.,* 12, 15, 1971.
159. Mulford, F. R. and Martens, D. C., Response of alfalfa to boron in fly ash, *Soil Sci. Soc. Am. Proc.,* 35, 296, 1971.
160. Martens, D. C., Schnappinger, M. G., Jr., and Zelanzy, L. W., The plant availability of potassium in fly ash, *Soil Sci. Soc. Am. Proc.,* 34, 453, 1970.
161. Doran, J. W. and Martens, D. C., Molybdenum availability as influenced by application of fly ash to soil, *J. Environ. Qual.,* 1, 186, 1972.
162. Page, A. L., Elseewi, A. A., and Straughan, I., Physical and chemical properties of fly ash from coal fired power plants with reference to environmental impacts, in *Residue Reviews 71,* Springer-Verlag, New York, 1979.
163. Swanson, V. E., Composition of Coal, Southwestern United States, U.S. Geological Survey, Southwest Energy Study, Coal Resources Work Group, Part II, 1972, 61.
164. Abel, K. H. and Rancitelli, L. A., Major, minor, and trace element composition of coal and fly ash, as determined by instrumental neutron activation analysis, *Adv. Chem. Ser.,* 141, 118, 1975.
165. Adams, L. M., Capp, J. P., and Gillmore, D. W., Coal mine spoil and refuse bank reclamation with power plant fly ash, 3rd Miner. Waste Util. Symp., Chicago, Ill., March 14-16, 1972, 105.
166. Adamkin, L. A., Relation between ash content and germanium content in coals, and its genetic significance, *Dokl. Akad. Nauk SSSR (Geochem.),* 192, 1353, 1970.
167. Alberts, J. J., Burger, J., Kalhorn, S., Seils, C., and Tisue, T., The relative availability of selected trace elements from coal fly ash and Lake Michigan sediment, *Proc. Int. Conf. Nucl. Methods Environ. Energy Res.,* 3, 379, 1977.
168. Aleksandrova, L. N., Ruzinov, L. P., and Starostina, K. M., Optimization of the leaching of germanium from ashes based on the layer combustion of coal by sodium hydroxide solutions, *Izv. Vyssh. Ucheb. Zaved., Tsvet. Met.,* 13, 146, 1970.
169. Anfimov, L. V., Distribution of some trace elements in the ash of coals from the Chelyabinsk basin, *Tr. Inst. Geol. Geokhim., Akad. Nauk SSSR, Ural. Filial,* 90, 64, 1971.
170. Antonijević, V., Pravica, M., Jovanović, M., Distribution of sulfur in ashes, *Hem. Ind.,* 33, 67, 1979.
171. Bangham, D. H., Ed., *Coal Ash: Chemical Composition as a Guide to Behaviour in Furnaces, Progress in Coal Science,* Butterworths, London, 1950.
172. Bickelhaupt, R. E., Effects of chemical composition on surface resistivity of fly ash, Prepared for Environmental Protection Agency, Contract 68-02-1303, NTIS PB-244 885, p. 41, August, 1975.
173. Bierman, A. H., McFarland, A. R., Pawley, J. B., Fisher, G. L., Prentice, B. A., Silberman, D., Odnov, J. M., and Ragaini, R. C., Size-Dependence of the physical and chemical properties of coal fly ash, *Am. Chem. Soc., Div. Fuel Chem., Prepr.,* 22, 1977; Symp. on the Prop. of Coal Ash, at Am. Chem. Soc. Chem. Inst. of Can. Jt. Conf., Montreal Quebec, Can., American Chemical Society, Washington, D.C., 1977, 149.
174. Biernacka, E., Content of trace elements in plants grown on recultivated coal and brown coal ash dumps, *Zesz. Probl. Postepow Nauk Roln.,* 179, 589, 1979.
175. Bochenin, V. I., Reduction in the error of determining the ash content of coal by the radioisotope method, *Solid Fuel Chem.,* 10, 145, 1976.
176. Bochenin, V. I., Shastov, V. P., and Zaitsev, N. I., Determination of the ash content of coal according to the scattered radiation of the iron-55 isotope, *Zaved. Lab.,* 36, 1071, 1970.
177. Bowden, D. N. and Roberts, H. S., Analysis and fusion characteristics of New Zealand coal ashes, *N.Z. J. Sci.,* 18, 119, 1975.
178. Borio, R. W. and Hensel, R. P., Coal-ash composition as related to high-temperature fireside corrosion and sulfur-oxides emission control, *J. Eng. Power,* 94, 142, 1972.

179. Borisenko, L. F., Delitsina, L. V., Melent'ev, G. B., Rossivskii, V. N., and Spitsyn, A. N., Utilization of coal ashes. Deposited Doc., , p. 17, No. VENITI 4051, 1976.
180. Borkowski, J., Krezel, R., Nowak, W., and Wysocki, W., Activation of brown coal ashes for plant cultivation, *Gorn. Odkrywkowe,* 18, 416, 1976.
181. Burek, R., Error in determination of ash content of coal by the method of x-ray transmission and x-ray and beta scattering, *Przegl. Gorn.,* 32, 504, 1976.
182. Burek, R. and Palica, M., Effect of variations in the iron and calcium contents on ash determination by absorption methods using 80-keV x-rays, *Nukleanika,* 14, 1011, 1969.
183. Corey, R. C., et al., Occurrence and determination of germanium in coal ash from power plants, U.S. Bureau of Mines, Bull., 575, 1959, 68.
184. Capp, J. P., A method for utilizing fly ash in surface mine spoil and job pile reclamation, U.S. Bureau of Mines, 1974, 8.
185. Carpenter, B. S. and Reimer, G. M., Fission track technique for uranium determination in coal and fly ash standard reference materials, Proc. Int. Conf. Nucl. Methods Environ. Res., 5T1/PUB/268, IAEA, Vienna, 1971.
186. Chou, K. S., Klem, W. A., Murtha, M. J., and Burnet, G., The lime-sinter process for production of alumina from fly ash, in Ash Utilization, Proc. 4th Int. Ash Util. Symp., St. Louis, Mo., 1976, MERC/SP-76/4, 433.
187. Cockrell, C. F. and Leonard, J. W., Characterization and Utilization Studies on Limestone modified fly ash, Coal Research Bureau Technical Report No. 60, School of Mines, West Virginia University, Morgantown, 1970, 20.
188. Coels, D. G., Ragaini, R. C., Ondov, J. M., Fisher, G. L., Silverman, D., and Prentice, B. A., Chemical studies of stack fly ash from a coal-fired power plant, *Environ. Sci. Technol.,* 13, 455, 1979.
189. Chu, T. Y. H., Ruane, R. J., Milligan, J. D., and Kim, B. R., Environmental evaluation of ash disposal at coal-fired power plants, *Energy Environ.,* 5, 565, 1978.
190. Dergachev, N. F., Kropp, L. I., Khar'kovskii, M. S., Loshkarev, Yu. H., Feshchenko, B. G., Zorin, V. A., and Teplitskii, O. M., Modernization of a wet ash collectors during combustion of fuel, with a high calcium oxide content in the ash, *Energetik,* 10, 11, 1973.
191. Dubinskii, Yu. N., Possibility of using ash of Kansk-Achinsk coals for production of calcium carbide and cyanamide, *Elektr. Stn.,* 7, 24, 1977.
192. Dreesen, D. R., Gladney, E. S., Owens, J. W., Perkins, B. L., Wienke, C. L., and Wangen, L. E., Comparison of levels of trace elements extracted from fly ash and levels found in effluent waters from a coal-fired power plant, *Environ. Sci. Technol.,* 11, 1017, 1977.
193. Dreesen, D. R., Wangen, L. E., Gladney, E. S., and Owens, J. W., Solubility of trace elements in coal fly ash, Department of Energy, Washington, D.C., *DOE Symp. Ser.,* 45, 240, 1978.
194. Dundr, C. J., Recirculation of fly ash as a means of increasing the concentration of germanium, *Strojirenstvi,* 11, 435, 1961.
195. Emlin, B. I., Man'ko, V. A., Nevskii, R. A., Druinskii, M. I., and Melikaev, N. P., Production of electrothermic ferrosilicoaluminium from secondary high-ash coals, Dnepropetr. Metall. Inst. Dnepropetrovsk USSR, *Stal.,* 4, 272, 1979.
196. Erdtmann, G., Determination of uranium and thorium in coal ash and power station precipitator ash and in bauxite and red sludge by activation analysis with epithermal neutrons, *Kerntechnik,* 18, 36, 1976.
197. Fieldner, A. C. and Selvig, W. A., Relation of ash composition to the uses of coal, *AIME Transition,* 74, 456, 1926.
198. Davis, J. E., Fly ash utilization — Status and prospects, Symp. on coal util., 3rd NCA/BCR Coal Conf. and Expo. Louisville, Ky., National Coal Association, Washington, D.C., 1976, 221.
199. Dent, J. K., England overview, Int. Ash Util. Symp., 4th Proc. St. Louis, Mo., Publ. ERDA, Morgantown Energy Res. Cent., MERC/SP-76/4, CONF-760322, W. Va. 14-26, 1976.
200. French, B. R., Fly ash formation in coal-fired boiler, M.S. thesis, Dept. of Chemical Engineering, Univ. of New Hampshire, 1976, 62.
201. Fuchs, W., Rare elements in German brown-coal ashes, *Ind. Eng. Chem.,* 27, 1099, 1935.
202. Furr, A. K., et al., National survey of elements and radioactivity in fly ashes, *Environ. Sci. Technol.,* 11, 1194, 1977.
203. Furr, A. K., Parkinson, T. F., Heffron, C. L., Reid, J. T., Haschek, W. M., Gutenmann, W. H., Pakkala, I. S., and Lisk, D. J., Elemental content of tissues of sheep fed rations containing coal fly ash, *J. Agric. Food Chem.,* 26, 1271, 1978.
204. Goncharov, V. F., Bezbakh, Zh. I., and Kutovai, P.M., Composition of volatile products during the coking of ore-coal mixtures, *Met. Koksokhim.,* 17, 66, 1970.
205. Goncharov, V. F., Bezbakh, Zh. I., Kutovai, P. M., Bim, A. M., Kovel'nitskaya, E. B., Glukhen'kii, A. G., and Ivanschenko, V. V., Behaviour of sulfur compounds in the coking or ore-coal mixtures, *Met. Koksokhim.,* 32, 65, 1972.

206. Gronhovd, G. H., Wagner, R. J., and Wittmaier, A. J., Comparison of ash fouling tendencies of high- and low-sodium lignite from a North Dakota mine, *Proc. Power Conf.*, 28, 632, 1966.
207. Gronhovd, G. H., Hark, A. E., and Paulson, L. E., Ash fouling studies of North Dakota lignite, in Technology and Use of Lignite, *U.S. Bur. Mines, Inf. Circ.*, 8376, 76, 1967.
208. Gronhovd, G. H., Beckering, W., and Tufte, P. H., Study of Factors Affecting Ash Deposition From Lignite and Other Coals, presented at ASME Winter Annual Meeting, Los Angeles, Calif., 1969, 10.
209. Gordon, B. A. and Nagarjan, V., Corrosion of Fe-10Al-Cr alloys by coal char, *Oxid. Met.*, 12, 313, 1978.
210. Gutenmann, W. H., Bache, C. A., Youngs, W. D., and Lisk, D. J., Selenium in fly ash, *Science*, 191, 966, 1976.
211. Gauger, A. W., The significance of ash softening temperature and ash composition in the utilization of coal. Procedures of the American Society of Testing Materials, 37, 376, 1937.
212. Jensen, G. A., Austin, G. T., Iron from coal minerals, *Ind. Eng. Chem., Process Des. Dev.*, 16, 44, 1977.
213. Hodgson, D. R. and Holliday, R., The agronomic properties of pulverized fuel ash, *Chem. Ind.*, 20, 785, 1966.
214. Hood, N., Mineral extraction and cellular concrete from fly ash, in Ash Utilization, Proc. 4th Int. Ash Util. Symp., St. Louis, Mo., MERC/SP-76/4, 1976, 380.
215. Horvath, D. J., Plants and foods of plant origin. Effects of sewage-recycling activities and land application of coal fly ash on the trace-element concentration in foods of plant origin, *Geochem. Environ.*, 3, 73, 1978.
216. Kawam, J., Coal ashes as raw materials for production of industrial metals, *Zesz. Nauk Politech. Slask. Chem.*, 50, 185, 1969.
217. Khrisanfova, A. I. and Sobolova, G. N., Spectrographic determination of coal ash composition, *Tr. Inst. Goryuch. Iskop.*, Moscow, 23, 16, 1967.
218. Kul'skaya, A. O. and Vdovenko, O. F., Spectrochemical determination of germanium, beryllium, and scandium in coal ash, *Khim. Fis. Khim. Spektral'n. Metody Issled. Rssd Redkikh Rasseyan. Elementov, Min. Geol. Okhrany Nedr. SSSR*, 135, 1961.
219. Kyle, J. J. J., On a vanadiferous lignite found in the Argentine Republic with analysis of the ash, *Chem. News*, 66, 211, 1892.
220. Kor, G. J. W., Desulfurization and sulfidation of coal and coal char. I. Desulfurization of coal and coal char at various temperatures and pressures. *Am. Chem. Soc., Div. Fuel Chem., Prepr.*, 22, 27, 1977.
221. Junkers, G., Treatment of fly ash from coal combustion and calcium sulfate from flue gas desulfurization installations, Ger. Offen., Pat. No. 2803764, 1979.
222. Leister, H., Direct reduction of iron minerals of rotary furnaces, Behaviour of ash from coal samples of Charqzeadas, *RS (Rio Grande do Sul) Met. ABM*, Ass. Brasil Metals, 30, 25, 1974.
223. Lebedov, I. K., Zavorin, A. S., Karyakin, S. K., and Privalikhin, G. K., Factional melting of ashes from the coal of Berezovo deposit in the Kansk-Achinsk basin, *Izv. Tomsk. Politekh. Inst.*, 283, 44, 1974.
224. Lipouch, B., Friedberger, R., Dundr, J., Zahradnik, L., Stovik, M., and Vondrakova, Z., Concentrating trace elements (germanium) in coal ash, Pat. No. 12131 (Polish).
225. Lustigova, M., Determination of manganese content in coal ash, *Sb. Pr. UVP*, 32, 149, 1976.
226. Lin, C. J., Liu, Y. A., Vives, D. L., Oak, M. J., Crow, F. E., and Huffman, E. L., Pilot-scale studies of sulfur and ash removal from coals, Institute of Electricians and Electronic Engineers, Piscataway, N. J., Trans Magn. MAG-12, 1976, 513.
227. Martirosyan, G. G. and Safaryan, A. M., Production of calcium hydrosilicates and white portland cement during complex processing of Ekibastuz coal ashes, *Kompleksn. Ispol'z. Miner. Syr'ya*, 6, 70, 1978.
228. Medvedev, G. V., Ruina, V. G., Nikol'skaya, L. N., Kuznetsova, A. A., and Sukhanova, E. I., Physicochemical and structural transformations in high-ash coal and Dzhezdy manganese ore during their simultaneous heating, from Ref. Zh. Metall., 1976. Abstr. No. 5A102, Tr. Khim.-metallurg. In-ta. AN Kaz. SSR, 27, 70-77, 1975.
229. Mikheikin, V. I. and Ziv., E. F., Use of mineralogical methods during the study of the composition of products of thermal treatment of coal ash and dust of iron ores, *Nauch. Tr., Gor. Nauch.-Issled. Proekt. Inst. Redkometsl. Prom*, 25, 129, 1971.
230. Mott, R. A. and Wheeler, R. V., The inherent ash of coal, *Fuel*, 6, 416, 1927.
231. Mukherjee, B. and Dutta, R., Germanium in Indian coal ash, *Sci. Culture*, 14, 538, 1949.
232. Michel, J. R. and Wilcoxon, L. S., Ash deposits on boiler surface from burning central Illinois coal, American Society of Mechanical Engineers, Paper No. 55-A-95, 1955, 17.

233. Lebedov, I. K., Razvalyaev, Yu. V., Levkovokii, A. V., Privalikhin, G. K., and Mosin, E. A., Effect of phase transformations in a bed on the hardening of ash deposits, *Vloyanie. Miner. Chasti Energ. Topl. Usloviya Rab. Parogeneratov, Mater. Vses. Konf.*, 2, 9, 1974.
234. Mikhailova, A. I. and Vlasov, N. A., Trace nutrient level in coals and carbon dioxide-humus fertilizers and their effect on the distribution of manganese, copper, and molybdenum in soil and plants, Mikroelem. Biosfere Primen. Ikh. Sel. Khoz. Med. Sib. Dal'nego Vostoka, Dokl. Sib. Konf., 3rd, 1971, 79.
235. Mikhailova, A. I., Vlasov, N. A., Aliulova, R. A., and Mutovina, T. K., Action of coal humic fertilizer on trace element levels in soil and plants, *Izv. Nauch. Issled. Inst. Nefte-Uglekhim. Sin. Irkutsk. Univ.*, 11, 97, 1969.
236. Mikhailova, A. I., Vlasov, N. A., and Egorova, T. M., Manganese levels in wheat straw and grain during application of coal humus fertilizers, *Izv. Nauch.-Issled. Inst. Nefte-Uglekhim. Sin. Irkutsk Univ.*, 11, 99, 1969.
237. Martens, D. C. and Beahm, B. R., Growth of plants in fly ash amended soils, in Ash Utilization, Proc. 4th Int. Ash Util. Symp., St. Louis, Mo., MERC/SP-76/4, 1976, 657.
238. Machin, J. S. and Wittera, J., Germanium in fly ash and its spectrochemical determination, Illinois State Geological Survey, Circular 216, 1956, 11.
239. Manz, O. E., Concrete Utilizing American Lignite Fly Ash, Paper presented at Economic Commission for Europe, Committee on Electric Power, Symp. on the Use of Ash, in particular in Production of Concrete and Prefabricated Construction Elements, Ankara, Turkey, 1970.
240. Maryland Environmental Service, Feasibility of returning flyash to Western Maryland for mine reclamation, Appalachian Regional Commission, Washington, D.C., NTIS PB-262-520, 1975, 26.
241. Murtha, M. J. and Burnet, G., Recovery of alumina from coal fly ash by high-temperature chlorination, *Proc. Iowa Acad. Sci.*, 83, 125, 1976.
242. Myers, J. E. and Immich, C. G., Natural radioactivity in Appalachian fly ashes, *Trans. Kan. Acad. Sci.*, 75, 174, 1972.
243. Nadkarni, R. A., Multielement analysis of coal and coal fly ash standards by instrumental neutron activation analysis, *Radiochem. Radioanal. Lett.*, 21, 161, 1975.
244. Naser, M. I., Basily, A. B., and Raafat, A. M., Spectrographic estimation of gallium and germanium in coal and coke, *Indian J. Technol.*, 12, 359, 1974.
245. Natesan, K. and Chopra, O. K., Corrosion behaviour of materials for coal-gasification applications, *Proc. Electrochem. Soc.*, 77-1, 493, 1976.
246. Natusch, D. F. S., Bauer, C. F., Matusiewicz, H., Evans, C. A., Baker, J., Loh, A., and Linton, R. W., Characterization of Trace elements in Fly Ash, Symp. Proc. 11, Int. Conf. on Heavy Metals in the Environment, Oct. 27-31, 1975.
247. Ness, H. M., Sondrel, E. A., and Tufte, P. H., Status of flue gas desulfurization using alkaline fly ash from Western coals, Symp. of flue gas desulfurization, New Orleans, PB-255 317, National Technical Information Service, Springfield, Va., 1976, 269.
248. Nichols, G. B. and Bickelhaupt, R. E., Electrostatic collection of fly ash from western coals: some special problems and the approach to their solution, Proc. Technol. and Use of Lignite, Grand Forks, N.D., 1975, 173.
249. Novitskii, N. V. and Ivanova, N. I., Rapid determination of the chemical composition of ashes, and deposits of energy-producing coals, *Khim. Tverd. Topl.*, 3, 60, 1972.
250. Ondov, J. M., Zoller, W. H., Olmez, I., Aras, N. K., Gordon, G. E., Rancitelli, L. A., Abel, K. H., Filby, R. H., Shah, K. R., and Ragaini, R. C., Elemental concentrations in the National Bureau of Standards, Environ. Coal and Fly Ash Standard Reference Materials, *Analyt. Chem.*, 47, 1102, 1975.
251. Perkova, R. I. and Pecherkin, L. A., Determination of germanium in coal ash, *Izv. Vyssn. Ucheb. Zaved., Gorn. Zh.*, 12, 9, 1969.
252. Popović, A., Noble metals in the ash of some coals of the Timok Basin, *Bull. Soc. Chim. Belgrade*, 19, 305, 1954.
253. Panin, V. I. and Shpirt, M. Ya., Quantitative distribution of silicon, aluminum, iron, and calcium compounds in fly-ash of burned powdered coal, *Khim. Tverd. Topl.*, 3, 81, 1971.
254. Papastefanou, C. and Charalambous, S., On the radioactivity of fly ashes from coal power plants, *Z. Naturforsch. A.*, 34A, 533, 1979.
255. Patterson, J. C., Henderlong, P. R., and Adams, L. M., Sintered fly ash as a soil modifier, *Proc. W. Va. Acad. Sci.*, 40, 151.
256. Rafter, T. A., Boron and Strontium in New Zealand coal ashes, *Nature (London)*, 155, 332, 1945.
257. Reiter, F. M., How sulfur content relates to ash fusion characteristics, *Power Eng.*, 59, 98, 1955.
258. Richardson, J. T., Thermomagnetic studies of iron compounds in coal char, *Fuel*, 51, 150, 1972.
259. Rees, W. J. and Sidrok, G. H., Plant nutrition on fly ash, *Plant Soil*, 8, 141, 1956.
260. Santoliquido, P. M., Use of inorganic ion exchangers in the neutron activation determination of arsenic in coal ash, *Radiochem. Radioanal. Lett.*, 15, 373, 1973.

261. Schubert, G., Production of iron concentrates from ashes, *Freiberg. Forschungsh. A.*, 506, 108, 1972.
262. Segnit, E. R., Corrosion of firebrick by brown coal ash, *J. Aust. Ceram. Soc.*, 4, 25, 1968.
263. Shou, J. K. and Pitts, W. S., Calculate coal liquefaction conversion by ash analysis, *Fuel*, 56, 343, 1977.
264. Shpirt, M. Ya., Coal ash benefication by a classification method, *Tr. Inst. Goryuch. Iskop.*, Moscow, 28, 90, 1972.
265. Sobinyakova, N. M. and Aleksandrova, L. N., Ion exchange method of complex treatment of power plant coal ashes. From Ref. Zh., Met., 1973, Abstr. No. 5G278, Miner. Syr'e, 23, 30, 1972.
266. Sobinyakova, N. M., Shmanenkov, I. V., Krainova, L. P., El'Khones, N. M., Starostina, K. M., and Aleksandrova, L. M., Use of an ion-exchange method during the complex processing of energy-producing coal ashes, Tr. Vses. Nauch.-Issled. Inst. Miner. Syr'ya, 23, 30, 1972.
267. Srivastava, S. C., Chatterjee, A., Santikari, A. K., Das, S. K., and Banerji, K. C., A radioisotopic method for nondestructive estimation of ash content in coal, *Natl. Symp. Isot. Appl. Ind.*, 519, 1977.
268. Svadkovskaya, F. E., Gruzinov, V. K., Gorokh, A. V., Pershina, F. R., Takenova, T. D., Reduction of oxides of lean chromium ore and ash from Ekibastuz coal, *Tr. Khim.-Met. Inst. Akad. Nauk Kaz. SSR*, 21, 77, 1973.
269. Szonntagh, J., Farady, L., and Janosi, A., Chromatographic determination of uranium in the ash of Hungarian coals, *Magy. Kem. Foly.*, 61, 312, 1955.
270. Sage, W. L. and McIlroy, J. B., Relationship of coal ash viscosity to chemical composition, *Combustion*, 31, 41, 1959.
271. Schobert, H. H., Petrochemistry of coal ash slags. II. Correlation of viscosity with composition and petrographic class, *Am. Chem. Soc., Div. Fuel Chem., Prepr.*, 22, 143, 1977.
272. Schobert, H. H., Barbie, D. L., Christensen, O. D., and Karner, F. R., Petrochemistry of coal ash slags, I. Formation of melilite and high temperature glass from a calcium-rich, silica-deficient slag, *Am. Chem. Soc., Div. Fuel Chem., Prepr.*, 22, 133, 1977.
273. Soloman, P. R. and Manzione, A. V., New method for sulphur concentration measurements in coal and char, *Fuel*, 56, 393, 1977.
274. Singh, N. and Mathur, S. B., Survey of indigenous coals fly ashes, and flue dusts as a potential source of germanium, *NML (Nat. Mettal. Lab., Jamshedpur, India), Tech. J.*, 15, 42, 1973.
275. Slates, R. V., Methods for analysis of trace elements in coal, coal fly ash, soil and plant samples, from ERDA Energy Res. Abstr. 1977, Abstr. No. 25780, Report, No. DP-1421, 21, 1976.
276. Straughan, I., Elseewi, A. A., and Page, A. L., Mobilization of selected trace elements in residues from coal combustion with special reference to fly ash, *Trace Subst. Environ. Health*, 12, 389, 1978.
277. Strojan, C. L. and Turnter, F. B., Trace elements and sulfur in soils and plants near the Mohave Generating Station in Southern Nevada, Jt. Conf. Sens. Environ. Pollut., Conf. Proc. 4th, 1978, 537.
278. Smith, A. C., The determination of trace elements in pulverized-fuel ash, *J. Appl. Chem.*, 8, 636, 1958.
279. Tomita, A., Mahajan, O. P., and Walker, P. L., Jr., Catalysis of char gasification by minerals, *Am. Chem. Soc., Div. Fuel Chem. Prepr.*, 22, 4, 1977.
280. Tufte, P. H. and Beckering, W., A proposed mechanism for ash fouling burning northern Great Plains lignite, Trans. of the ASME, *J. Eng. Power Series A*, 97, 407, 1975.
281. Weaver, J. N., Neutron activation analysis of trace elements in coal, fly ash and fuel oils, *Anal. Methods Coal Coal Proc.*, 1, 377, 1978.
282. Weclewska, M., Polarographic determination of trace metals in coal ash, *Prace Glownego Inst. Gornictwa, Komun. (Katowice) Ser.*, 242, 1, 1960.
283. Wen, C. Y., Sears, J. T., and Galli, A. F., Role of the carbon-carbon dioxide reaction in gasification of coal and char. From Energy Res. Abstr. No. 46210, 1978.
284. White, H. J., Effects of fly ash characteristics on collector performance, *Air Rep.* 5, 37, 1955.
285. Wijatno, H., Aluminum recovery from coal fly ash by high temperature chlorination, From Energy Res. Abstr. 1978, Abstr. No. 20903, Report, 88, 1977.
286. Wochok, Z. S., Fail, J. L. and Hosmer, M., Analysis of plant growth in fly ash amended soil, in Ash Utilization, Proc. Fourth Internat. Ash Utilization Symp., St. Louis, Mo., MERC/SP-76/4, 1976, 642.
287. Zagorodnyuk, A. V., Magunov, R. L., and Stasenko, I. V., Phase analysis of ashes for germanium compounds, *Zavod. Lab.*, 39, 1060, 1973.
288. Zhukova, T. S., Sheludyakov, L. N., Rogozhkina, S. F., Crystallization of high-alumina iron-containing melts based on ash from Ekibastuz deposit coals, *Izv. Akad. Nauk Kaz. SSR, Ser. Khim.*, 3, 5, 1979.
289. Rose, F., Effect of the volatile constituents of coal on the decomposition behaviour of iron ore during reduction in the rotary kiln, *Arch. Eisenhuettenwes.*, 47, 465, 1976.

290. Bland, A. E., Robl, T. L., and Rose, J. G., Evaluation of interseam and coal cleaning effects on the chemical variability of past and present Kentucky coal refuse, *Soc. Mining Eng., AIME Trans.*, 262, 331, 1977.
291. Bland, A. E., Robl, T. L., and Jerry, G., Kentucky coal preparation plant refuse characterization and uses. Univ. Ky. Inst. Min. Miner Res. Rep. IMMR n. 21, Ky.Coal Refuse Disposal and Util. Semin., 2nd, Proc. Pineville, May 20-21, 1976, Publ. by Univ. of Kentucky, Off. of Res. and Eng. Serv., Lexington, 1976, 21.
292. Bishop, C. and Simon, N. R., Univ. Ky. Inst. Min. Miner Res. Rep. IMMR n 21, Kentucky Coal Refuse Disposal and Utility Seminar, 2nd Proc. Pineville, May 20-21, 1976. University of Kentucky, Off. of Res. and Eng. Serv., Lexington, 1976, 61-67.
293. Faber, J. H., Babcock, A. W., and Spencer, J. D., Eds., Proceedings 4th International Ash Utilization Symposium, St. Louis, 1976, ERDA Report, CONF-760322.
294. Markon, G., World Overview, Conf-760322, 27, 1976.
295. Plank, C. O. and Martens, D. C., Amelioration of soils with fly ash, *J. Soil Water Conserv.*, 28, 177, 1973.
296. Pursglove, J., Address at the first Ash Utilization Symposium at Pittsburgh in 1967.
297. Meyers, R. A., *Coal Desulfurization,* Marcel Dekker, New York, 1977.
298. Condry, L. Z., Recovery of alumina from coal refuse — An Annotated bibliography, Coal Res. Tech. Report No. 130, College of Mineral and Energy Resources, West Virginia University, Morgantown, 1976, 8.
299. Condry, L. Z., Possible effluents from coal desulfurization processes, West Virginia University Coal Research Bureau School of Mines, Tech. Rep. 121, 1976, 3.
300. Cole, W. A., Grosh, W. A., and Stehlik, C. J., Iowa coals as a surface of sulfur, *U.S. Bur. Mines, Rep. Invest.*, 5362, 23, 1957.
301. Kestner, W. M., Leonard, J. W., and Wilson, E. B., Reduction of sulfur from steam coal by magnetic methods, *Min. Cong. J.*, 53, 70, 1967.
302. Leonard, J. W., Holland, C. T., and Syed, E. U., Unusual methods of sulfur removal coal: A Survey, Coal Res. Bureau Technical Report, No. 30, School of Mines, West Virginia University, Morgantown, 1967, 12.
303. Cavallaro, J. A. and VanEck, O. J., Preparation characteristics and desulfurization potential of Iowa coals, U.S. Bur. Mines, Rep. Invest., 7830, 32, 1973.
304. Chung, K. E., Products from coal minerals, Office of Coal Research, R&D Report No. 53, Development of a Process for Producing an Ashless, Low-Sulfur Fuel from Coal, VI, Part 3, Contract No. 14-01-0001-495, NTIS PB-237 764, 1974, 40.
305. Meyers, R. A., Removal of pyritic sulfur from coal using solutions containing ferric ions, U.S. Patent 3917465, 1975.
306. Meyers, R. A., Removal of pyritic sulfur from coal, U.S. Patent 3926575, 1975.
307. Leonard, J. W., Potential for further lowering sulfur in intensively cleaned coal at mines, Coal Research Bureau Technical Report, College of Mineral and Energy Resources, West Virginia University, Morgantown, 1975, 3.
308. Stewart, O. W., Shou, J. K., and Jackson, E. V., Clean coal combustion through desulfurization and blending, Univ. Ky Inst. Min. Miner. Rep. Res. IMMR, 19-PD14-76 Energy Resour. Conf. 4th, Univ. of Ky., Lexington, Jan. 7-8, 1976, 63.
309. Attar, A., Corcoran, Am. H., and Gibson, G. S., Transformation of sulfur functional groups during pyrolysis of coal, *Am. Chem. Soc., Div. Fuel Chem., Prepr.*, 21, 1976; Paper presented at Am. Chem. Soc. Natl. Meet., 172nd, San Francisco, Calif., 106, 1976.
310. Ganguli, P. S., Hsu, G. C., Gavalas, G. R., and Kalfayan, S. H., Desulfurization of coal by chlorinolysis, Am. Chem. Soc. Div. Fuel Chem. Prepr. 21, 1976, Pap. presented at Am. Chem. Soc. Natl. Meet. 172nd, San Francisco, Calif., 118, 1976.
311. Ponder, W. H., Stern, R. D., and McGlamery, G. G., SO_2-Control Methods Compared, *Oil Gas J.* 74, 60, 1976.
312. Stambaugh, E. P., Levy, A., Giammar, R. D., and Sekhar, K. C., Hydrothermal coal desulfurization with combustion results, in Energy and the Environment (L. Theodore, ed.), Proc. 4th Nat. Conf. on Energy and Envir. Cincinnati, 1976.
313. Stambaugh, E. P., Miller, J. F., Tam, S. S., Chauhan, S. P., Feldman, H. F., Carlton, H. E., Foster, J. F., Nack, N., and Oxley, J. H., Coal desulfurization by the battelle hydrothermal coal process, IEEE Power Eng. Soc. Text of a Paper From the Summer Meet., Portland, Oreg., Institute Electricians and Electronic Engineers, Piscataway, New Jersey, Pap A 76, 1976.
314. Stambaugh, E. P., Miller, J. F., Tam, S. S., Chauhan, S. P., Feldman, H. F., Carlton, H. E., Nack, N., and Oxley, J. H., Clean fuels from coal, Coal Process Technol. 3, American Institute of Chemical Engineers, New York, 1977, 1.

315. Stambaugh, E. P., Miller, J. F., Tam, S. S., Chauhan, S. P., Feldman, H. F., Carlton, H. E., Nack, N., and Oxley, J. H., The battelle hydrothermal coal desulfurization process, *Proc. Air Pollut., Ind. Hyg. Conf. Air Qual., Manage. Electr. Power Ind.*, 12th, 1976, 211.
316. Friedman, S. and Warzinski, P. P., ASME Pap n 76-WA/APC-2 for Meet. Dec. 5-10, 1976, Pittsburgh, Chemical Cleaning of Coal, 4.
317. Cohen, E., Application of a superconducting magnet system to the cleaning and desulphurization of coal, *IEEE Trans. Magn.* MAG-12, 503, 1976.
318. Fine, H. A., Lowry, M., Power, L. F., and Geiger, G. H., Proposed process for the desulfurization of finely divided coal by flash roasting and magnetic separation, *IEEE Trans. Magn.* MAG-12, 523-527, 1976.
319. Hendel, F. J. and Winkler, J., Desulfurization of coal-oil mixtures by attrition grinding with activated iron powder, Monogr. on Alternate Fuel Resources Based on Pap. Presented at the Symp. on Alternate Fuel Resources, Santa Maria, Calif. 1976, Publ. by West Period. Co. (AIAA v 20), North Hollywood, 1976, 357.
320. Vasan, S. and Willett, H. P., Economics of desulfurization of coal-gas vs. flue-gas desulfurization. Symp of Coal Util., 3rd Natl. Coal Assoc./Bitum Coal Res., Inc. Coal Conf. and Expo 3, Paper, Louisville, Ky., NCA, Washington, D.C., 1976, 111.
321. Murray, H. H., Magnetic desulfurization of some Illinois basin coals, *Am. Chem. Soc., Div. Fuel Chem., Prepr.*, 22, 1977; Symp. on Desulfurization of Coal Char, New Orleans, La., 1977, 106.
322. Boateng, D. A. D. and Phillips, C. R., Desulfurization of coal by flotation of coal in a single-stage process, *Sep. Sc.*, 12m, 71, 1977.
323. Friedman, S., LaCount, R. B., and Warzinski, R. P., Oxidative desulfurization of coal, *Am. Chem. Soc., Div. Fuel Chem., Prepr.*, 22, 1977; at 173rd Natl. Meet., Symp. on Desulfurization of Coal and Coal Char, New Orleans, La., March 21-25, 1977, 100.
324. Wilson, D.C., Dry table-pyrite removal from coal, *Am. Chem. Soc., Div. Fuel Chem., Prepr.*, 22, 1977; at 173rd Natl. Meet. Symp. on desulfurization of coal and coal char, New Orleans, La., Mar. 21-25, 1977.
325. Lin, C. J. and Liu, Y. A., Desulfurization of coals by high-intensity high-gradient magnetic separation (HIGMS): Conceptual Process Design and Cost Estimation, *Am. Chem. Soc., Div. Fuel Chem., Prepr.*, 22, 1977; at 173rd Natl. Meet., Symp. on Desulfurization of Coal Char, New Orleans, La., 1977, 93.
326. Sareen, S. S., Sulfur removal from coals; ammonia/oxygen system, *Am. Chem. Soc., Div. Fuel Chem., Prepr.*, 22, 1977; at 173rd Natl. Meet., Symp. on Desulfurization of Coal Char, New Orleans, La., 1977, 93.
327. Hamersma, J. W., Kraft, M. L., and Meyers, R. A., Applicability of the Meyers process for desulfurization of U.S. Coal (A survey of 35 coals), *Am. Chem. Soc., Div. Fuel Chem., Prepr.*, 22, 1977.
328. Fleming, D. K., Smith, R. D., and Aquino, M. R. Y., Hydrodesulfurization of coals, *Am. Chem. Soc., Div. Fuel Chem., Prepr.*, 22, 1977; at 173rd Natl. Meet., Symp. on Desulfurization of Coal and Coal Char., New Orleans, La., Mar. 21-25, 1977, 45.
329. Haldipur, G. B. and Wheelock, T. D., Desulfurization of coal in a fluidized bed reactor, *Am. Chem. Soc., Div. Fuel Chem., Prepr.*, 22, 1977.
330. Huang, E. T. K. and Pusifer, A. H., Coal desulfurization during gaseous treatment, Am. Chem. Soc. Div. Fuel Chem. Prepr. 22, No. 7, 1977.
331. Evans, J. M., *Alternatives to Stack-Gas Scrubbing, Coal Process Technology*, American Institute of Chemical Engineers (CEP Tech. Man.), New York, 1977, 47.
332. Ratcliffe, C. T. and Pop, G., Chemical reduction of sulfur dioxide to free sulfur with lignite and coal. I. Steady state reaction Chemistry and interaction of volatile components, *Fuel*, 59, 237, 1980.
333. Guruz, K. and Celebi, S., Sulfur removal by pyrolysis of Turkish lignites, 1497832, *Fuel*, 58, 893, 1979.
334. Morrison, R. E., Power Plant Ash: A New Mineral Resource, in CONF-760322, 200, 1976.
335. Mitchell, D. R., The recovery of pyrite from coal-mine refuse, *Trans. Am. Inst. Mining Metall. Eng., Coal Div.*, 157, 141, 1944.
336. Sun, S. C. and Savage, K. L., Flotation Recovery of Pyrite from Bituminous Coal Refuse, *Trans. Soc. Mining Eng. AIME*, 241, 377, 1968.
337. Ergun, S. and Bean, E. H., Magnetic separation of pyrite from coals, *U.S. Bur. Mines, Rep. Invest.* 7181, 25, 1968.
338. Jensen, G. A., The kinetics of gasification of carbon contained in coal minerals at atmospheric pressure, *I & EC Process Design Dev.*, 14, 308, 1975.
339. Kelley, K. K., Shomate, C. H., Young, F. E., Naylor, B. F., Salo, A. E., and Huffman, E. H., Thermodynamic Properties of Ammonium and Potassium Alums and Related Substances, with reference to Extraction of Alumina from Clay and Alunite, U.S. Bureau of Mines, Tech. Paper 688, 1946.

340. Grim, R. E., Machin, J. S., and Bradley, W. S., Amenability of Various Types of Clay Minerals to Alumina Extraction by the Lime Sinter and Lime-Soda Sinter Processes, Div. of State Geological Survey, Bull. 65, Urbana, Illinois, 1945.
341. Cavin, D. C., A Study of Iron and Aluminum Recovery from Power Plant Fly Ash, M.S. thesis, Ch. E. Dept., Iowa State University, Ames, 1973.
342. Funnell, J. E. and Curtice, D. K., Characterization Studies of Fly Ash From the Four Corners Generating Station, Final Report, SWRI Project No. 07-2940, Southwest Research Institute, San Antonio, Texas, 1971.
343. Kealy, D., Backer, R. R., Atkins, L., Busch, R., Those waste banks could be sources for fuel, alumina, *Coal Mining and Processing,* August, 1976, 46.
344. Powell, A. R., Extraction of germanium and gallium, *Chem. and Ind.,* 40, 1225, 1954.
345. Lever, F. M., Germanium and gallium, *Metal Ind.,* 80, 3, 1952.
346. Crawley, R. H. A., Sources of Germanium in Great Britain, *Nature (London),* 175, 291, 1955.
347. Reynolds, F. M., The occurrence of vanadium, chromium and other unusual elements in certain coals, *J. Soc. Chem. Ind.,* 67, 341, 1948.
348. Fletcher, M. F., The Concentration and Location of Germanium in Some British Coals, Great Britain, Nat. Coal Board. Sci. Dept., Central Res. Establishment, Rept. 1232, December 1954.
349. Beard, W. J., Germanium, Min. Res. I.C. No. 12, Mines Branch, Ottawa, Canada, 10, 1955.
350. Williams, A. E., Germanium From Coal, *Iron Coal Trades Rev.,* 163, 29, 1951.
351. Chirnside, R. C. and Cluley, H. J., Germanium from coal. A British source of Germanium for use in crystal valves, *G.E.C. J.,* 19, 94, 1952.
352. Losev, B. I., Mel'nikova, A. N., and El'piner, I. E., Halogenation and extraction of germanium from coal in a field of ultrasonic waves, *Izvest. Akad. Nauk S.S.S.R.,* Otdel. Tekh. Nauk, No. 9, 90, 1957, Chem. Abs. 52, p. 5790, 1958.
353. Losev, V. I. and Nikiforova, T. S., Germanium in coal, *Zhur. Priklad. Khim.,* 33, 730, 1960.
354. Bylyna, E. A., Losev, B. I., and Troyanskaya, M. A., Extraction of Germanium from coal with gamma-irradiated carbon tetrachloride, *Izvest. Akad. Nauk SSSR, Otdel. Tekh. Nauk* 4, 124, 1958.; Chem. Abstr., 52, 17004, 1958.
355. Pogrebitskii, E. O., Regularity in distribution of Ge in the donets basin coals, *Zapiski Leningrad, Gron. Inst.,* 35, 87, 1959.
356. Bunina, M. V., The coal field of Ubagansk in the coal basin of Turgaisk, Trudy Ob'edinennoi Kustanaiskoi Nauch Sessii Posvyashennoi Problemam Turgaisk, Regional. *Ekon. Kompleksa,* 2, 290, 1958, Chem. Abstr., 54, p. 24179, 1960.
357. Takas, P. and Horvath, A., Possibilities of producing germanium and gallium from the by-products of the coal industry, *Kahaszati Lapok,* 7, 1959.
358. Zahradnik, L., Germanium in the products of direct coal combustion and its extractability by muriatic acid, *Chem. Prum.,* 9, 1959.
359. Hampel, A. C., Ed., *Germanium, Rare Metals Handbook,* Reinhold, New York, 1954.
360. Menkovskii, M. A. and Aleksandrova, A. N., The Selection of Conditions for Ashing Coal for Germanium Determination, *Zavod. Lab.,* 29, 1963.
361. Miller, R. L. and Gill, J. R., Uranium from coal, *Scientific American,* 191, 36, 1954.
362. George, D., Comparative characterization of coal shale and petroleum liquids, *Am. Chem. Soc., Div. Pet. Chem., Prepr.,* 22, 1977; Anal. Chem. of Tar Sands and Oil Shale, New Orleans, La., 1977, 785.
363. Erving, R. A. et al., Recovery of uranium from North Dakota lignites, Report EMI-237, June 1, 1955.
364. Pitmon, G. R., Peter, M. A., Gardner, H. E., Winslow, G., and Shimin, R. L., Pilot Plant Testing of Dakota Lignites, U.S. Atomic Energy Commission Report WIN-81, 1957.
365. Breger, I. A. and Deul, M., The organic geochemistry of uranium. Contribution to the geology of uranium and thorium by the United States Geological Survey and Atomic Energy Commission for the United Nations International Conference on Peaceful Uses of Atomic Energy, Geneva, Switzerland, U.S. Geol. Surv. Profess. Pap., 300, 505, 1955.
366. Breger, I. A., Deul, M., and Meyrowitz, R., Geochemistry and Mineralogy of a Uraniferous Subbituminous coal. *Econ. Geol.,* 50, 610, 1955.
367. Breger, I. A., Meyrowitz, R., and Duel, M., Effects of destructive distillation on the uranium associated with selected naturally occuring carbonaceous substances, *Science,* 120, 310, 1954.
368. Breger, I. A., Meyrowitz, R., and Warr, J. J., Extraction of uranium from the red desert coal of Wyoming, *U.S. Geological Survey TEI,* 372, 21, 1953.
369. Breger, I. A. and Schopf, J. M., Germanium and uranium in coalfield wood from upper devonian black shale, *Geochim. Cosmochim. Acta,* 7, 287, 1955.
370. Porter, E. S. and Petrov, H. G., Recovery of uranium from lignites, *Min. Eng.,* 1004, 1957.

371. Woody, R. J., George, D. R., Petrow, H., Breymann, J. B., and Porter, E. S., Laboratory Investigation of Dakota Lignites, U.S. Atomic Energy Commission Report WIN-54, 1957.
372. Blake, C. A., Brown, K. B., and Coleman, C. F., The extraction and recovery of uranium (and vanadium) from acid liquors with di(2-ethylhexyl)phosphoric acid and some other organophosphorous acids, Oak Ridge National Laboratories, ORNL-1903, May 13, 1955.
373. Crouse, D. J. and Brown, K. B., Amine extraction processes of uranium recovery from sulfate liquors, Vol. I, Oak Ridge National Laboratories, ORNL-1959, May 27, 1954.
374. Crouse, D. J., Brown, K. B., Arnold, W. D., Moore, J. B., and Lowrie, R. S., Progress report on uranium extraction with organonitrogen compounds, Oak Ridge National Laboratories, ORNL-2099, May 14, 1956.
375. Grinstead, R. R., Shaw, K. G., and Long, R. S., Solvent extraction of uranium from acid leach slurries and solutions, Proceedings of the International Conference on the Peaceful Uses of Atomic Energy, 8, 71, 1956.
376. Smith, K. R., The coal-uranium breeder: uranium from coal, *Energy*, 2, 171, 1977.
377. Josa, J. M., Merino, J. L., and Villoria, Y. A., Lignitos radioactivos Espanoles, naturaleza y solubilizacion del urania, in *Proc. of Low-Grade Uranium Ores*, International Atomic Energy Agency, Vienna, 1967, 157.
378. Hayes, J. R., Bed Combustion, *Ind. Res. Dev.*, January 1980, 44.
379. Jung, K. and Sanmore, B. R., Fluidized bed combustion of wet brown coal, *Fuel*, 59, 74, 1980.
380. Braunstein, H. M. and Copenhaver, E. D., Environmental Health and Control Aspects of Coal Conversion: An Information Overview, Oak Ridge National Lab., Tenn. ORNL/EIS-94, 1977, 538.
381. Kasper, S., Coal conversion chemistry, see it in common terms, *Ind. Res. Dev.*, Jan. 1981, 164.
382. Glenn, R. A., Basu, A., Wolfarth, J. S., and Katz, M., Coal hydrogenation process studies. I. On the role of coal sulphur in the hydrogenation of coal, *Fuel*, 29, 149, 1950.
383. Given, P.H., et al., Dependence of coal liquefaction behaviour on coal characteristics. I. Vitrinite-Rich Sample; II. Role of Petrographic Composition, *Fuel*, 54, 34, 1975.
384. Given, P. H., et al., The relation of coal characteristics to coal liquefaction. Pennsylvania State University, sponsored by National Science Foundation, RANN Program, Grant GI-38974, NTIS PB 239-261, 1974, 83.
385. Filby, R. H., Shah, K. R., Hunt, M. L., Khalil, R. S., and Sautter, C. A., Solvent refined coal (SRC) Process, Trace elements NTIS, Report FE-496-T-17, National Technical Information Service, Springfield, Va., 1978.
386. Filby, R. H., Shah, K. R., and Sautter, C. A., Trace elements in the solvent refined coal process, U.S. Environmental Protection Agency, Office Research and Development, Report No. EPA/600/7-78/063, 1978, 266.
387. Filby, R. H., Shah, K. R., and Sautter, C. A., A study of trace element distribution in the solvent refined coal (SRC) process using neutron activation analysis, *J. Radioanal. Chem.*, 37, 693, 1977.
388. Filby, R. H., Shah, K. R., and Yaghmaie, F., The nature of metals in petroleum fuels and coal-derived synfuels, Ash Deposits Corros., Impurities Combust. Gases, 51, 1978.
389. Coleman, W. M., Perfetti, P., Dorn, H. C., and Taylor, L. T., Trace element distribution in various solvent refined coal fractions as a function of the feed coal, *Fuel*, 57, 612, 1978.
390. Wakeley, L. D., Davis, A., Jenkins, R. G., Mitchell, F. D., and Walker, P. L., The nature of solids accumulated during solvent refining of coal, *Fuel*, 58, 379, 1979.
391. Saxby, J. D., Atomic H/C ratios and the generation of oil from coals and kerogens, *Fuel*, 59, 305, 1980.
392. Miller, J. K., Christianson, G. A., Schock, M. R., Morrison, W. W., and Billings, M. I., Trace element effects of energy conversion facilities: a phase one final report to the Old West Regional Commission, Report NP-23774, 204, 1977.
393. Austin, L. G., Mechanical and comminutice properties of coal, Materials problems and research opportunities, in *Coal Conversion*, Vol. 2, Staehle, R.W., Ed., Ohio State University, Columbus, NTIS PE-248 081, April 1974, 49.
394. Babu, S. P., et al., Suitability of West Virginia coals to coal conversion processes, Coal-Geology Bulletin, No. 1, West Virginia Geological and Economic Survey, December 1973, 32.
395. Clavenna, L. R., Recovery of alkali metal compounds for use in catalytic coal conversion processes. Ger. Offen. (270778), Patent 2802825, 1978.
396. Condry, L. Z. and McClung, J. D., Potential waste by-products from selected coal conversion processes, Coal Research Bureau Technical Report, 113, 1975, 25.
397. Cox, J. L., Catalysts for coal conversion, Clean Fuels from 2nd Coal Symp., Illinois Institute of Technology, Chicago, June 23-27, 1975, including Selected Papers from 1st Symp., Illinois Institute of Technology, Chicago, Published by Institute of Gas Technology, Chicago, Ill., 271, 1975.
398. Dailey, N. S., Process effluents: quantities and control technologies in environmental, health and control aspects of coal conversion, Braunstein, H. M., Copenhever, E. D., and Pfuderer, H. A., Eds., Report ORNL/EIS-94, Oak Ridge National Laboratories, Oak Ridge, Tenn., 1977.

399. Davis, A., Spackman, W., and Given, P. H., The influence of the properties of coals in their conversion into clean fuels, *Symposium on the Role of Technology in the Energy Crisis,* American Chemical Society, Washington, D.C., 1974, 461.
400. Dooley, J. E. and Thompson, C. J., Analysis of liquids from coal conversion process, *Am. Chem. Soc., Div. Fuel Chem., Prepr.,* 21, 1976; Symp. on Coal Liq. 172nd, Natl. Meet., San Francisco, Calif., 1976, 243.
401. Eisenberg, W. C. and Solomon, I. J., Formation of benzene polycarboxylic acids in the oxygen oxidation of solid by-products from coal conversion process (CCP) in an alkaline medium, *Fuel,* 56, 181, 1977.
402. Leonard, J. W., Holland, C. T., and Syed, E. U., Unusual methods of sulfur removal coal: A Survey, Coal Res. Bureau Technical Report, No. 30, School of Mines, West Virginia University, Morgantown, 1967, 12.
403. Ensminger, J. T., Coal: Origin, classification, and physical and chemical properties, in Environmental, health and control aspects of coal conversion, Braunstein, H. M., Copenhover, E. D., and Pfuderer, H. A., Eds., Report ORNL/EIS-94, Oak Ridge National Laboratory, Oak Ridge, Tenn., 1977.
404. French, J., Fairchild, G. A., Wilson, W. E., Nadler, J. C., Knapp, K. T., Turner, B., Samfield, M., Smith, J. R., and Jefcoat, J. A., Preliminary Assessment of the Current Impact Resulting from the Conversion from Oil or Gas to Coal Fuel in Large Utility Boilers, U.S. Environmental Protection Agency, Washington, D.C., 1975.
405. Given, P. H., Problems in the chemistry and structure of coals as related to pollutants from conversion processes, Symp. Proceedings, Environmental Aspects of Fuel Conversion Technology, EPA-650/2-74-118, St. Louis, Mo., 1974, 27.
406. Goeken, R. J., Coal conversion building toward commercial development, Proc. Rocky Mt. Coal Mining Institutes 72nd Reg. Meet., Vail, Colo., 1976, 32.
407. Herbes, S. E., Southworth, G. R., and Gehrs, C. W., Organic Contaminants in Aqueous Coal Conversion Effluents, Environmental Consequences and Research Priorities, Proc. 10th Annual Conf. on Trace Substances in Environmental Health, Columbia, Missouri, 1976.
408. Hicklen, W. L., The construction of coal conversion vessels, *Energy Common.,* 2, 133, 1976.
409. Hildebrand, S. G., Cushman, R. M., and Carter, J. A., The potential toxicity and bioaccumulation in aquatic systems of trace elements present in aqueous coal conversion effluents, *Trace Subst. Environ. Health,* 10, 305, 1976.
410. Hill, G. R., Coal conversion — A partial answer, *EPRI J.,* 2, 22, 1977.
411. Hodgson, R. L., Hydroconversion of coal with combinations of catalysts, S. African Patent 69 05181, 1970.
412. Hoffman, J. R., Long, W. T., and Williams, R. L., Demonstration of reliability methodology using two coal conversion plant models, *Natl. Bur. Stand., Spec. Publ.,* 468, 1977.
413. Howard-Smith, I. and Werner, G. J., Coal conversion technology, Coal Convers. Technol. Publ. by Noyes Data Corp. Park Ridge, N.Y., 1976, 133.
414. Jahnig, C. E., Evaluation of pollution control in fossil fuel conversion processes, Gasification, Section 5, BI-GAS Process, EPA-650/2-74-009-g, NTIS PB-234 694, 1975, 65.
415. Jahnig, C. E., Evaluation of Pollution control in fossil fuel conversion process, Gasification, Section 6, HYGAS Process, EPA-650/2-74-009-h, NTIS PB-247 225, 1975, 53.
416. Jahnig, C. E., Evaluation of pollution control in fossil fuel conversion processes, Gasification: Section 7, U-Gas Process, EPA-650/2-74-009-i, NTIS PB-247 226, 1975, 3.
417. Katzman, H., A research and development program for catalysis in coal conversion process, Report No. EPRI 207-0-0, NTIS PB 242-412, 1974, 229.
418. Koester, P. A. and Zieger, W. H., Analysis for radionuclides in SRC and coal combustion samples, from Gov. Rep. Announce Index, U.S., 1979, Report, 33, 1978.
419. Koppenhaal, D. W. and Manahan, S. E., Hazardous chemicals from coal conversion processes, *Environ. Sci. Technol.,* 10, 1104, 1976.
420. Leas, A.M., Intergrated coal conversion, U.S. Patent No. 4132627, 1979.
421. Lee, B. S., Hygas process achieves 92% coal conversion, *Oil Gas J.,* 75, 74, 1977.
422. Magee, E. M., Evaluation of pollution control in fossil fuel conversion processes, Final Report, EPA-600/2-72-101, NTIS PB-255 842, 1976, 292.
423. Petrakis, L. and Grandy, D. W., Free radicals in coals and coal conversion. II. Effect of liquefaction processing conditions on the formation and quenching of coal free radicals, *Fuel,* 59, 227, 1980.
424. Samans, C. H. and Hulsizer, W. R., Current progress in materials development for coal conversion, *J. Eng. Mater. Technol.,* 99, 372, 1977.
425. Spackman, W., Davis, A., Walker, P. L., Lovell, H. L., Essenhigh, R. H., Vastola, F. J., and Given, P. H., Characteristics of American coals in relation to their conversion into clean energy fuels, NTIS, Report FE-2030-7, National Technical Information Service, Springfield, Va., 1977.

426. Staehle, R. W., Critical materials problems in coal conversion, Natl. Bur. Stand. Spec. Publ., 468, 1977; Prov. of Failures in Coal Convers, Proc. of Meet. of Mech. Failures Prev. Group, 24th, Columbus, Ohio, 1976, 50.
427. Stone, L. K., Emissions from coal conversion processes, *Chem. Eng. Prog.,* 72, 52, 1976.
428. Williams, J. M., Henslry, W. K., Wewerka, E. M., Wanek, P. L., and Olsen, J. D., Trace element distribution in several coal conversion processes: an exchange program between the Los Alamos Scientific Laboratory and National Coal Board of England, From Energy Res. Abstr. 1978, Abstr. No. 32594, Report, 1978, 26.
429. Anderson, L. L., Wood, R. E., and Wiser, W. H., Clean liquid energy from coal, *Trans. Soc. Min. Eng. AIME,* 260, 318, 1976.
430. Appell, H. R., Costeam-Low-Rank coal liquefaction: an updated analysis, *Energy,* 1, 24, 1976.
431. Brunson, R. J., Liquefaction of calcium-containing subbituminous coals and coals of lower rank, U.S. Patent 416460.
432. Carlson, F. B., Yardumian, L. H., and Atwood, M. T., Toscoal process — coal liquefaction and char production. Clean Fuels from 2nd Coal Symp., Paper Illinois Inst. of Technol. Chicago, from 1st Coal Symp. Illinois Inst. of Technol., Sept. 10-14, 1973, Publ. by Inst of Gas Technol. Chicago, 1975, 495.
433. Charanjit, R., Appell, H. R., and Illig, E. G., Liquefaction of Western subbituminous coals with synthesis gas. *Am. Chem. Soc., Div. Pet. Chem., Prepr.,* 22, 1977; Gen. Pap. Am. Chem. Soc. Meet., Chicago, Ill. Aug. 28-Sep. 2, 1977, 1108.
434. Clarke, J. W. and Rantell, T. D., Filtration in coal liquefaction. Influence of filtration conditions in nonhydrogenated systems, *Fuel,* 59, 35, 1980.
435. Fan, L. T., Miyanami, K., and Fan, L. S., Model for the extractive coal liquefaction-l. Modeling and Simulation of the Nenken Coal Liquefaction Process, Kansas State Univ. Inst. Syst. Des Optim. Rep. 71, 1976, 33.
436. Farcasiu, M., Fractionation and structural characterization of coal liquids, *Fuel,* 56, 9, 1976.
437. Frankel, H. E. and Ogren, J. R., Reliability problems in coal gasification and liquefaction, Natl. Bur. Stand. Spec. Publ. No. 468, Apr. 1977; Prev. of Failures in Coal Convers, Proc. of 24th Meet. of Mech. Failures Prev. Group, Columbus, 1976, 3.
438. Friedman, S., Akhtar, S., and Yavorsky, P. M., Overview of coal liquefaction projects, Technol. and Use of Lignite, Proc. of Symp., Grand Forks, N.D., Publ. by ERDA, Office of Public Affairs, Technical Information Center (GFERC/IC 752), Oak Ridge, Tenn., 1975, 202.
439. Fu, Y. C. and Batchelder, R. F., Catalytic liquefaction of coal, *Am. Chem. Soc., Div. Fuel Chem., Prepr.,* 21, 1976; Symp. on Coal Liq., 172nd, Natl. Meet., San Francisco, Calif, 1976, 78.
440. Given, P. H., Inorganic species in coal and their potential catalytic effects on liquefaction processes, Appendix in A Research and Development Program for Catalysis in Coal Conversion, NTIS No. PB 242-412, A-122-A-140, National Technical Information Service, Springfield, Va., 1974.
441. Gleim, W. K. T., Liquefaction of coal without catalyst using selective hydrogenation, *Am. Chem. Soc., Div. Fuel Chem., Prepr.,* 21, 1976; Symp. on Coal Liq., 172nd Natl. Meet., San Francisco, Calif., 1976, 91.
442. Gorbaty, M. L., Taunton, J. W., Liquefaction for calcium-containing sub-bituminous and low-grade coals, Ger. Offen. Patent 2807203, 1978.
443. Gorin, E., Kulik, C. J., and Lebowitz, H. E., Deashing of coal liquefaction products via partial deasphalting. I. Hydrogen donor extraction effluents, *Ind. Eng. Chem. Process Des. Dev.,* 16, 95, 1977.
444. Gorin, E., Kulik, C. J., and Lebowitz, H. E., Deashing of coal liquefaction products via partial deasphalting. II. Hydrogenation and hydroextraction effluents, *Ind. Eng. Chem. Process Des. Dev.,* 16, 102, 1977.
445. Granaf, B., Mineral matter effects in coal liquefaction, First Quarterly Report, October 1-December 31, 1978, Report SAND-79-0306, 1978.
446. Granoff, B., Thomas, M. G., Mineral matter effects in coal liquefaction. I. Autoclave Speening Study, *Am. Chem. Soc., Div. Fuel Chem., Prepr.,* 22, 1977; presented at Am. Chem. Soc. Natl Meet., 174th, Chicago, Ill., Am. Chem. Soc., Washington, D.C., 1977, 183.
447. Guin, J. A., Tarrer, A. R., Pitts, W. S., and Prather, J. W., Kinetics and solubility of hydrogen in coal liquefaction reactions, *Am. Chem. Div., Fuel Chem., Prepr.,* 21, 1976; Symp. on Coal Liq. 172nd, Natl. Meet., 1976, San Francisco, Calif., 1976, 170.
448. Holloway, P. H. and Kramer, D. K., Surface analysis of a deactivated coal liquefaction catalyst, From Energy Res. Abstr. No. 16417, 1978., Report SAND-77-1389, 1977, 37.
449. Holy, N., Nalesnik, T., and McClanahan, S., Coal liquefaction with soluble transition-metal complexes, *Fuel,* 56, 47, 1977.
450. Jahnig, C. E., Evaluation of pollution control in fossil fuel conversion processes, Liquefaction: Section 3, H-Coal Process, EPA-650/2-74-009-m, NTIS PB-249 847, 1975, 60.

451. Jahnig, C. E. and Magee, E. M., Evaluation of pollution control in fossil fuel conversion processes, Gasification Section I: CO_2 Acceptor Process, EPA-650/2-74-009-d, NTIS PB-241 141, 1974, 60.
452. Kimura, H., Conversion into fluid phase and direct liquefaction of coal, *J. Fuel Soc. Jpn.*, 55, 792, 1976.
453. Knudson, C. L., Schiller, J. E., and Rudd, A. L., Temperature effects on coal liquefaction, Rates of depolymerization and product quality as determined by gel permeation chromatography, *Am. Chem. Soc., Div. Fuel Chem.*, Preprint 22, 1977; *Prod. of Liq. Fuels Ill.*, American Chemical Society, Washington, D.C., 1977, 49.
454. Lang, L. E., Leas, R. L., and Leas, A. M., Coal liquefaction bottoms conversion by coking and gasification, U.S. Patent 4060478, 1977.
455. Lee, J. M., Tarrer, A. R., Guin, J. A., and Prather, J. A., Selectivity of coal minerals as catalysts in coal liquefaction and hydrodesulfurization, *Am. Chem. Soc., Div. Fuel. Chem., Prepr.*, 22, 1977.
456. Lee, J. M., VanBrackle, H. F., Lo, Y. L., Tarrer, A. R., and Guin, J. A., Coal minerals catalysis in liquefaction, *Coal Process. Technol.*, 4, 1, 1978.
457. Magee, E. M., Jahnig, C. E., and Kalfaldelis, C. D., Environmental influence of coal liquefaction, Coal Process Technol., 3, (CEP Tech. Man.), American Institute of Chemical Engineers, New York, 1977, 16.
458. Mayo, F. R., Chemistry of coal liquefaction, *Am. Chem. Soc., Div. Fuel Chem., Prepr.*, 22, 1977; presented at Symp. on the Org. Chem. of Coal, Chicago, Ill., 1977, 103.
459. Migut, K. and Kasper, S., Costs of solid-liquids separation methods in coal liquefaction, *Am. Chem. Soc., Div. Fuel Chem., Prepr.*, 22, 1977; Comp. Econ. of Coal Convers. Processes, Symp. Presented at 174th Am. Chem. Soc. Natl. Meet., Chicago, Ill., Aug. 29-Sep. 2, 1977, 51.
460. Moore, R. H., Martin, E. C., Cox, J. L., Elliott, D. C., Coal liquefaction by aromatic interchange with phenol and catalytic hydrogenolysis, *Am. Chem. Soc., Div. Pet. Chem., Prepr.*, 22, 1977.
461. Mori, K., Taniuchi, M., Yaguchi, S., Coal liquefaction, Japan Kokai Tokkyo Koho, Patent 78-106604, 1978.
462. Morita, M., Future of coal liquefaction techniques by direct catalytic hydrogenation, *J. Fuel Soc. Jpn.*, 55, 831, 1976.
463. Morita, M. and Sato, S., Liquefying coal and lignite by hydrogenation and thermal cracking, Japan Kokai Patent 76-13803, 1976.
464. Nakako, Y., Nada, J., Motonaga, K., Osawa, T., Yokota, S., and Kakunai, H., Liquefaction of coal, Japan Kokai Tokkyo Koho, Patent 78-126005, 1978.
465. Nakane, T., Yokota, S., and Taniuchi, M., Coal liquefaction catalyst, Japan Kokai Patent 77145386, 1977.
466. Nedoshivin, Yu. N., Demborskaya, E. A., Koshevnik, Yu. A., and Nikiforova, T. S., NMR study of some products from the liquefaction of coal, *Khim. Tverd. Topl. Moscow*, 5, 51, 1978.
467. Newman, J. O. H., Akhtar, S., and Yavorsky, P. M. Coagulation and filtration of solids from liquefied coal of synthoil process, *Am. Chem. Soc., Div. Fuel Chem., Prepr.*, 21, 1976; Symp. on coal Liq., 172nd Natl. Meet. San Francisco, Calif., 1976, 109.
468. Oberg, C. L., Falk, A. Y., Hood, G. A., and Gray, J. A., Coal liquefaction under high-mass flux and short-residence time conditions, *Am. Chem. Soc., Div. Fuel Chem., Prepr.*, 22, 1977; at 173rd Natl. Meet., Symp. on Chem. from Coal, New Orleans, La., 1977, 185.
469. Oder, R. R., Magnetic desulfurization of liquefied coals, Conceptual, Process design and cost determination, IEEE Trans. Magn., MAG-12, Institute of Electricians and Electronic Engineers, Piscataway, New Jersey, 1976, 532.
470. O'Hara, J. B., Coal liquefaction, *Hydrocarbon Process*, 55, 221, 1976.
471. O'Hara, J. B., Becker, E. D., Jentz, N. E., and Harding, T., Petrochemical feedstocks from coal, *Chem. Eng. Prog.*, 73, 64, 1977.
472. O'Hara, J. B., Lochmann, W. J., and Jentz, N. E., Materials challenges of coal liquefaction, *Chem. Eng.*, 84, 147, 1977.
473. Ohshima, M., Kashima, K., Liquefaction of coal by high-pressure hydrogenation over an iron-sulfur or iron hydroxide-sulfur catalyst, Japan Kokai, Patent 75-38702, 1975.
474. Reber, S. A., Nadkarni, R. M., Hyde, R. W., Schutte, A. H., and Stickles, R. P., Liquefy coal to petrochemical feed, *Hydrocarbon Process*, 55, 229, 1976.
475. Weintraub, M., Weiss, M. J., Akhtar, S., and Yavorsky, P. M., Filterability of a coal-derived liquid, *Am. Chem. Soc., Div. Fuel Chem., Prepr.*, 21, 1976; Symp. on Coal Liq. 172nd, Natl. Meet., San Francisco, Calif., 1976, 43.
476. Korosi, A., Woebcke, H. N., and Virk, P. S., Pyrolysis of a Hydrogenated coal liquid, *Am. Chem. Soc., Div. Fuel Chem., Prepr.*, 21, 1976.
477. Krichko, A. A., Zamanov, V. V., Makar'ev, S. V., Davydov, V. P., Titova, T. A., and Yulin, M. K., Liquid products from coal, *Khim. Tverd. Topl. (Moscow)*, 3, 144, 1977.

478. Karn, F. S., Brown, F. R., and Sharkey, A. G., Aging characteristics of coal liquids, *Am. Chem. Soc., Div. Fuel Chem., Prepr.,* 22, 1977.
479. Kershaw, J. R., Fluorescence spectroscopy in the characterization of coal derived liquids, *Fuel,* 57, 299, 1978.
480. Reuther, J., Kinetics of heterogeneously catalyzed coal hydroliquefaction, *Ind. Eng. Chem. Process Des. Dev.,* 16, 249, 1977.
481. Sakabe, T., Industrialization and studies of direct liquefaction of coal, *J. Fuel Soc. Jpn.,* 53, 808, 1976.
482. Schiller, J. E., Analysis of solvent-refined coal, Recycle solvents, and coal liquefaction products, *Am. Chem. Soc., Div. Pet. Chem., Prepr.,* 22, 1977; Anal. Chem. of Tar Sands and Oil Shale, New Orleans, La., 1977, 638.
483. Schweighardt, F. D., Retsofsky, H. L., and Friedel, R. A., Chromatographic and NMR analysis of coal liquefaction products, *Fuel,* 55, 131, 1976.
484. Schweighardt, F. K., Retsofsky, H. L., and Raymond, R., Asphaltenes from coal liquefaction, *Am. Chem. Soc., Div. Fuel Chem., Prepr.,* 21, 1976; Pap. presented at 172nd Am. Chem. Soc. Natl. Meet., San Francisco, Calif., 1976, 27.
485. Schweighardt, F. K., White, C. M., Friedman, S., and Schultz, J. L., Heteroatom species in coal liquefaction products; *Am. Chem. Soc., Div. Fuel Chem., Prepr.,* 22, 1977; Symp. on the Org. Chem. of Coal, presented at 174th Am. Chem. Soc. Natl. Meet., Chicago, Ill., 1977, 124.
486. Severson, D. E., Souby, A. M., and Baker, G. G., Continuous liquefaction of lignite, a process development unit, *Am. Chem. Soc., Div. Fuel Chem., Prepr.,* 22, 1977; Prod. of Liq. Fuels from Coal, presented at 174th Am. Chem. Soc. Natl. Meet., Chicago, Ill., 1977, 161.
487. Sharkey, A. G., Schultz, J. L., Schmidt, C. E., and Friedel, R. A., Mass spectrometric analysis of stream from coal gasification and liquefaction processes, PERC/RI-75/5, U.S. Energy Research and Development Administration, 1975.
488. Silvestri, A. J. and Naro, P. A., Liquefaction of coal, U.S. Patent 3923634, 1975.
489. Stanulonis, J. J., Gates, B. C., and Olson, J. H., Catalyst aging in a process for liquefaction and hydrodesulfurization of coal, *AIChE, J.,* 22, 576, 1976.
490. Stein, R. T., Voltz, S. E., and Callen, R. B., Coal liquefaction, *Ind. Eng. Chem. Prod. Res. Dev.,* 16, 61, 1977.
491. Tarrer, A. R. and Guin, J. A., Effects of coal minerals on reaction rates during coal liquefaction, *Am. Chem. Soc., Div. Fuel Chem., Prepr.,* 21, 1976.
492. Tarrer, A. R., Guin, J. A., Pitts, W. S., Hanley, J. P., Prather, J. W., and Styles, G. A., Effect of coal minerals on reaction rates during coal liquefaction, *Am. Chem. Soc., Div. Fuel Chem., Prepr.,* 21, 1976; Symp. on Coal Liq., 172nd, Natl. Meet., San Francisco, Calif., 1976, 59.
493. Tateno, M., Komi, T., Magitaka, W., Coal liquefaction catalyst, Japan Kokai, Patent 78-73486, 1978.
494. Wett, T., Coal future source of oil and gas, *Oil Gas J.,* 76, 517, 1977.
495. Whitehurst, D. D. and Mitchell, T. O., Short contact time coal liquefaction-1. Techniques and product distributions, *Am. Chem. Soc., Div. Fuel Chem., Prepr.,* 21, 1976; Symp. on Coal Liq., 172nd Natl. Meet., San Francisco, Calif., 1976, 127.
496. Winnicki, J., Occurrence and manner of binding of some rare elements in Polish coals, *Pr. Nauk. Inst. Chem. Niorg. Met. Pierwiastkow Rzadkich Politech. Wroclaw,* 18, 45, 1973.
497. Yarzab, R. F., Given, P. H., Spackman, W., and Davis, A., Dependence of coal liquefaction behaviour on coal characteristics. IV. Cluster analyses for characteristics of 105 coals, *Fuel,* 59, 81, 1980.
498. Yu-Ping, H. and Yen, T. F., Evaluation of coal liquefaction efficiency based on various rands, *Energy Source,* 3, 39, 1976.
499. Agarwal, J. C. and Ahner, W. D., Conversion of coal into a fuel gas, Ger. Offen. (130375) P Patent 2443740.
500. Agosta, J., Coal to gas test facility, Proc. Ill Min. Inst. Annu. Meet., 84th, Springfield, Ill. Oct 21-22, 1976, Sponsored by Ill. Min. Inst., 87-94, 1976.
501. Ahn, Y. K. and Bolez, C. A., Regional and feedstock effects on economics of integrated coal gasification/power plant systems, *Am. Chem. Soc., Div. Fuel Chem., Prepr.,* 22, 56, 1977.
502. Arora, J. L., Burnham, K. B., and Tsaros, C. L., High-and low-btu gas from Montana subbituminous coal, *Am. Chem. Soc., Div. Fuel Chem., Prepr.,* 22, 1977; Comp. Econ. of Coal Convers. Processes, Symp. Presented at 174th Am. Chem. Soc. Natl. Meet., Chicago, Ill., Aug. 29-Sep. 2, 1977, 72.
503. Ashworth, R. A., Vyas, K. C., and Bonamer, D. G., Gas from coal for iron ore pelletizing, *Proc. Ironmaking Conf.,* 37, 282, 1978.
504. Ashworth, R. A., Vyas, K. C., and Bonamer, D. G., Gas from coal for iron ore pelletizing, *I & SN,* 6, 24, 1979.

505. Attari, A., Fate of trace constituents of coal during gasification, from Govt. Rep. Announce. 1973, 73(20), 66, U.S. National Technical Information Service, PB Rep., No. 223001/9, 1973, 39.
506. Attari, A., Mensinger, M., and Pau, J. C., Fate of trace element constituents of coal during gasification, Part II. Institute of Gas Technology, paper presented at 169th meeting, Am. Chem. Soc., Div. Fuel Chem., Philadelphia, Pa., April 6-11, 1975, 15.
507. Attari, A., Pau, J. C., and Mensinger, M., Fate of trace and minor constituents of coal during gasification, From. Gov. Rep. Announce. Index (U.S.) 1977, 77(23), 279. U.S. NTIS, PB Rep., 46 pp, No. PB-270913, 1976.
508. Babcock-Hitachi, K. K., Removal of hydrogen sulfide from coal gasification gases, Otani, Yoshinori, Japan, Kokai, (060478) P Patent 78 37582.
509. Backer, D. F. and Murthy, B. N., Feasibility of reducing fuel gas clean-up needs: Phase I. Survey of the effects of gasification process conditions on the entrainment of impurities in the fuel gas, Gilbert Associates, Inc., NTIS FE-1236-15, 49 p., June 20, 1976.
510. Baria, D. N., A survey of trace elements in North Dakota, lignite and effluent streams from combustion and gasification facilities, The Engineering Experiment Station, University of North Dakota, Grand Forks, May 1975, 64.
511. Bloom, R., Jr., Illinois coil gasification group project incorporation the cogas process, Proc. of Synth. Pipeline Gas Symp., 8th, Chicago, Ill., October 18-20, 1976, 465.
512. Botts, W. V., Kohl, A. L., and Trilling, C. A., Low-BTU gasification of coal by Atomics International's Molten Salt Process, *Proc. 11th Intersoc. Energy Convers. Eng. Conf.*, American Institute of Chemical Engineers, New York, 1976, 1, 280.
513. Brewer, R. E. and Ryerson, L. H. Production of High-Hydrogen Water Gas from Younger Cokes, Effects of Catalysts, *Ind. Eng. Chem.*, 27, 1047, 1935.
514. Buskies, U. and Summers, F., Coal gasification for the production of ammonia and direct reduced iron, *Erdoel Kohle, Erdgas, Petrochem.*, 31, 474, 1978.
515. Carinci, G. G. and Meissner, D. C., Use of coal gasification in the direct reduction Midrex process, Uses Carbon Sider, Absteciment Technol., Trab., Congr. ILAFA-Carbon, 305, 1976.
516. Chambers, H. F., Mirna, J. A., and Yavorsky, P. M., Free fall dilute-phase hydrogasification of coal, *Energy and the Environ., Proc. 4th Natl. Conf.*, American Institute of Chemical Engineers, Dayton, 1976, 19.
517. Chauhan, S. P., Feldmann, H. F., Nack, H., Stambaugh, E. P., and Oxley, J. H., Phase I summary report on a novel approach to coal gasification using chemically incorporated catalyst, From ERDA Energy Res. Abstr. 1977., Abstr. No. 15654, Report, p. 134, No. BMI-1953, 1976.
518. Chernenkov, I. I., Shafir, G. S., and Gavrilova, A. A., Gasification of coal in a fluidized bed under pressure with the high-temperature purification of the gas in a pilot plant of the Moscow Coke and Gas Plant, *Khim. Tverd. Topl. (Moscow)*, 4, 137, 1978.
519. Chow, T. J. and Earl, J. L., Lead Isotopes in North American coals, *Science*, 176, 510, 1972.
520. Coates, R. L., High rate coal gasification, *Coal Process Technology*, American Institute of Chemical Engineers (CEP Tech. Man.), New York, 1977, 89.
521. Detman, R. F., Preliminary economic comparison six processes for pipeline gas from coal, Proc. of Synth. Pipeline gas Symp., 8th, Chicago, Ill., Sponsored by Am. Assoc. (Cat. No. L51176), 515, 1976.
522. Ellman, R. C. and Johnson, B. C., Slagging fixed-bed gasification at the Grand Forks Energy Research Center, Synth. Pipeline Gas Symp., 8th, Chicago, Ill., 497, 1976.
523. Epperly, W. R. and Stegel, H. M., Catalytic coal gasification for SNG production, Proc. 11th Intersoc. Energy Convers. Eng. Conf., American Institute of Chemical Engineers, New York, 1, 1976, 249.
524. Estep-Barnes, P. A. and Kovach, J. J., Chemical and mineralogical characterization of core samples from underground coal gasification sites in Wyoming and West Virginia, MERC/RI -75/2, 1975, 16.
525. Exxon Research and Engineering Co., Coal gasification with chemically incorporated catalysts, Oak Ridge National Laboratory, Oak Ridge, Tenn., ORNL/FE, 1977, 1.
526. Feldmann, H. F., Chauhan, S. P., Longanbach, J. R., Hissong, D. W., Conkle, H. N., Curren, L. M., and Jenkins, D. M., Novel approach to coal gasification using chemically incorporated calcium oxide (Phase II). From Energy Res. Abstr. No. BMI-1986, Report, 1977, 318.
527. Feldmann, H. F., Chauhan, S. P., Choi, P., and Conkle, H. N., Gasification of calcium oxide catalyzed coal, *Proc. Intersoc. Energy Convers. Eng. Conf.*, 14, 856, 1979.
528. Ferretti, E. J., Industrial fuel gas from coal — its technology and economics, *3rd Symp. on Coal Util., NCA/BCR (Natl. Coal Assoc/Bittum Coal Res. Inc.) Coal Conf. and Expo. 3*, NCA, Washington, D.C., 1976, 80.
529. Ferretti, E. J. and Kasper, S., Coal gasification update, *Energy Commun.*, 3, 203, 1977.
530. Fischer, D. D., Monitoring of emissions from an in situ coal gasification experiment, Proc. Second Annual Underground Coal Gasification Symposium, Morgantown, West Virginia, NTIS MERC/SP-76/3, 1976, 242.

531. **Flowers, Ab.,** Status of processes on the USA for the production of SNG from coal, 13th World Gas Conf. Int. Gas Union, London, 1, 1976, Tech. Program, Pap IGU/B4-76, 1976, 19.
532. **Forney, A. J., Haynes, W. P., Gasior, S. J., and Kenny, R. F.,** Effect of additives upon the gasification of coal in the synthene gasifier, *Am. Chem. Soc., Div. Fuel Chem., Prepr.,* 19, 111, 1974.
533. **Forney, A. J., Haynes, W. P., Gasior, S. J., Kornusky, R. M., Schmidt, D. E., and Sharkey, A. G.,** Trace elements and major component balances around the synthane PDU gasifier, U.S. Environ. Prot. Agency, Off. Res. Dev. EPA 000076 EPA-600/2-76-149, Symp. Proc. Environ. Aspects Fuel Convers. Technol., II, 1975, 67.
534. **Fornoni, L.,** Continuous method and apparatus for the treatment of solid materials in general, and in particular, for the gasification of coal and shale and for the reduction of iron ore, Ger. Offen., Patent 2749605, 1978.
535. **Fornoni, L.,** Gasification of coal and direct reduction of iron ores, Braz. Pedido PI, Patent 77-02321, 1978.
536. **Frank, M. E., Sherwin, M. B., Blum, D. B., and Mendick, R. L.,** Liquid phase methanation-shift PDU results and pilot status, Proc. of Synth. Pipeline Gas Symp., 8th, Chicago, Ill., Oct. 18-20, 1976, Sponsored by Am. Gas Assoc. (Cat. No. L51176), 159, 1976.
537. **Franke, F. H.,** Gasification of brown coal to synthesis and reducing gases, *Haus Tech.-Vortragsveroeff.,* 405, 60, 1978.
538. **Furman, A. H.,** Pressurized feed for coal gasification, *Coal Process Technology,* (CEP Tech. Man.) American Institute of Chemical Engineers, New York, 1977, 60.
539. **Gasior, S. J., Lett, R. G., Strakey, J. P., and Haynes, W. P.,** Major, minor and trace element balances for the synthane PDU gasifier, Illinois, *Am. Chem. Soc., Div. Fuel Chem., Prepr.,* 23, 88, 1978.
540. **Vasyutinskii, N. A.,** Influence of chlorine and chlorides on the gasification of carbon, *Solid Fuel Chem.,* 10, 148, 1976.
541. **Hahn, R. and Heuttinger, K. J.,** Kinetics of iron-catalyzed hydrogen pressure gasification of coal, *Chem. Eng. Tech.,* 50, 954, 1978.
542. **Harada, M., Hayashi, K.,** Coal gasification apparatus, Japan Kokai Tokyo Koho, Patent 7913505, 1979.
543. **Harada, M., Taoda, K.,** Coal gasification apparatus for the molten salt process, Japan Tokyo Koho, Patent 78-31643, 1978.
544. **Hausen, L. D., Phillips, L. R., Mangelson, N. F., and Lee, M. L.,** Analytical study of the effluents from a high temperature entrained flow gasifier, *Fuel,* 80, 323, 1980.
545. **Hill, V. L. and Howes, M. A. H.,** Metallic corrosion in coal gasification pilot plants, *Mater. Perform.,* 17, 22, 1978.
546. **Hirschfeld, F.,** What's holding up coal gasification, *Mech. Eng.,* 99, 32, 1977.
547. **Howes, M. A. H.,** Selection of materials used in coal gasification plants, *Natl. Bur. Stand. Spec. Publ.* 468, 1977.
548. **Illinois Institute of Technology,** IITRI experiments with molten iron coal gasification, From Energy Res. Abstr. 2, 1977., Report, 1976 (NP-22188), 24.
549. **International Nickel Co., Inc.,** Development and evaluation of high chromium weld deposits overlays to protect less alloyed substrates from corrosion in a coal gasification atmosphere, From Energy Res. Abstr. No. 7207, 1979, Report No. FE-2621-4, 1978, 14.
550. **Ioffe, V. G., Klimov, B. P., and Klimov, L. P.,** Prospects for producing hot reducing gas from coal, *Izv. Vyssh. Uchebn. Zaved., Chern. Metall.,* 2, 114, 1979.
551. **Jahnig, C. E.,** Evaluation of pollution control in fossil fuel conversion processes, Gasification, Section 8, Winkler Process, EPA-650/2-74-009-j, NTIS PB-249 846, 1975, 42.
552. **Jensen, G. A.,** The kinetics of gasification of carbon contained in coal minerals at atmospheric pressure, *I & EC Process Design Dev.,* 14, 308, 1975.
553. **Jones, C. H. and Donohue, J. M.,** Comparative evaluation of high- and low-temperature gas cleaning for coal gasification-combined cycle power systems, From ERDA Energy Res. Abstr. No. 42363, 1977.
554. **Juentgen, H. and Van Heek, K. H.,** Technical scale gas generator for steam gasification of coal using nuclear heat, *Nucl. Technol,* 35 581, 1977.
555. **Kaku, H. and Meguri, N.,** Coal gasification apparatus, Japan Kokai Tokkyo Koho, Patent 79 83903, 1979.
556. **Kamody, J. F. and McGurl, G. V.,** Koppers-Totzek Gasification for electric power generation, ASME Pap. 76-JPGC-Fu-1 for Meet., 12, 1976.
557. **Karnavas, J. A., LaRosa, P. J., and Pelczarski, E. A.,** Coal gasification, Two-stage coal combustion process, *Chem. Eng. Progr.,* 69, 54, 1973.
558. **Kata, H. and Takamoto, S.,** High-btu gases from gasification of heavy oils and coal with oxygen released from transition metal oxide catalysts, Japan Kokai, Patent 85602, 1975.

559. Kim, C. S., Baddour, R. F., Howard, J. B., and Meissner, H. P., Calcium carbide production from calcium oxide and coal or hydrocarbons in a rotating-arc reactor, *Ind. Eng. Chem. Process Des. Dev.*, 18, 323, 1979.
560. Kimura, S. and Takahashi, T., Hydrogen sulfide removal from coal gas, Japan Kokai Tokkyo Koho, Patent 79 39394, 1979.
561. Kindig, J. K. and Turner, R. L., Improving coal, U.S. Patent 3938966, 1976.
562. Kindig, J. K. and Turner, R. L., Removal of impurities from coal, U.S. Patent 4098584, 1978.
563. King, B. A., Coal gasification — A new alternative in clean energy production, Monogr. on Alternate Fuel Resour. Based on Pap. Presented at Symp., Santa Maria, Calif., Publ. by West Period Co., North Hollywood, Calif., 1976, 331.
564. Knueppel, H., Brotzmann, K., and Fassbinder, G. H., Desulfurization of coal by gasification in an iron bath reactor, Ger. Offen. Patent 2520584, 1976.
565. Koppenaal, D. W., Trace element studies on coal gasification process streams, *Diss. Abstr.*, Int. B. 1979, 40, 201, p. 219, 1978.
566. Kornosky, R. M., Gasior, S. J., and Strukey, J. P., Gasification of Iowa coals in the synthane PDU gasifier, *Am. Chem. Soc., Div. Fuel Chem., Prepr.*, 22, 197, 1977.
567. Krieb, K. H., KDV process (High Pressure Coal Gasification) for Power Generation, 13th World Gas Conf., London, 1976, Tech. Program Pap. IGU/B3, 1976, 9.
568. Kwon, J. T., Recovery of calcium hydroxide and hydrogen sulfide for coal gasification ashes, Ger. Offen, Patent 2400451, 1974.
569. Ledent, P., La gazeification souterraine du carbon, *Ind. Miner.*, 59, 81, 1977.
570. Lee, B. S., Current development of the hygas program, Proc. of 8th Synth. Pipeline Gas Symp., Chicago, Ill. 1976.
571. Leonhard, H. F., MacDonald, S., Direct reduction with gas from the Koppers-Totzek coal gasification process. *Ironmaking Conf. Proc.*, 35, 422, 1976.
572. Lewis, R., Synthane gasification proves responsive, *Oil Gas J.*, 75, 81, 1977.
573. Lewis, R., Synthane coal to gas pilot plant, *Natl. Bur. Stand., Spec. Publ.*, 468, 1977.
574. Luthy, R. G., Vassilidu, P., and Carter, M. J., Experimental analysis of the leading characteristics of residual hygas coal gasification solids, Report FE-2496-28, National Technical Information Service, Springfield, Va., 1978.
575. Manokhin, A. P. and Chernyak, A. S., Germanium and gallium behaviour during industrial semi-coking and gasification of long-flame coals, *Nauch. Tr., Irkutsk. Gos. Nauch. Issled. Inst. Redk. Tsvet. Metal.*, 19, 219, 1968.
576. Massey, M. J., Nakles, D. V., Forney, A. J., and Haynes, W. P., Effluents from synthane gasification of lignite, *Coal Process Technology*, American Institute of Chemical Engineers, 1977, 53.
577. Mei, J. S. and Keddy, E. S., Heat pipes for fluid-bed gasification of coal: metallurgical condition of heat pipes after tests in process environment, *Proc. 11th Intersoc. Energy Convers. Eng. Conf.*, American Institute of Chemical Engineers, 1, 1976, 883.
578. Miles, J. M., Status of the Bi-Gas program-1. Pilot plant activities, Proc. of 8th Synth. Pipeline Gal. Symp. Chicago, Ill., 1976 (Cat.No. L51176), 77.
579. Morel, W. C., Economic comparison of coal feeding systems in coal gasification-lock hopper vs. slurry, *Am. Chem. Soc., Div. Fuel Chem., Prepr.*, 22, 1977.
580. Morel, W. C. and Yim, Y. J., Economics of producing methanol from coal by entrained and fluidized-bed gasifiers, *Am. Chem. Soc., Div. Fuel Chem., Prepr.*, 22, 1977.
581. Muralidhara, H. S. and Sears, J. T., Effect of calcium on gasification, *Coal Process. Technol.*, 4, 22, 1978.
582. Murray, R. H., Industrial fuel gas from coal, *Iron Steel Eng.*, 53, 25, 1976.
583. Nakamura, T. and Iwasaki, Y., Coal gasification, Japan Kokai, Patent 76-02728, 1976.
584. Nakamura, T. and Iwasaki, Y., Gasification of solid fuels, Ger. Offen. Patent 2521080, 1975.
585. Nakamura, Y. and Tokumitsu, N., Solid fuel gasification using a molten metal bath, Japan Kokai Tokkyo Koho, Patent 79-47707, 1979.
586. Neavel, R. C. and Lang, R. J., Gasification of coal, Ger. Offen. Patent 2854908, 1979.
587. Nemeth, E. J., Use of reducing gas by coal gasification for direct iron ore reduction, U.S. Patent 3853538, 1974.
588. Nishimoto, Y., Mizumoto, Y., and Imai, T., Gasification of solid carbon compound, Japan Kokai, Patent 76127102, 1976.
589. Onozawa, M., Coal gasification with molten slag and iron, Japan Kokai, Patent 77-41606, 1977.
590. Onozawa, M., Coal gasification apparatus, Japan Kokai, Patent 77-41605, 1977.
591. Otto, K., Bartosiewicz, L., and Shelef, M., Effects of calcium, strontium, and barium as catalysts and sulfur scavengers in the steam gasification of coal chars, *Fuel*, 58, 565, 1979.
592. Oxley, J. H., The reactions of sulfur during the gasification of coal, Thesis, College of Engineering and Science, Carnegie Institute of Technology, Pittsburgh, Pennsylvania, 1956, 286.

593. Page, G. C., Fate of pollutions in industrial gasifiers, U.S. Environ. Prot. Agency, Off. Res. Dev., EPA, 1978, 191.
594. Parsons, R. C. and Yang, K., Upgrading coal gasification products, U.S. Patent 3823775, 1974.
595. Rao, M. J., Ramacharyulu, M., and Vaidyeswaren, R., Methanation of synthesis gas from coal, *Chem. Ind. Dev.*, 11, 15, 1977.
596. Rath, L. K., Margaritis, P. J., Shah, R. D., Cherish, P., and Salvador, L. A., Operation of the Westinghouse coal gasification PDU, *Energy and the Environ. Proc. of the 4th Natl. Conf.*, American Institute of Chemical Engineers, Dayton, 1976, 28.
597. Raymon, N. S. and Sadler, L. Y., Refractory lining materials for coal gasifiers, *U.S. Bur. Mines, Inf. Circ.*, 8721, 1976, 25.
598. Samuel, W. A. and Grandy, G. A., Application of coal gasification for the production of iron ore pellets, Proc. 50th Annu. Meet. Min. Sect. AIME, Paper No. 19, 1977, 31.
599. Sasaki, A., Taoda, K., Takeuchi, Y., and Yanagi, M., Gasification or thermal decomposition of coal, Japan Kokai Tokkyo Koho, Patent 78-102907, 1978.
600. Sather, N. F., Swift, W. M., Jones, J. R., Beckner, J. L., Addington, J. H., and Wilburn, R. L., Potential trace element emissions from the gasification of Illinois coals, From ERDA Energy Res. Abstr. 1976, 18519, Report No ANL-75-XX-1, 1975, 22.
601. Scaefer, A. O., The AGA-ERDA-MPC program on materials for the gasification of coal, NBS Spec. Publ., 468, 1977, 80.
602. Shah, R. D., Margaritis, P. J., Rath, L. K., Cherish, P., and Salvador, L. A., Operation of the Westinghouse coal gasification process development unit, *Proc. 11th Intersoc. Energy Convers. Eng. Conf.*, Paper 769047, American Institute of Chemical Engineers, New York, 1976, 249.
603. Somerville, M. H., Elder, J. L., and Todd, R. G., Trace elements: analysis of their potential impact from a coal gasification facility, Technol. Use Lignite, 1978, 285.
604. Stanton, G. C. and Ellingson, W. A., In-service infrared thermal imaging of coal-gasification plant components, *Proc. of the Bienn. Infrared Ing. Exch.*, AGA Corp., Secaucus, New York, 1977, 119.
605. Starkovich, J. A. and Blumenthal, J. L., Hydrogen production by catalytic coal gasification, U.S. Patent 4069304, 1978.
606. Schrider, L. A., Brandenburg, C. F., Fischer, D. D., Boyd, R. M., and Campbell, G. G., Technol. and Use of Lignite, Proc. Symp., Grand Forks, N.D., May 14-15, 1975, ERDA, Off. of Public Aff., Tech. Inf. Cent. (GFERC/IC-75/2), Oak Ridge, Tenn., 1975, 254.
607. Cavagnaro, D. M., Ed., NTISearch NTIS/PS-76/0407/7ENS, Search period covered 1964-1976, National Technical Information Service, Springfield, Va.
608. Fisher, D. D., King, S. B., and Humphrey, A. E., Report of the Successful Development of Underground Coal Gasification at Hanna, Wyoming, *Am. Chem. Soc., Div. Fuel Chem., Prepr.*, 22, 1977; *Symp. on In Situ Process of Coal*, American Chemical Society, Washington, D.C., 1977, 49.
609. Fleming, K. D., Evaluation of factors that affect the genesis and disposition of constituents in coal gasification, Sampling Strategy and Charact. of Potential Emiss. from Synfuel Prod., Symp./Workshop, Proc., Austin, Tex. ERDA (CONF-760602), Washington, D.C., 1976, Available from NTIS, Springfield, Va., 1976, 7.
610. Hommert, P. J. and Beard, S. G., Descriptions of reverse combustion linkage and forward gasification during underground coal gasification, *Am. Chem. Soc., Div. Fuel Chem., Prepr.*, 22, 1977.
611. Roehl, A. A., Brown, R. A. S., and Jenson, E. J., Underground coal gasification field test in Alberta-1976, *Am. Chem. Soc., Div. Fuel Chem., Prepr.*, 22, 1977; *Symp. on In Situ Process of Coal* American Chemical Society, Washington, D.C., 1977, 92.
612. Schrider, L. A., Brandenburg, C. F., Fisher, D. D., Boyd, R. M., and Campbell, G. G., Outlook for underground coal gasification, *Erdoel Kohle Erdgas Petrochem. Brennst. Chem.*, 29, 409, 1976.
613. Schrider, L. A. and Whieldon, C. E., Underground coal gasification, A status report, *J. Pet. Technol.*, 29, 1179, 1977.
614. North American Coal Corporation, 1962 Annual Report, Cleveland, Ohio, 1963.
615. Statistical data on the uranium industry, U.S. Energy and Research Administration, Grand Junction Office, Colorado, 600-100(77), January 1, 1977.
616. Uranium resources, production and demand, Joint report of the Nuclear Energy Agency of the Organization for Economic Co-operation and Development (OECD) and the International Atomic Energy Agency, 1975.
617. Anon., Calsinter process recovers aluminum, *Ind. Res. Dev.*, January, 1980, p. 92.

Chapter 2

ENVIRONMENTAL CONSIDERATIONS

I. TRACE CONTAMINANTS FROM COAL-FIRED POWER PLANT

The quantities of pollutants entering the environment as the result of coal combustion increase with the steady growth in amount of coal being utilized. In this chapter we shall discuss environmental contamination by trace elements in coal.

Electric power generation requires large tonnages of varying quality coals; therefore it is not possible to ignore trace elements present in the coal at very low levels. This is particularly true since some are known to be harmful to life at relatively low concentrations. Trace element mobilization resulting from coal use is discussed in detail by Zubovic et al.[1] They have estimated the amounts of 18 elements that potentially could be mobilized as a result of coal production (see Table 1). The coal production tonnages are those reported by Wood and Simon[2] for the years 1978 and 2000. Trace-element data used in these estimates are based on the average content of these elements in U.S. coals. The authors feel that world estimates based on these averages are valid because U.S. coals were deposited in a variety of geologic environments during different geologic ages and represent coal ranks ranging from lignite to low volatile bituminous coal. Thus, one could expect that the average U.S. coal would closely approximate the composition of the average world coal.

Data shown in Table 1 are based on the following figures for coal production tonnages[2]: U.S., 1978 — 650×10^6 tons; U.S., 2000 — 1.600×10^6 tons; World, 1978 — 3.600×10^6 tons; World, 2000 — 9.300×10^6 tons; and cumulative production tonnage 1979 to 2000 is estimated to be 151.000×10^6 tons.

The most important data in Table 1 are in the last column where cumulative amounts of each of the elements mobilized are shown. During this period, about 1.7×10^9 tons of sulfur would be mobilized. In the U.S., policies which will reduce sulfur emissions to the atmosphere as a result of coal combustion are being devised. These policies should be considered on a world-wide basis if the incidence of acid rains is to be reduced. This phenomena is already a problem in some areas; with the increased use of coal it could become a world-wide problem. The cumulative mobilization for other elements listed in Table 1 indicate ranges from 14,000 short tons of mercury to 9,700,000 tons of fluorine. Both of these elements are volatile and are almost totally emitted to the atmosphere during coal combustion. Although the amount of mercury mobilized is lowest, it is one of the most toxic of the elements listed. The form or chemical combination of fluorine emitted during coal combustion is not known, nor have the effects of fluorine been fully evaluated. There is considerable concern in the U.S. directed toward fluorine emissions from aluminum reduction plants, and strict controls are in effect concerning such emissions. However, no great concern is being voiced over fluorine emissions from power plants. A potential hazard can result when an aluminum reduction plant and a major power plant are located in the same area.[1]

Other elements which are potentially toxic, volatile, and found in significant quantities in coal are As, Se, and Sb. Most of the other listed elements are nonvolatile and would be retained in the bottom ash and fly ash. These nonvolatile elements are readily leachable from these combustion products and their disposal poses a problem if ground water contaminations are to be prevented. Many of these elements are leachable from coal conversion residues and attain concentrations in leachates above recommended levels for human consumption.

Table 1
AMOUNTS OF ELEMENTS MOBILIZED AS A RESULT OF COAL PRODUCTION IN SHORT TONS[1]

Element	U.S. 1978	U.S. 2000	World 1978	World 2000	Cumulative 1979-2000
As	3,300	8,400	19,000	47,000	770,000
Be	910	2,300	5,100	130,000	210,000
Cd	91	230	510	1,300	21,000
Co	2,270	5,800	13,000	33,000	530,000
Cr	6,110	15,000	34,000	87,000	1,400,000
Cu	7,800	20,000	43,000	110,000	1,800,000
F	42,000	110,000	230,000	590,000	9,700,000
Hg	59	150	328	843	14,000
Li	5,400	14,000	30,000	77,000	1,300,000
Mn	16,000	41,000	91,000	230,000	3,800,000
Mo	1,200	3,000	6,500	17,000	270,000
Ni	4,700	12,000	26,000	68,000	1,100,000
Pb	4,600	12,000	25,000	65,000	1,000,000
Sb	380	950	2,100	5,400	88,000
Se	1,100	2,800	6,200	16,000	250,000
U	850	2,100	4,700	12,000	200,000
V	9,800	25,000	54,000	140,000	2,300,000
Zn	9,100	23,000	51,000	130,000	2,100,000
S	7.34×10^6	18.6×10^6	40.9×10^6	105×10^6	$1,695 \times 10^6$

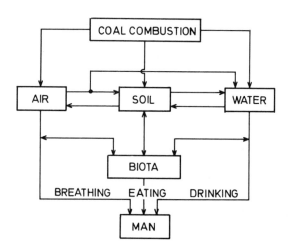

FIGURE 1. Schematic diagram of pathways of trace contaminants from coal combustion to ecosystem and to man (source: ERDA[235]).

Before the effects of coal-derived trace contaminants on man and ecosystems can be evaluated, it is necessary to estimate the transport, transformation, and bioaccumulation of these contaminants as they move from the power plant to the receptor (Figure 1). In addition to determining routes of transport and transformation, rates for these processes also must be determined before one can predict environmental fate and levels of exposure. Information on the physical and chemical characteristics (e.g., particle size, solubility, and valence state) of materials released to the environment is a vital prerequisite to a study of transport processes.[3] Therefore one should study air pollution, soil pollution, and water pollution as a result of coal combustion. One has to keep in mind that the variability in coal composition is large, and the mobilization of trace elements from its use will largely depend on the coal, mining methods, type of transport, and energy or industrial conversion process.

Of primary concern to the world community are the volatile elements (S, As, F, Hg, and Se) which can move across country and continental boundaries to produce undesirable effects throughout the world unless adequate control mechanisms are adopted. Other trace elements could have local environmental impacts.[1]

A. Air Pollution

Air pollution as a result of coal combustion is the most obvious type of environmental pollution. Emissions from coal combustion have been studied in several instances. The amount of trace elements emitted from a coal-fired power plant will depend mainly on the initial trace element concentrations in the coal, but also on combustion engineering and pollution control at the plant. Different combustion processes will yield different primary emissions.

To be able to calculate the emission of elements in the atmosphere it is necessary to know the fate of trace elements during combustion. This has been discussed in some detail in chapter 1 of this volume, where the measurements of material balance in power plants were described. It has been shown that a partitioning of elements between slag and fly ash occurs. The process is not understood in detail but it is quite clear that elements (such as As, Cd, Hg, Pb, Sb, Se, Zn, and halogens), having low boiling points (below the normal combustion temperature, 1300 to 1600°C), are volatilized in connection with combustion. Some of the elements (such as Hg, Se, and halogens) remain wholly or partially in the vapor phase in the stack gases. Other elements, shortly after combustion, adsorb or condense onto surrounding particles in the flue gases. As a result of this, the finer particles, having a larger surface area per mass unit than the coarser ones, will have higher concentrations of the volatile elements than the coarser particles. This enrichment process is of great importance for the emissions, since the finer particles are more difficult to remove from the stack gases with conventional particle precipitators.[4]

The degree of volatilization depends not only on the temperature but also to some extent on the coal matrix. For example, for coals with a high concentration of Ca, more As will be retained as arsenite or orsenate in the ash than for a coal with a low Ca concentration.

Boulding[5] gives the following equation for calculating trace element emissions:

$$E_p = \frac{8.76 \, E_c \, K \, h_r \, I}{h_c} \, f \qquad (1)$$

where

E_p = amount emitted (tons/year)
E_c = concentration of the element in coal ($\mu g/m^3$)
K = plant capacity (MW)
h_r = plant heat rate (Btu/kWh)
I = plant load factor
h_c = heat content of the coal (Btu/ton)
f = emission factor = mass flow of the element emitted/mass flow of the element with coal

Table 2
EMISSION FACTORS FOR TRACE ELEMENTS OBTAINED IN MASS BALANCE STUDIES AT COAL-FIRED STEAM PLANTS

Element	Allen Steam Plant[a] (Klein et al.[9-12]) Precipitation efficiency: 96.5%	99.5%	Allen Steam Plant (Bolton et al.[6-8])	Valmont Power Station[b] (Kaakinen and Jordan[14]) Total fly ash removal efficiency: ~95%	Calculated U.S. mean values[c] (Klein et al.[9-12])
As	3.8	3.1	0.64—2.1		1.7
Ba		0.31	0.15		0.57
Be					
Br		~100			92
Ca	0.38	0.16	0.20—0.32		0.40
Cd	1.7	2.9	2.2		2.8
Cl		~97			97
Co	0.81	0.47	0.89—2.7		0.62
Cr		1.2	0.22—2.3		1.0
Cu			1.3	7.9	
F					
Fe	1.0	0.38	0.76—2.9	4.6	0.39
Hg	(60)	90—			89
In					
Mn	1.2	0.40	0.77—1.5		0.51
Mo			0.16	15.7	
Na	0.95	0.39	1.1		0.41
Ni			1.5		
Pb	2.2		0.8		3.5
Sb		27			2.1
Se		12.5	35 5.3	12.6	14.2
Sr				5.0	
Ti	0.86	0.54	0.23 0.56		0.62
Tl					
V	1.7	0.95	0.46 0.54		0.90
Zn	1.7	2.9	1.5	12.7	1.8

Note: Mass flow of element emitted in percent of the mass flow of the element with coal.

[a] Allen Steam Plant: Cyclone-fed boiler, high efficient electrostatic precipitation. 1000 MW(e): Coal consumption 365 tons/hr.
[b] Valmont Power Station: Pulverized coal-fired boiler. Mechanical dust collector followed by electrostatic precipitator and wet scrubber in parallel. 1000 MW(e): Coal consumption 465 tons/hr.
[c] Calculated as a mean for U.S. (13% of the coal burned in cyclone-fed boilers and stokers, 87% burned as pulverized coal. High-efficient (99.5%) electrostatic precipitation. Element partitioning between slag and fly ash in accordance with T. A. Allen Steam Plant measurements).

The emission factor (f) differs from element to element and also from plant to plant. It should therefore be determined empirically for each element by mass balance studies at each plant.[4]

Emission factors (mass flow of the element emitted in % of the mass flow of the element with coal) obtained in mass balance measurements at the T. A. Allen Steam Plant[6-8,9-12] and Valmont Power Station[13,14] are shown in Table 2.

Table 3 shows the emission estimates calculated by Loevblad[4] for a 1000-MW cyclone-fed plant with high-efficient (99.5%) electrostatic precipitation, made on the basis of the emission factors obtained at the T. A. Allen Steam Plant.[9-12] The coal consumption per hour is estimated at 365 tons. The emission factor for Hg is assumed to

Table 3
ESTIMATED EMISSION OF TRACE ELEMENTS AT A 1000 MW(e) CYCLONE-FED POWER PLANT IN G/HR.[4]

	Norway	England	U.S.(A)	U.S.(B)	Ger.(A)	Ger.(B)	Pol.(A)	Pol.(B)	Sov.(A)	Sov.(B)	Sov.(C)	Australia
As	23	170	72	9.0	33	8.2	22	8.2	19	36	14	29
Ba	380	140	100	84	110	120	170	200	330	180	280	370
Br	<730	36·10³	8.4·10³	4.4·10³	5.8·10³	2.9·10³	4.7·10³	3.7·10³	5.1·10³	1.8·10³	1.8·10³	<730
Ca	1300	880	880	303	590	1200	2400	3200	3100	1600	260	140
Cd	<1.0	<1.0	<1.0	<1.0	<1.0	<1.0	<1.0	<1.0	<1.0	<1.0	<1.0	3.2
Cl	0.11·10⁵	13·10⁵	5.5·10⁵	2.6·10⁵	1.8·10⁵	2.9·10⁵	4.4·10⁵	4.7·10⁵	4.4·10⁵	0.53·10⁵	0.11·10⁵	0.28·10⁵
Co	1.0	11	4.5	23	8.5	11	5.2	7.0	7.0	11	5.9	31
Cr	15	110	98	33	84	130	55	70	64	150	64	55
F	20·10³	27·10³	14·10³	6.2·10³	16·10³	15·10³	26·10³	12·10³	21·10³	19·10³	30·10³	23·10³
Fe	8.6·10³	8.4·10³	12·10³	2.0·10³	9.2·10³	3.1·10³	12·10³	13·10³	8.1·10³	37·10³	20·10³	2.8·10³
Hg	22	40	22	5.1	40	37	16	23	58	37	24	<5.1
Mn	<15	69	24	<15	320	35	180	200	230	53	110	29
Na	2.0·10³	2.6·10³	360	585	460	780	1.4·10³	1.4·10³	1.7·10³	3.8·10³	1.2·10³	85
Pb	<10²	<10²	<10²	<10²	<10²	<10²	<10²	140	<10²	<10²	<10²	610
Sb	1.8	150	28	8.0	150	91	77	80	95	47	14	240
Se	71	64	170	86	46	86	39	86	71	56	<23	27
Ti	280	1600	1600	<200	<390	740	670	670	1800	2100	1200	2200
V	20	140	56	21	84	120	31	50	84	95	45	73
Zn	3.2	160	33	12	100	140	260	360	170	250	190	1200

Note: 1000 MW(e) cyclone-fed boilers, high-efficient electrostatic precipitator (99.5% total efficiency); coal consumption ~365 ton/hr.

be 90% and for F 100%. The emissions of Be, Cu, In, Mo, Ni, and Sr are not estimated since no corresponding emission factors are available for these elements.

The trace element emissions from coal combustion given in Table 3 are calculated for a power plant with extremely good particle pollution abatement equipment and are probably hard to reduce further with conventional abatement techniques and without special pretreatment of the coal. However, a wet scrubber for sulfur dioxide removal in addition to the electrostatic precipitator will probably decrease the emissions of some of the trace elements, at least the emissions of halogens. The emissions of elements such as F, Hg, Sb, and Se from a 1000-MW coal-fired plant are considerable and of the same order of magnitude as those from some industrial processes.[4]

The environmental impact of emissions from coal combustion is of course not only related to the quantity emitted but also to the chemical and physical state of the emitted elements. Fly ash from coal combustion consists largely of relatively insoluble aluminum silicate glass particles. The major part of the elements contained in these particles is therefore not easily available for reactions in the environment. It appears, however, as if elements on the surface layer of particles would be quite readily soluble. This is of importance when considering that some of the elements enriched on the particle surfaces are toxic.

Loevblad[4] has produced an interesting comparison of emissions from coal and oil combustion based on the heat value of the respective fuel. In preparing this comparison, Loevblad[4] has assumed percentage emission at combustion coal to be same as reported by Bolton et al.[6-8] (see Table 2). Coal composition is taken as a mean value of 12 different coal samples. Trace element composition of fuel oil was assumed to be the mean value of 9 different oils and it was assumed that 100% of the element present is emitted at combustion. The results are shown in Table 4. There are some significant differences. Fuel oils have large emissions of V and Cl, while coals have much larger emission of Fe, Mn, and Sb. (For similar comparison see also Table 21.)

Particle size distribution of fly ash and flue-gas desulfurization (FGD) sludge have been discussed by several authors. Results obtained by Page et al.[15] are shown in Table 5. The results for FGD sludge are those reported by Bern[16] and are shown in Table 6.

Ragaini and Ondov[17-21] have performed a study to identify the trace element emissions and to correlate them by particle size from coal-fired power plants using electrostatic precipitators (ESP) or a scrubber unit as an emission control system. Samples of coal, ESP ash, bottom ash, and scrubber slurries were taken from 2 western U.S. mine-mouth power plants and analyzed for 27 elements by neutron activation analysis (NAA). Particle-size distribution parameters for stack fly ash were obtained by a scanning electron microscope (SEM). The results of the analysis of stack fly ash indicated that the elements could be classified into three groups:

1. Group I consisted of elements that have inorganic species with a boiling point of sublimation temperature at or below the combustion temperature of a coal-fired power plant (1500 to 1600°C).
2. Group II consisted of elements contained in small particle size fly ash that have no reasonable inorganic species with a boiling or subliming point at temperatures less than or equal to 1500°C.
3. Group III consisted of nonvolatile mineral phases found in coal with little or no small fly ash particle association.

Comparison of the two emission control systems resulted in the following findings: depending on the volatility of the element and a corresponding predominance of small fly ash particles, emission of Se, Ba, Sb, As, W, and U were greater for the scrubber

Table 4
COMPARISON OF EMISSION
FROM COAL AND OIL
COMBUSTION NORMALIZED
TO g/kWh · 10^6 (Loevblad[4])

Element	Coal (g/kWh·10^6)	Oil (g/kWh·10^6)
As	13	6.4
Br	2.2	10
Cd	<0.3	<0.9
Cl	120	1200
Co	3.6	33
Cr	27	2.7
Fe	3800	400
Hg	9.3	0.20
Mn	38	6.4
Pb	52	65
Sb	28	0.26
Se	24	9
V	23	5700
Zn	83	83

Table 5
PARTICLE SIZE
DISTRIBUTION OF FLY ASH[15]

Particle diameter (μm)	Mass fraction (%)
Wet sedimentation	
>50	32.5
2—50	63.2
<2	4.3

Table 6
PARTICLE SIZE
DISTRIBUTION OF FLUE-GAS
DESULFURIZATION

Particle diameter μm	Mass fraction (%)
Untreated	
74—2000	2
2—74	95
<2	3
1:1 FGD sludge/fly ash	
74—2000	1
2—74	94
<2	5

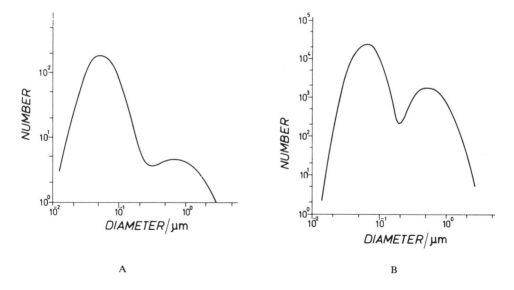

FIGURE 2. Number-size distribution of particles collected on a back-up filter of a 2-hr sample (A). Number-size distribution of particles collected on a whole filter sample (B). Data are from Ragaini and Ondov.[17]

systems than ESP units; generally the elements classified as Group I were found in emissions from the scrubber units in the range of 2 to 7 times greater than in emissions from ESP units; Group III elements Fe, Al, and Se were emitted in greater quantities per unit power output from ESP than from scrubber units; and quantities of emissions of submicron fly ash particles from ESP units were less than 10% of the total emissions, while in the scrubber units 95% of the fly ash mass containing insoluble elements was of submicron particle size.

The size distribution of particles in the fly ash is shown in Figure 2.[17] In Figure 2B the number-size distribution observed for the whole-filter sample is shown. The distribution observed is clearly bimodal and can be resolved into two distinct distributions having number median diameters at 0.065 and 0.565 μm. Corresponding M_{50} values for the two distributions are 0.088 and 1.04 μm; respective σ_g are 1.37 and 1.57. These parameters represent the distribution of total particulates. Values of the M_{50} for the large-particle distribution can also be obtained from impactor data.

Based on the back-up filter (Figure 2A), the M_{50} of small-particle distribution is 0.013 μm with σ_g of 1.37 determined for the whole-filter sample. Due to the possibility of distortion of the back-up filter distribution, such as by collection of some of the small particles on impactor stages, the whole-filter M_{50} and σ_g are considered the more accurate parameters to describe the total mass distribution.

In this study, particle size distribution parameters for stack aerosols observed were obtained from scanning electron microscope (SEM) observation of particles collected on the membrane filters and on impactor stages. In the former case, number-size distributions were obtained by counting particles in discrete size ranges, after solvent suspension and sonic redispersion, with the aid of an SEM-Qiantiment system. Since the number distributions observed were log normal, both mass and surface median diameters (M_{50} and S_{50}, respectively) were calculated from the observed number median diameter (N_{50}) and geometric standard deviation (σ_g). In the latter case, values of the M_{50} and σ for the aerosol were obtained from log probability plots of the percent cumulative mass vs. N_{50}. Table 7 lists the relative amounts of elements observed on small- and large-particle distributions in stack fly ash, along with distribution para-

Table 7
RELATIVE AMOUNTS OF ELEMENTS
OBSERVED ON SMALL AND LARGE PARTICLE
DISTRIBUTIONS IN FLY ASH[17]

	Small distribution ($0.08\ \mu m < m_{50} < 0.09\ \mu m$)	Large distribution		
	Fraction (%) ± $^a\sigma$	Fraction (%)	$M_{50}(\mu m)$	σg
Group I				
I	62 ± 8	38 ± 12	0.93	2.00
Ba	55 ± 9	45 ± 11	0.92	1.93
Br	62 ± 29	38 ± 12	0.91	2.46
As	57 ± 7	43 ± 12	0.93	1.93
Sb	49 ± 2	51 ± 5	1.05	1.98
Mo	44 ± 8	56 ± 9	1.03	2.03
U	43 ± 7	57 ± 6	ND	ND
Hg	40 ± 9	60 ± 5	1.01	1.94
V	31 ± 6	69 ± 9	0.95	1.96
Ga	29 ± 7	71 ± 15	1.00	1.85
Se	27 ± 3	73 ± 9	1.11	1.89
Zn	18 ± 2	82 ± 10	1.01	2.03
Cl	11 ± 9	89 ± 45	0.85	2.47
Group II				
Co	20 ± 3	80 ± 9	1.13	1.87
Cr	20 ± 2	80 ± 6	1.10	1.84
Fe	10 ± 3	90 ± 6	1.06	1.86
Mn	8 ± 4	92 ± 12	1.09	1.87
Ti	6 ± 3	94 ± 5	1.11	1.87
Group III				
Sm	<9	91 ± 5	1.20	1.83
Ta	<6	94 ± 8	ND	ND
Th	<6	94 ± 8	1.17	1.85
Sc	<5	95 ± 5	1.10	1.79
Mg	<14	86 ± 26	ND	ND
Na	<5	95 ± 11	1.15	1.77
In	<6	94 ± 24	1.13	1.78
Al	<6	94 ± 12	1.20	1.93
La	<6	94 ± 9	1.24	1.87

Note: $^a\sigma$ represents the dispersion of individual measurements or analytical uncertainty, whichever is larger.

meters determined for the large particle mode. In Group I and Group II are listed elements with significant small-particle association. Elements with little or no small-particle association are listed in Group III, correlated as a function of impactor stage; having correlation coefficients greater than or equal to 0.94. In Group I all of the elements except V have inorganic species with boiling point or sublimation temperature at or below the combustion temperature of a coal-fired power plant, typically 1500 to 1600°C. In Group II, all of the elements listed have no reasonable inorganic species boiling or subliming at temperatures < 1550°C.

At combustion temperatures, inorganic forms (except for some volatile chlorides) of U, V, Co, Cr, Fe, Mn, and Ti are essentially nonvolatile. However, significant small-particle association is observed. Mn, Fe, and Cr are major components of stainless steel which is used in many parts in contact with the gas stream. Abrasion may be responsible for some of the small-particle component observed for these elements. Elements in Group III (i.e., rare earths, Sc, Al, Mg, and Na) are associated with nonvolatile mineral phases found in coal, such as clay minerals. Particle size distribution observed for these elements is similar to the total mass distribution.

A similar study was reported by Natusch and Wallace.[22] In their measurements, concentrations of 14 minor and trace elements were determined for various size fractions of fly ash samples collected from 8 power plants within the U.S. A volatilization-adsorption mechanism was proposed relating elemental concentration to particle diameter. Samples were categorized according to fly ash retained in the precipitation system of the plant and airborne fly ash leaving the plant. Material was separated into a wide range of particle size fractions. Elemental analyses were performed by mass and emission spectroscopy, X-ray fluorescence, AA, polarography, and colorimetry. The concentration of several elements was particle-size dependent (e.g., Pb, Cd, Se, Ni, Cr, and Zn), while the concentrations of others exhibited no such dependency (e.g., K, Ca, Ti, Co, Cu, Sn, and Bi). Beryllium and Mn showed sharp concentration increases in the submicron size range. Trace elements were concentrated in the smallest fly ash particles by initial volatilization in a high-combustion zone followed by subsequent recondensation onto the large surface area of the smaller particles. Existing collection devices, although efficient for bulk emission reduction in the form of large particles, preferentially permit emission of the most toxic smaller particles.

Billings et al.[23] indicate that about 90% of the mercury in coal burned in a pulverized-coal furnace appears as vapor in the flue gas. Bolton et al.[6] and Gordon et al.[24] reported several such elements occurring in greater concentrations in fly ash as compared to bottom ash. Thus, the process of coal combustion releases trace elements to the atmosphere as vapors and particles, and these particles have relatively greater concentrations of certain trace elements than the feed coal or the collected fly ash.

Toca[25] has examined the particulate effluent from a coal-fired boiler to determine the particle size relationship between Cd and Pb concentration in the effluent particles. Tests utilized effluent particles from a southern Illinois (3/4 in.) bituminous coal fired in a Detroit Stoker spreader-type traveling grate. The different effluent particle size ranges were analyzed for amounts of: C, Fe, Mg, Si, Ca, Be, Al, Mn, B, Cu, Pb, Zn, Ge, Na, Ni, Ti, and Mo as determined by emission spectroscopy; and for Br, Ga, Sb, Cs, Co, Sc, Nb, Zr, Cl, Cd, Hg, Ci, Ba, La, K, Eu, Se, Tb, and Cl by neutron activation analysis. Quantitative determinations of Pb and Cd were made by atomic absorption spectroscopy. Both Pb and Cd were found in higher concentrations in the small particles. The highest concentrations of Pb and Cd in the particulate effluent were found in those particles less than 5 μm, which are respirable by humans.

Emissions from some plants in the U.S. have been studied in some detail. For example, Gladney and Owens,[26,27] Oglesby,[28] and Gordon et al.[24] have followed in their study 37 elements from the feed coal through the Chalk Point, Md., coal-fired power plant, and the plant was found not to be a major source for the enrichment of some trace elements on urban air particulates. Samples of coal, bottom slag, economizer ash, fly ash, in-stack suspended particulates, and ambient aerosol in the vicinity of the power plant were analyzed by various neutron activation analysis techniques. The enrichment factor of each element was based on its enrichment or depletion relative to Al, as compared to the elemental ratio in the crust of the earth. Though the power plant increased the total particulate loading in the surrounding atmosphere, it did not act as a strong source for the highly enriched trace elements observed in urban air (Sb,

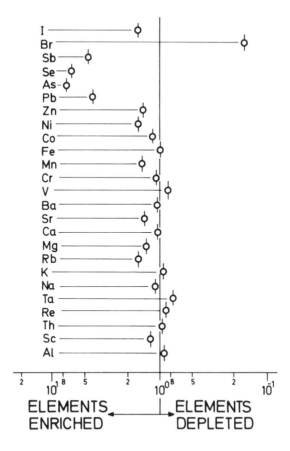

FIGURE 3. Enrichment factors of various elements on suspended stack particles with respect to concentrations in the coal.[26]

As, Zn, Se, In, V, and Pb). Elements were consistently enriched, or depleted, in various ash fractions throughout the plant. Bromine, I, and Hg leave the plant in the vapor phase. Elements enriched on particles carried beyond the precipitator are Hg, As, Se, Sb, and Pb; but most elements show little, if any, enrichment over their concentration in coal during their passage through the power plant.

The values for enrichment of an element in the suspended fly ash relative to its concentration in the coal are shown in Figure 3.

Magee et al.[29] have studied three Northern Great Plains plants. At each of the plants, all of the incoming and outgoing streams were sampled periodically over a time interval of 2 days. The flue gas dusts were sampled utilizing a wet electrostatic precipitator sampler. Collection efficiency for this sampler was reported as 99% when compared to the EPA filter method. A materials balance approach for 27 elements was used to characterize the effluents around the power plants.

In discussing the results, it was noted that the amount of an element which exists at a steam-electric power plant in each of the various ash streams depends on several factors, including the following:

1. Elemental concentration in the coal
2. Boiler configuration and firing conditions
3. Flue gas emission control devices
4. Properties of the element and its compounds

Ash distribution among exiting ash streams varied with the plants. The tangentially fired boilers at Plants A and B each produced about 22% bottom ash, while the Plant C cyclone boiler produced about 63% bottom ash. The venturi scrubbers at Plant A and the electrostatic precipitator at Plant B showed collection efficiencies of 99.6 and 99.1%, respectively. The mechanical collector at Plant C had approximately 65% collection efficiency. Thus, Plants A, B, and C had total ash percents in the flue gas of 0.3, 0.7, and 12.9%, respectively.

This study agreed with other studies discussed in its recognition that elements are partitioned into three distinct groups with respect to their distribution in the ash fractions. Enrichment was noted in the flue gas plus fly ash at all three plants for the following elements: S, Hg, Cl, Sb, F, Se, V, Pb, Mo, Ni, B, Zn, Cd, Cr, Cu, Co, U, As, and Ag. Elements which were approximately equally distributed in the bottom ash and fly ash included Ba, Be, Fe, Al, Ca, Ti, Mn, and Mg. Some elements enriched in the flue gas (S, Hg, and Cl) were primarily discharged to the atmosphere in the vaporous phase.

Davison et al.[30] have characterized the ash that passed the collector at the Midwestern Power Plant. The ash from the collector hoppers was size-differentiated physically in the laboratory by sieving and by aerodynamic separation. The fly ash passing the collector was sampled *in situ* using an Andersen stack sampler. However, since the researcher failed to use a backup filter in this sampler, the particles less than 0.5 μm in diameter were not collected.

The following conclusions were derived from this study:

1. The concentration of trace elements in the ash is dependent on particle size. Generally, increasing concentrations are correlated with decreasing particle size.
2. There is a definite enrichment of certain elements in the smallest particles emitted from a power plant. These elements include lead, thallium, antimony, cadmium, selenium, arsenic, nickel, chromium, zinc, and sulfur.
3. The highest concentrations of the trace constituents occur in particulates in the 0.5 to 10.0 μm diameter range; the size of particulate that can be inhaled and deposited in the pulmonary region of the respiratory system.
4. Presently available emission control devices for fine particulates are less effective for removing particulates in the size range that contains the most toxic elements.

Lee et al.[31] have studied the emission from the Illinois Power Plant which had an electrostatic precipitator (ESP) with a rated efficiency of 97.7%. The separation of the particles into size classes was accomplished by means of an in-stack cascade impactor which utilized aluminum discs as collection surfaces. Inlet and outlet samples from the ESP were analyzed by neutron activation for Fe, V, Cr, Ni, Mn, Pb, Sb, Cd, Zn, and Se.

It was reported that, of the elements emitted from the electrostatic precipitator, those concentrated in particulates in the submicron diameter range were antimony, chromium, lead, selenium, and zinc. Nickel was more greatly enriched in particulates in the 5- to 10-μm diameter range. The method of analysis for all of the elements, except selenium, was graphite furnace atomic absorption spectrophotometry. Selenium was determined by neutron activation analysis. Potential sources of error were felt to be related to problems with the sampling methodology, such as particle reentrainment, calibration inaccuracies, and wall loss effects. Wall losses were believed to be the most serious error, since they can range from 30 to 50% of the total amount collected.

Gowherd et al.[32] have studied the impact on the environment made by the Widows Creek Power Plant, which is a 125-MW, tangentially fired boiler equipped with a mechanical fly ash collector. The purpose of the project was to quantify the potentially hazardous pollutants in the waste stream of a typical, coal-fired utility boiler. In this study, 22 trace elements, nitrates, sulfates, polycyclic organic compounds, and polychlorinated biphenyls were identified as potentially hazardous air pollutants resulting from the combustion of coal.

The analytical results showed that, with the exception of antimony, barium, beryllium, manganese, tellurium, titanium, and vanadium, there was a tendency for potentially hazardous pollutants to progressively concentrate in the ash streams further downstream from the boiler. Fine particles were enriched with most of the trace metals. Beryllium, cadmium, copper, and zinc exhibited the greatest degree of enrichment in these particles. Another finding was that organic compounds found in the coal, such as polycyclic organic material (POM), were approximately evenly distributed between the bottom ash and the collector ash during one test period. The polychlorinated biphenyls (PCBs) were apparently formed during the combustion of the coal. None of these were found in the coal, but they were found in all of the ash streams.

Linton[33] has performed tests upon fly ash samples from two coal-fired power plants burning midwestern bituminous coals which indicate that a number of minor and trace elements are found to be preferentially concentrated on the fly ash particle surfaces. Elemental analysis was performed to determine the presence and concentration of 19 matrix (Al, Fe, and Si) and trace elements. Primary analysis for the surface concentration of element types was determined by ion microprobe mass spectrometry. Corroborative determinations were made by Auger microprobe and electron microprobe analyses. Bulk concentrations of elements contained in the fly ash were determined by spark-source mass spectrometry. It was found that Be, C, Ca, Cr, K, Li, Na, P, Pb, Si, Tl, V, and Zn are preferentially concentrated on the fly ash particle surfaces. The conclusion is drawn that because of this phenomenon, the highest minor and trace element concentrations occur in those smaller particles with high surface-to-mass ratios. These smaller particles will more readily pass through control devices, remain suspended in the atmosphere for long periods, and will deposit in the innermost regions of the human respiratory tract when inhaled.

There are several surveys of literature on this problem. For example, Smith and Gruber[34] made a literature survey of over 300 references covering the chemical and physical properties of coal, coal combustion theory, coal utilization, and coal combustion smoke emissions, particulate emissions, and gaseous emissions. General chemical and mineralogical information was tabulated in the survey. The relative frequency of the occurrence of minerals was presented in tabular form for 43 minerals comprising 7 groups: clay and shale, sulfides and sulfates, carbonates, oxides and hydroxides, phosphates, silicates, and salts. Illite and seracite were designated as the two most frequently occurring minerals. Of the 43 mineral types, 17 were described as occurring in coal in only extremely rare occasions. The typical compositional limits of coal ash were presented for SiO_2, Al_2O_3, Fe_2O_3, CaO, MgO, TiO_2, Na_2O, K_2O, and SO_3. The Cl content of seven selected U.S. coals was also tabulated. Ranges in the composition of fly ash were reported for C, Fe_2O_3, MgO, CaO, Al_2O_3, SO_3, TiO_2, CO_3, SiO_2, P_2O_5, K_2O, and Na_2O.

Winchester and Nifong[35] made a survey of literature to evaluate whether air pollution along the southwestern shore of Lake Michigan could be a significant source of trace element contamination in Lake Michigan. Data were tabulated for the estimated amount of emissions per year for 30 trace elements from coal burning sources, coke manufacture, fuel oil combustion, iron and steel manufacture, cement manufacture, and transportation sources. The trace elements most commonly emitted from iron and steel operations were Fe, Mn, Cu, and Zn; coal, coke, and fuel oil combustion were mainly responsible for As, Cr, Sn, Ti, Ni, and V emissions. Accumulation of data on air and water pollutants, and meteorological information, indicate that air pollution could be an important source of certain trace elements contaminating the waters of Lake Michigan.

B. Atmospheric Deposition and Soil Pollution

In this section we shall discuss transport phenomena associated with particulate and gaseous emissions to the atmosphere and with disposal of sludge, fly ash, bottom ash, and scrubber liquors to ash basins and landfills. The result of all these activities will be soil pollution, which will be discussed in some detail.

Within approximately 10 km of an effluent source, current transport modeling techniques rely primarily on a Gaussian plume formulation or on numerical solutions to the gradient-transport diffusion equation. Both approaches require some degree of *a priori* specification of the local wind velocity and turbulence levels; this may or may not be available. Gradient-transport models require detailed information and a computer to solve the equations. The Gaussian plume equation has, therefore, become popular because of its simplicity and because empirical information on the requisite dispersion parameters has been collected and summarized.[36-38] Gifford[39] reviewed this literature and recommended expressions for the dispersion parameters for particular uses. Properly used, this model can provide average concentration estimates to within a factor of 2 or 3.[3]

Most of the data from which published dispersion parameters were deduced were taken over open, flat, homogeneous terrain of moderate aerodynamic roughness and under atmospheric conditions ranging from unstable to moderately stable. Concentration calculations will, therefore, be reliable only under similar conditions.

For distances greater than 10 km one cannot assume, in general, that power plant site measurements of wind direction, speed, and atmospheric stability characteristics are going to persist either in space or in time.[3] In order to use these data to calculate the path of a pollutant plume, it has been necessary to develop various interpolation schemes (in time and space). These range from straightforward interpolation schemes with distance (from point of concern to wind observation) weighting functions,[40] to allowances for impenetrable topography,[41] to rigorous mass conservation schemes where the data on winds, topography, and top of mixed layer are fitted by a statistical technique to the equations of momentum conservation.[42]

Several case studies to relate the emissions from fossil fuel combustion to the trace element content of surrounding soil and plants have been conducted. Klein and Russell[12] reported enrichment of Ag, Cd, Co, Cr, Cu, Fe, Hg, Ni, Ti, and Zn in soil and Cd, Fe, Ni, and Zn in plants collected around a 650-MW power plant near Holland, Mich. Enrichment was confirmed to the upper 2 cm of soil and correlates well with wind patterns and the metal content of coal for each element except mercury, which was only slightly enriched.

The Oak Ridge-Allen Plant study also investigated major and minor elements[6] in soil and moss (Dicranum) samples collected along a 40-mi transect through the plant. The concentrations of most major elements (Al, Ca, Fe, Mg, Mn, Na, and Ti) in soil varied by approximately a factor of 2 over the sampled areas. The concentrations of some minor elements in the upper 1 cm of soil also varied considerably; whereas for others, the concentrations varied only slightly. Soil samples collected under trees generally contained higher concentrations of trace elements than samples collected in open areas. The analytical data for soil indicate no accumulation of trace elements that could be attributed to the power plant. Trace element concentrations in moss were uniformly lower than in the surrounding soil.

Klein and Russell[12] found that soils around a coal-burning facility (650 MW, 90% precipitator efficiency) were enriched in silver, cadmium, cobalt, chromium, copper, iron, mercury, nickel, titanium, and zinc. They also found that plant materials (native grass, maple leaves, and pine needles) were enriched in cadmium, iron, nickel, and zinc. Soil enrichment, except for mercury, was correlated with wind patterns and metal content of coal. Bradford et al.,[43] however, concluded that 4 years of operation at the Mojave Generating Station resulted in no measurable contamination of either soil or vegetation in the region surrounding the facility.

Poelstra et al.[44] studied the behavior of mercury compounds in 15 different European soils. Mercury and its compounds were strongly adsorbed in the top layer of soil. Leaching of mercury is minimal. Evaporation of dimethyl mercury from soil occurs at a high rate, but for most other mercury compounds the evaporation rate is probably low.

Peyton and McIntosh[45] reported elevated levels of cadmium, zinc, and lead in soil from a borrow pit and in sediment from a small pond, both near the Indiana Toll Road. Cadmium-lead-zinc ratios suggest that steel-making processes contribute to the metal content of the borrow pit; whereas, automobile exhaust emissions are responsible for the high lead content of the pond sediments. Initial AAS data indicate a correlation between percent organic matter and heavy metal content.

Benenati[46] used AAS to study the concentrations and distributions of zinc, lead, cadmium, and arsenic in soil, vegetation, and water resources sampled within a 7-mi radius of the zinc smelter in Blackwell, Okla. Within a pasture adjacent to the smelter, he found that decreased productivity, altered floristic composition, and lowered species diversity correlated with increased proximity to the smelter and increased levels of metals in both soil and vegetation.

Anderson and Smith[47] examined concentrations of mercury in the surface 2 cm of soil within 20 km of a 1200-MW coal-burning facility. Soils sampled in the northerly directions (prevailing downwind) contained higher concentrations than those sampled in the southerly directions. Despite some increased concentrations of mercury, however, the projected soil enrichments of the element over a 35-year period are not critical.

A mass deposition model was formulated by Jurinak et al.[48] for particulates emitted from a hypothetical 3000-MW coal-burning facility situated in a semiarid environment. Some of the basic assumptions of this model are emission rate, 1.05 tons/hr; deposition velocity, 0.03 to 0.1 m/sec; particle size distribution for particles > 10 μm that was 3% of 25 μm radius, 5% of 12.5 μm radius, and 10% of 7.5 μm radius. Deposition rates using this model are calculated for the winter, spring, summer, and fall seasons. The changes in pattern of deposition with season were modeled to reflect changes in the wind speed and direction and atmospheric stabilities. Deposition was found to be highest in summer and lowest in fall. Total particulate mass deposition ranged from 10 to 50 kg/km^{-2}/month^{-1}. The model also predicted that fallout over a 50-year period would be insignificant with regard to the total amount of zinc, chromium, lead, and cadmium originating from the stack and deposited in the final environmental sink for the region. Fallout, however, would contribute 65% of the mercury loading of this sink.

Based on the maximum deposition rate (50 kg/km^{-2}/month^{-1}) found by Jurinak et al.[48] and concentration of trace elements in the < 3 μm fraction, the concentration of fly ash-derived trace elements in plants and soils was established by Page et al.[15]

These estimates show measurable, but small, enrichments in plants or soils in trace elements originating from stack emissions. Lyon[49] and Vaughan et al.[50] arrived at similar conclusions using different approaches to estimated fallout.

Horton and Dorselt[51] have studied the environmental impact of trace elements emitted by the Savannah River Plant. The study[52] is aimed to develop new and improved data on the movement and effects of trace metals in the terrestrial ecosystem surrounding this power plant.

The deposition of trace elements by rain has been discussed by several authors.

Precipitation scavenging of particles and gases has been measured in the neighborhood of coal-fired power plants, and the experimental results are fairly well modeled theoretically, at least for estimates of total mass scavenged. Details of the removal as a function of particle size, however, remain in a primitive state. Nevertheless, useful predictions of atmospheric concentrations and deposition rates can be made if heavy reliance is placed on available experimental data. For sites downwind from the coal-burning power plant where trace elements or their complexes reach the soil surface by wet or dry deposition from the atmosphere, the most important determinants of their environmental fate are likely to be sorption and desorption on soil particles at the soil surface and the subsequent mobility of these trace elements as a result of soil erosion caused by both wind and water. A few key experiments should be conducted on the sorption-desorption processes associated with the trace elements that stem from fly ash material.[3]

The general principles of chemistry, physics, and biology utilized in mathematical modeling are relatively well established for dealing with trace contaminants at or below the soil surface. Simulations and reasonable predictions of the subsequent movement, retention, and transformation of trace contaminants in particular situations are possible, provided the relevant environmental parameters have been measured.

Klein et al.[9-11] have estimated the amounts of elements that annually reach the oceans in U.S. rivers. They used a river runoff value of 1.485×10^{15} ℓ/year for the U.S. and a weighted mean suspended load of 602 mg/ℓ. Elemental concentrations for both dissolved and particulate matter were taken from the literature. Results from these calculations are shown in Table 8.

The effect of power plant effluents on soil is discussed also by Roffman et al.[53] According to these authors, there is no evidence of accumulation of trace elements caused by deposition of particulate matter from the power plant (see Table 9). Analysis and comparison of floral and faunal samples collected in areas of high potential impact and of low potential "control" areas showed no conclusive evidence of significant increase in concentrations of trace elements. The samples were analyzed for 34 trace elements, and while the variability of the amounts present was high for some elements, no significant increases were noted in the impact areas. Visual observations of vegetation damage due to trace element emissions in gaseous or particulate forms were conducted and have indicated no apparent damage.

Concentrations of trace elements in grasses and two browse species found near the Four Corners Generation Power Plant are listed in Table 10.[53] The study concluded that concentrations of potentially harmful trace elements are reported to be low in the Four Corners region as compared to the entire U.S. Comparison of trace element values after 10 years of operation of the Four Corners Generation Power Plant does not indicate significant changes.

This is contrary to the figures published in the report:[54] "Trace Element Geochemistry of Coal Resources — Development Related to Environmental Quality and Health". In this report, the trace-element contamination of vegetation arising from emission from coal-fired power plant was estimated. The emission of trace elements is based on work by Page et al.,[15] Jurinak et al.,[48] Ondov et al.,[55] and assuming 3000-MW power plant using western U.S. coal equipped with an electrostatic precipitator of 99% efficiency. The concentrations deposited on vegetation are calculated assuming

Table 8
ANNUAL U.S. DISCHARGE OF TRACE
ELEMENTS FROM COAL COMBUSTION[9]

Element	Flows (10³ tons/year)		Annual flow as % of weathering mobilization
	Discharged slag and fly ash	Weathering mobilization	
Al	3640	64000	5.7
As	1.72	8	21
Ba	22.7	480	4.7
Br	0.1	34	0.3
Ca	1510	34000	4.4
Cd	0.16	0.6	27
Co	1	7.3	13.7
Cr	6.2	90	6.9
Cs	0.38	5.4	7
Cu	2.9	28	10
Fe	3780	35000	10.8
Hg	0.005	0.12	4
K	536	16000	3.4
Mg	420	11000	3.8
Mn	11.7	770	1.5
Mo	3.1	2.7	115
Na	243	15000	1.6
Ni	5.6	36	15.6
Pb	1.7	13	13
Rb	5.3	90	5.9
Sb	0.17	5	3.4
Sc	0.77	6.3	12.2
Se	0.66	0.5	132
Si	8020	30500	2.6
Sr	8	370	2.2
Th	0.73	11	6.6
Ti	176	4500	3.9
U	0.76	0.95	80
V	9.9	90	11
Zn	26.8	75	36

100% canopy, yield 3770 kg of dry matter per hectare, 4-month exposure, particulates deposited to be 3 μm, and all particulates deposited remain with harvested crop. The results obtained are shown as microgram of element per gram of dry matter in Table 11. Definite studies of the pathways and rates of transfer of trace contaminants in aquatic and terrestrial ecosystems are of the highest significance. These studies, however, cannot be adequately designed, performed, or evaluated without knowledge of the chemical species and transformations of contaminants in the ecosystem. For example, changes occur when fly ash particulates react at the water-sediment interface of anaerobic lakes, resulting in transformations of the molecular form and solubility of many compounds.[56] Effective study of contaminant transport in ecosystems depends on the ability to characterize and measure the forms of contaminants at the ambient concentrations.

Recently much effort has been invested into studies of different aspects of environmental pollution as a result of coal production and use. Some of the resulting reports and papers were presented in this chapter, most of them are only listed: see References 57-184.

Table 9
TRACE ELEMENTS IN SOILS AROUND FOUR CORNERS GENERATION POWER PLANT, U.S.[53]

	Surface soils[a]		Subsurface soils (15 cm)[a]	
Element	Control	Impact	Control	Impact
Arsenic	4.20	4.13	4.63	4.60
Beryllium	2	2	2	2
Cadmium	0.153	0.263	0.227	0.157
Chromium	10	13	10	13.33
Copper	10.93	11.60	6.87	13.07
Fluorine	148.33	174.67	220.00	177.67
Mercury	0.02	0.02	0.02	0.02
Magnesium	8,300	8,700	8,700	>10,000
Manganese	366.67	200.00	233.33	166.67
Nickel	8.33	8.33	8.33	10
Lead	10.17	5.00	10.50	21.00
Antimony	1.11	0.20	0.17	0.15
Selenium	0.20	0.23	0.23	0.23
Vanadium	23.33	16.67	23.33	20
Zinc	16.67	10	20	10
Aluminum	>10,000	>10,000	>10,000	>10,000
Silver	>1	>1	>1	>1
Boron	16.67	26.67	16.67	23.33
Bismuth	16.67	10.00	16.67	10.00
Calcium	>10,000	>10,000	>10,000	>10,000
Cobalt	30.00	33.33	33.33	26.67
Iron	>10,000	>10,000	>10,000	>10,000
Potassium	>10,000	>10,000	>10,000	>10,000
Molybdenum	<3	<3	<3	<3
Sodium	⩾10,000	⩾10,000	⩾10,000	⩾10,000
Niobium	<10	<10	<10	<10
Silicon	⩾10,000	⩾10,000	⩾10,000	⩾10,000
Tin	<5	<5	<5	<5
Titanium	933.33	933.33	933.33	1,000
Zirconium	83.33	116.67	90.00	120.00
Barium	1,266.00	1,600.00	1,233.33	1,566.67
Galium	13.33	13.33	16.67	13.33
Germanium	7.67	10	7.67	10
Lithium	53.33	60.00	56.67	63.33
Strontium	200	200	233.33	233.33
Scandium	10	13.33	10	13.33

[a] Concentrations in ppm.

II. COAL CLEANING

In this chapter we shall discuss some of the measures done in order to reduce harmful effects of coal use. Most of the work done in this area is related to the reduction of sulfur compounds emission from power plants.

Table 12 shows estimated world-wide sulfur compound emission. These figures have increased since then. Coal combustion comprises about 67% of man-made sulfur pollution and about 23% of all sulfur air emissions. At first glance, coal combustion would thus appear to be only a small part of the sulfur compound air emission problem. However, coal combustion sources, i.e., electric utilities, industrial boilers and commercial boilers, are heavily concentrated in and around the industrial cities of the

Table 10
TRACE ELEMENT CONTENT OF TWO GRASSES AND TWO BROWSE SPECIES FOUND NEAR THE FOUR CORNERS GENERATION POWER PLANT[53]

	Indian ricegrass		Galleta		Ephedra		Fourwing saltbush	
	Control	Impact	Control	Impact	Control	Impact	Control	Impact
As	1.15	0.75	36.00	36.50	0.18	0.44	0.15	0.11
Be	0.1	0.1	0.1	0.1	0.1	0.1	0.1	0.1
Cd	0.2	0.2	0.1	0.11	0.1	0.1	0.2	0.2
Cr		9.00	6.00	6.00	22	3	23	3
Cu	10.93	8.74	9.65	4.91	2.85	10.0	0.67	4.66
F	30.50	17.00	9.50	6.50	4	6	3.5	3
Hg	0.02	0.027	0.035	0.015	0.04	0.04	0.01	0.01
Mg	830.00	860.00	810.00	875.00	550	550	1,985	1,950
Mn	132.50	105.00	75.00	70.00	45	55	60	30
Ni	7.00	4.67	4	4	2	2	4.5	2
Pb	1.66	1.04	0.965	1.85	0.67	0.69	0.425	0.77
Sb	0.32	0.52	0.43	0.39	0.26	0.70	0.48	0.38
Se	0.21	0.19	0.075	0.09	0.12	0.09	0.74	0.16
V	14.50	12.67	4.50	3.50	2	3	2	2
Zn	17.00	11.00	18	8.50	11	11	13.5	6
Al	6,750	5,943	8,050	7,100	280	540	175	130
Ag	1	1	3	3	1	1	1	2
B	107.50	15	22.5	20	10	10	30	35
Bi	3	3	3	3	3	3	3	3
Ca	10,000	10,000	10,000	10,000	10,000	10,000	10,000	10,000
Co	ND<3	ND<3	ND<3	ND<3	ND<3	ND3	ND<3	ND<3
K	500	500	500	500	500	500	500	500
Mo	18.5	2.67	2	2	2	2	2	2
Na	300	300	300	300	300	300	300	300
Nb	ND<3	ND<3	ND<3	ND<3	ND<3	ND<3	ND<3	ND<3
Si	10,000	10,000	10,000	10,000	10,000	10,000	4,750	3,300
Ti	182.50	155.00	187.50	130.00	33	45	25	30
Zr	26.00	68.33	56.50	17.00	11	14	5.5	6
Ba	42.50	46.67	22.50	17.00	25	25	17.5	13
Ga	ND<10	ND<15	ND<10	ND<10	ND<5	ND<5	ND<7	ND<7
Ge	ND<3	ND<3	ND<3	ND<3	ND<3	ND<3	ND<3	ND<3
Li	1.75	1.67	1.0	1.0	0.5	0.5	3.0	1.0
Sr	50.00	55.00	85.00	66.00	220	160	105.5	130
P	182.50	208.33	537.50	602.50	165	270	335	325

Note: Concentrations in ppm.

world which also, of course, have the highest population density. By contrast, the natural marine and terrestrial emissions are dispersed from the outset over a wide geophysical area.

To contend with this problem, some governments and local authorities have established performance standards for power plants. For example, in the U.S., the Federal Government has established a performance standard for all new coal-fired plants greater than 25 MW of 1.2 lbs $SO_2/10^6$ Btu of coal heat input. For most eastern U.S. coals, this corresponds to a maximum allowable coal sulfur content of approximately 0.8% and about 0.6% for western U.S. coals.

Pollution of the atmosphere arising from coal combustion can be combatted in a number of ways including: flue gas desulfurization by scrubbing, the building of tall stacks to disperse the pollutant over a wider area, conversion of coal to low sulfur

Table 11
ESTIMATED TRACE ELEMENT
CONCENTRATION OF VEGETATION
BY 3000-MW COAL POWER PLANT[54]

Element	Conc. deposited on vegetation	Typical conc. for vegetation
As	0.07	0.4
Cd	0.002	0.2
Cr	0.034	1.5
Cu	0.74	10
Pb	0.15	3
Mo	0.027	1
Se	0.107	0.2
Zn	0.30	25
Sb	0.011	0.06
Be	0.006	0.03
Co	0.011	1.0
Ga	0.096	1.2
Ni	0.021	5
Th	0.016	0.05
U	0.016	0.04
V	0.18	1

Table 12
ESTIMATED WORLDWIDE SULFUR
COMPOUND AIR EMISSIONS[185]

Compound	Source	Emissions as sulfur[a]
SO_2	Coal combustion	51
	Petroleum refining	3
	Petroleum combustion	11
	Smelting of ores	8
H_2S	Industrial emissions	3
	Marine emissions	30
	Terrestrial emissions	70
SO_4 aerosols	Marine emissions	44
Total air emissions		220

[a] Tons per year × 10^6.

(and low ash) fuel oil, and coal cleaning. The technology for these approaches is in varying stages of development, from laboratory and pilot scale in the case of liquefaction, and demonstration plant development for flue gas scrubbing, to a fully developed and mature industry in the case of physical cleaning of coal.

Many power plants in the U.S. are using special equipment to remove SO_2 from the flue gases. The wet-scrubbing method is currently preferred, in which a slurry of lime or limestone is contacted with the flue gas. By 1985, the estimated annual production of FGD sludge will amount to 37 million tons originating from 42,535 MW capacity having FGD systems in operation.[16]

A more interesting approach is a pilot-plant at the Pennsylvania Electric Company Seward Generating Station which recently demonstrated the technical feasibility of extracting sulfur dioxide gas from the flue gases and converting it to usable sulfuric acid. Such an arrangement could recover 90% of the sulfur in the coal. Also, the resulting sulfuric acid produced from this process might find use as a leaching agent to beneficiate coal ash as part of a coal-associated mineral processing complex at power plants for the production of titanium, silica, iron, and alumina.[186] This idea is further developed in the chapter on element recovery from coal ash.

An extensive survey of the sulfur reduction potential of the coals of the U.S. was performed by Deurbrouck at the U.S. Bureau of Mines.[187] Coals were selected from all major coal-producing regions of the U.S. The average coal contained 2.05% pyritic sulfur and 3.23% total sulfur. Only about 7% of the as-mined coals were in the 0.8% sulfur range needed to meet the U.S. Standards for New Stationary Sources.

In the U.S., an extensive air quality control program is under way and organized by Electrical Power Research Institute (EPRI). The research is being conducted on the technical aspects of removing SO_2 by injecting a dry sorbent into the flue gas head of the fabric filter (baghouse) typically used for fly ash removal. If effective, this dry sorbent injection process would enable integrated, continuous removal of particulate matter and SO_2 from the flue gas stream. Compounds of interest for use as dry sorbents are nahcolite (naturally occurring sodium bicarbonate, $NaHCO_3$) and trona (a naturally occurring mixture of sodium bicarbonate and sodium carbonate).[188]

The primary advantages of SO_2 control by dry sorbent injection are the lower capital costs associated with removing particulate matter and SO_2 in a single system and the higher availability levels and reduced maintenance requirements resulting from system simplicity. Also, energy and water consumption are lower than in a conventional wet scrubbing system. Disadvantages include limited levels of SO_2 removal, potentially limited availability and high cost of dry sorbents, lack of operating experience, and disposal of the spent sodium sorbent.

Previous investigations have confirmed that calcium- and magnesium-based sorbents are relatively ineffective in removing SO_2, while alkaline materials containing sodium are attractive as dry sorbents. Also, injection of sorbent into higher-temperature flue gas generally results in better SO_2 removal, although detailed information on the kinetics of the heterogeneous reactions is not available. Most dry scrubbing studies have focused on nahcolite as the sodium-based dry sorbent. Only limited data are available on the effectiveness of trona as a dry scrubbing agent.[188]

In order to characterize dry removal of SO_2 with nahcolite and trona in conjunction with a baghouse, a laboratory scale study was conducted.[188,234] The experimental system, which fired a western coal with a sulfur content of 0.45%, had a flue gas flow rate of 0.38 m³/sec at the baghouse inlet. The experiments investigated the effects of the following parameters on the dry removal process: sorbent type (nahcolite, trona, commercial sodium bicarbonate), sorbent utilization, sorbent particle size, baghouse temperature, air-to-cloth ratio, cleaning cycle time, sorbent injection schedule (continuous, batch), flue gas temperature at injection point (up to 427°C), predecomposition of the nahcolite, and inlet SO_2 level.

With the nahcolite sorbent, the reaction time required to achieve maximum steady-state SO_2 removal at a baghouse inlet temperature of 123°C was about 40 min. A stoichiometric quantity of nahcolite yielded an average SO_2 removal level of 50% for a 90-min test period; the maximum SO_2 removal was 67%. A sorbent ratio, SR (actual sodium-to-sulfur ratio normalized by the sodium-to-sulfur ratio required to form sodium sulfate), of 1.5 was required for an average SO_2 removal level of 70%. The average removal level at an SR of 0.9 was increased from 42% to 66% by precoating the bags with the nahcolite sorbent.

Injecting the nahcolite into high-temperature (up to 427°C) combustion products increased the level of SO_2 removal; maximum removal occurred at 288°C. Using stoichiometric amounts (SR = 1), the SO₂ removal with injection into the duct at 550°F was 80% compared with 67% with injection at the baghouse inlet (132°C). SO₂ removal decreased to 60% as the duct temperature was increased to 800°F. Varying the baghouse air-to-cloth ratio from 0.6 to 1.2 m/min had no effect on SO₂ removal.

Trona was also effective in removing SO₂ from the combustion products, although the level of removal was not as great — 40% at an SR of 1, with an SR of 3 required for 70% removal. The trona reacted rapidly with the SO₂ on injection; thus SO₂ removal was essentially independent of cleaning cycle time. At an SR of 2, injecting the trona at elevated flue gas temperatures enhanced SO₂ removal to a level of 70 to 78%, compared with 50 to 62% removal when the sorbent was injected at the baghouse inlet. Both of these dry sorbents also showed some removal of NO; at an SR of 1, nahcolite removed 12% of the NO and trona 5%.[188]

Physical cleaning of coal to remove ash along with a part of the pyrite content has been in commercial practice for a number of years. Because the continuous and nonselective mining machines of today result in a substantial impurity content in coal, the majority of utility fuel is now cleaned prior to combustion. Coal cleaning (or washing) is the separation of raw coal into ash-rich and ash-lean fractions by virtue of the relatively high density and hardness of ash in relation to the hydrocarbon coal matrix. This is done in commercial practice through first crushing to liberate some pyrite, then passing the coal through screens, jigs, dense-medium processes, tables, etc. The behavior of coal in mechanical cleaning plants is approximated in the laboratory through float-sink testing of crushed coal in liquids of various density.

The most widely used washing techniques (some 75% of the total commercial production) utilize density separation methods (see Table 13).

To simulate these processes on a laboratory scale, investigators use static vats of organic solvents with varying densities. The float portion of a sample in a solvent of some given density is scooped out and the sink fraction is drained and subjected to the next solvent mixture. When analyzed, these float-sink samples provide a histogram which depicts the behavior of the property measured as the density of the sample (washing medium) changes (see for example paper by Schultz et al.[190]).

A report by Cavallaro et al.[87] presents the results of a washability study showing the trace element contents of various specific gravity fractions for ten coal samples collected from various coal-producing regions of the U.S. Reliable analytical methods were developed to determine cadmium, chromium, copper, fluorine, mercury, manganese, nickel, and lead in the whole coals and the various specific gravity fractions of the coals. The material balances for the 8 trace elements for the 10 coals ranged from 85 to 115% with an average of 99% and a 95% confidence interval of ±3%. The data from the analytical determinations on the washed coals are plotted as washability curves so that the quantity and quality of the clean coal products can be obtained at the desired specific gravity of separation. Generally, the data showed that most of the trace elements of interest concentrated in the heavier specific gravity fractions of the coal, indicating that they are associated with mineral matter; removal of this material should result in significant trace element reductions, ranging up to 88%. Some of the data obtained by Cavallaro et al.[87] are shown in Table 14.

Capes et al.[191] have examined the rejection of trace metals from coals during the process of fine grinding followed by selective oil agglomeration of the carbonaceous constituents. Six different steam and coking coals, mainly from Canadian and U.S. sources, were treated. The level of most heavy elements was generally low but, in light of the large throughputs of power plants, the potential environmental hazard may be significant, especially for those elements which are volatile or may form volatile compounds when fired. Many of the trace metals were substantially removed during ag-

Table 13
COAL CLEANING METHODS[189]

Washer type	Fraction of coal cleaned by equipment type (%)			
	1942	1952	1962	1972
Jigs				
Dense-medium processes	8.8	13.8	25.2	31.4
Concentrating tables	2.2	1.6	11.7	13.7
Flotation	—	—	1.5	4.4
Pneumatic	14.2	8.2	6.9	4.0
Classifiers	7.4	8.5	2.1	1.0
Launderers	13.1	5.2	2.2	1.0

glomerative beneficiation of the coal. Some which are apparently in organic association in the coal remained in the agglomerated product.

A summary of results obtained by Capes et al.[191] is presented in Table 15. It is evident, from the average values of the ratio of metal rejection to overall ash rejection, that many of the trace elements are concentrated in the tailings and are thus removed by fine grinding and oil agglomeration. A number of elements, however, are apparently associated with the organic constituents of coal and thus report with the agglomerated product. Oil agglomeration is hence not a satisfactory separation method for these constituents. Most prominent among these materials are barium, beryllium, boron, germanium, mercury, selenium, titanium, and zirconium. There is no apparent reason to explain the association of these particular elements with the carbonaceous fraction in coal, although the results are consistent with previous work.

Ford et al.[192] have made an evaluation of the effect of coal cleaning on trace element removal. We shall describe their work in some detail.

Eight geographically diverse bituminous coals, used basically for power generation, were examined to determine the suitability of a concentrating table for deep-cleaning coals at a topsize of 30-mesh. In this study, the eight raw feed coals and the separated fractions of zones A, B, C, D, and E for each of the eight were analyzed for trace element content. The end of the concentrating table corresponds to zone E. This zone should contain refuse material high in pyrite content. The next area, zone D, should contain nonpyrite refuse material. At the end of the table, zone A should contain the cleanest coal with zone B also containing clean coal, but less clean than in zone A. Zone C, being in the middle, should contain an expectedly intermediate type of material — mostly coal, but with a relatively high ash content. In most cases, the zone C material is combined with the zone A and B material as the cleaned coal while the zones D and E material are the refuse.

The trace element content of the raw coals used in this study is shown in Table 16. The summary of the results obtained for trace elements removal is shown in Table 17. Even using the rough coal cleaning as described in this report, significant reductions resulted in many of the trace elements reported to be pollutants when released during coal utilization. As shown in the data in Table 17, arsenic and manganese were removed with the greatest consistency; beryllium and fluorine were removed the least, with beryllium actually enriched in half of these coals. The average percent removed, by trace element and by coal, was 24%. Of particular significance was the amount of coal actually reporting to the refuse fractions, zones D and E, as given in the "Weight percent" column. These amounts can generally be related to the amounts of trace elements removed. Qualitatively, a constituent removed at a lower value than that for

Table 14
WASHABILITY ANALYSES SHOWING THE LEVELS OF TRACE ELEMENTS IN THE SAMPLE CRUSHED TO 14-MESH TOP SIZE[87]

Pittsburgh coalbed, Pennsylvania

Product	Weight (%)	Direct (ppm)									Cumulative								Weight (%)
		Cd	Cr	Cu	F	Hg	Mn	Ni	Pb		Cd	Cr	Cu	F	Hg	Mn	Ni	Pb	
Float-1.30	59.4	0.03	11	4.6	17	0.08	2.8	7.4	1.7		0.03	11	4.6	17	0.08	2.8	7.4	1.7	59.4
1.30-1.40	29.3	0.09	19	6.7	33	0.09	5.9	10	3.9		0.05	14	5.2	22	0.08	3.8	8.2	2.4	88.7
1.40-1.60	5.9	0.35	31	19	81	0.28	19	30	0.10		8.7	3.1							4.8
Sink-1.60	5.4	0.39	43	43	125	1.7	150	30	26		0.09	16	8.1	31	0.18	13	9.8	4.3	100.0
Head Sample	—	—	—	—	—	—	—	—	—		0.09	16	9.0	35	0.19	11	11	4.3	100.0

Waynesburg coalbed, Ohio

Product	Weight (%)	Direct (ppm)									Cumulative (ppm)								Weight (%)
		Cd	Cr	Cu	F	Hg	Mn	Ni	Pb		Cd	Cr	Cu	F	Hg	Mn	Ni	Pb	
Float-1.30	23.4	0.14	15	6.1	27	0.13	4.3	8.1	2.1		0.14	15	6.1	27	0.13	4.3	8.1	2.1	23.4
1.30-1.40	40.7	0.06	18	5.6	53	0.07	8.2	9.6	2.4		0.09	17	5.8	44	0.09	6.8	9.0	2.3	64.1
1.40-1.60	20.6	0.15	24	10	113	0.15	20	12	5.6		0.10	19	6.8	60	0.10	10	9.8	3.1	84.7
Sink-1.60	15.3	0.36	30	41	146	0.61	66	41	26		0.14	20	12	73	0.18	19	15	6.6	100.0
Head sample	—	—	—	—	—	—	—	—	—		0.14	21	11	78	0.18	18	16	6.7	100.0

Table 15
REJECTIONS OF ELEMENTS FROM COAL BY
GRINDING FOLLOWED BY OIL AGGLOMERATION[191]

	Average concentrations (ppm)			% Element rejected/ (% ash rejected) (SD)
Element	Feed coal	Agglomerates	Tailings	
Aluminum	15,000	10,000	100,000	0.88
Antimony	1.3	0.7	7	0.68
Arsenic	72	15	270	1.11(0.18)
Barium	40	38	100	0.42(0.21)
Beryllium	0.5	0.7	0.03	0.02
Boron	21	18.8	32	0.49(0.26)
Cadmium	0.3	0	2.0	1.23
Calcium	2200	1300	4800	0.65
Chromium	652	186	2837	0.90(0.19)
Copper	301	171	1103	0.79(0.15)
Germanium	14	14	0	0
Iron	49,000	15,700	193,500	0.89(0.15)
Lead	75	33	242	0.90(0.25)
Magnesium	250	125	1500	0.88
Manganese	275	55	935	0.97(0.14)
Mercury	0.49	0.41	1.2	0.51(0.34)
Molybdenum	183	43	667	1.01(0.24)
Nickel	59	24	204	0.88(0.17)
Selenium	4	3	20	0.44
Sodium	50	25	500	1.48
Tin	65	20	170	0.90
Titanium	425	400	710	0.33
Vanadium	150	61	410	0.87(0.28)
Zinc	284	144	1150	0.84(0.29)
Zirconium	25	25	100	0.42

the weight of coal reporting to the refuse does not reflect a significant reduction of that particular constituent. The data in Table 18 show that the averages of the percent of the major constituents removed were 25%, or about the same as found for the trace elements. The calorific value, as evidenced by the negative sign, was increased in proportion to the percent of ash removed.

One should mention the EPRI coal-cleaning test facility (EPRI Report RP 1400, see also Yeager[188] 1980). The aim of the facility is to develop superior flow sheets for steam coal, verifying flow sheet design before installation, verifying and developing new processes, and testing various unit operations. The test facility is capable of processing 20 tons/hr of 3/3 in x 0 coal through a number of flow sheets. Its unit processes include heavy-media cyclones, water-only cyclones, tables, and flotation cells. Auxiliary equipment includes classifying cyclones, centrifuges, and thickeners. The design of this facility is so flexible that almost any commonly used flow sheet can be set up by reconnecting various units.

The test facility has been designed to enable easy sampling, and it will be instrumented and automated to allow continuous monitoring of various process parameters. These features will ensure data reliability and will permit a better understanding of the effects of process variables on product quality. It is expected that during the first 4 or 5 years of operation a large coal-cleaning data base will be developed to assist the industry in designing plants on a more scientific basis.[188]

In this chapter we should also mention a more recent technology to minimize SO_2

Table 16
ANALYTICAL DATA FOR COALS USED IN THE STUDY BY FORD ET AL.[192]

Constituent	\multicolumn{8}{c}{Coal Sample Number}	8 coals mean
	2047	2856	2860	2881	2889	2900	2926	2928	
Arsenic[a]	5.63	49.6	19.4	37.7	4.99	59.6	3.71	58.7	29.9
Beryllium[a]	2.55	1.45	1.71	1.78	2.68	1.85	1.24	2.56	1.98
Chromium[a]	9.04	28.2	39.7	28.5	13.1	21.4	11.8	40.2	24.0
Copper[a]	10.4	32.8	26.3	24.4	21.2	21.0	19.7	52.0	26.0
Fluorine[a]	46.8	137	88.4	107	175	76.2	90.8	105	103
Lead[a]	6.75	18.1	29.8	14.1	47.0	14.7	100	19.3	31.2
Manganese[a]	56.0	31.7	45.2	18.9	93.0	21.2	108	31.2	50.7
Mercury[a]	574	597	517	618	262	382	165	262	422
Nickel[a]	15.9	22.8	28.5	28.3	13.5	22.6	15.6	32.4	22.5
Selenium[a]	2.08	6.64	4.80	4.92	2.03	1.91	2.10	6.62	3.89
Vanadium[a]	17.1	50.9	63.9	49.9	25.0	34.9	19.4	87.1	43.5
Zinc[a]	424	33.1	31.6	35.1	32.9	26.9	30.9	22.2	79.6
Silicon[b]	1.56	6.00	6.16	5.11	2.40	3.15	2.63	6.50	4.19
Aluminum[b]	0.69	2.92	3.44	2.91	0.67	1.98	1.20	4.08	2.24
Iron[b]	3.39	2.75	3.45	1.50	2.54	2.46	3.01	1.36	2.56
Calcium[b]	0.10	0.26	0.16	0.11	1.92	0.07	1.32	0.17	0.51
Ash[b]	10.1	24.0	26.4	19.9	13.3	15.0	14.8	25.3	18.6
Volatile matter[b]	45.7	19.8	18.0	25.0	41.3	34.2	38.6	31.4	31.8
Fixed carbon[b]	44.2	56.2	55.6	55.1	45.4	50.8	46.6	43.3	49.7
Sulfur, total[b]	5.48	2.88	4.18	1.86	4.21	3.36	4.02	1.96	3.49
Sulfur, pyritic[b]	4.66	2.42	3.64	1.42	2.96	2.78	3.23	1.42	2.82
Sulfur, organic[b]	0.80	0.46	0.52	0.42	1.17	0.54	0.78	0.54	0.65
Calorific Value (Btu/lb)	12,772	11,604	11,193	12,294	12,882	12,576	12,826	11,044	12,150

[a] Parts per million.
[b] Percent.

Table 17
SUMMARY OF % TRACE ELEMENT REMOVED IN EXPERIMENTS BY FORD ET AL.[192]

	Coal sample number								Mean, by element
	2847	2856	2860	2881	2889	2900	2926	2928	
%[a]	4	16	28	16	7	3	3	18	12
Arsenic	21	67	67	59	15	39	11	45	41
Beryllium	−3	10	17	8	−4	−1	−2	23	6
Cadmium	21	—	—	—	—	—	—	—	21
Chromium	4	22	40	26	15	1	2	26	17
Copper	3	44	56	25	12	8	8	30	23
Fluorine	−1	9	14	12	6	1	−1	21	8
Lead	28	53	63	37	8	19	16	34	32
Manganese	16	53	76	58	39	9	24	18	37
Mercury	23	64	68	44	3	15	4	20	30
Nickel	8	36	55	20	21	8	3	23	22
Selenium	9	61	55	39	19	10	2	29	28
Vanadium	−1	14	39	25	10		1	15	13
Zinc	43	41	39	34	13	28	17	46	33
Mean, by coal	13	40	49	32	13	11	7	28	

[a] Weight percent in refuse fraction.

Table 18
SUMMARY OF % MAJOR CONSTITUENTS REMOVED IN EXPERIMENTS BY FORD ET AL.[192]

	Coal sample number								Mean, by constituent
	2847	2856	2860	2881	2889	2900	2926	2928	
%[a]	4	16	28	16	7	3	3	18	12
Ash	25	35	63	40	33	7	8	53	33
Total sulfur	24	56	45	26	9	18	7	11	25
Pyritic sulfur	29	67	56	37	15	21	10	17	32
Calorific value	−2	−12	−31	−14	−7	−2	−2	−19	−11
Silicon	18	30	67	42	32	3	2	57	31
Aluminum	7	27	61	35	22	3	3	55	27
Iron	36	65	58	40	24	22	12	26	35
Calcium	10	23	31	27	57	14	29	35	28
Mean, by coal	18	36	44	29	23	11	9	29	

[a] Weight percent in refuse fraction.

emission from coal combustion which uses fluidized-bed combustion boilers where crushed coal is burned on a bed of inert ash and crushed limestone. The lime reacts with SO_2 released during the combustion of the coal to form sulfates. The resulting fluidized-bed boiler (FBB) waste is a granular solid material consisting of $CaSO_4$, unreacted CaO, other metal oxides, and fly ash (see also chapter 1 of this volume).

III. WASTE MANAGEMENT

The collected residues from coal-fired electric power plants are disposed of or recycled on land. However, there is also a significant amount of liquid discharge into streams as well as releases in the effluents from coal refuse and mineral wastes. In the U.S., the coal industry has been under constant pressure by environmental legislation to clean up liquid discharge into streams. Consequently, rather large sludge ponds have developed which consist of fine coal, fine soil, and water. These ponds are generally formed behind large piles of coarse coal waste and in conjunction therewith formulate the often-criticized coal waste impoundments.

Ash disposal is not a new problem. However, government regulations are affecting disposal operations. For many years disposal was simply handled through large bulk sites, usually located near the source of waste material. Now, there are two possibilities:

1. To undertake an extensive research of waste product uses and recycling operations
2. To formulate more creative and efficient uses of existing on-site disposal operations and methods

Present statistics indicate only 20% of all ash generated is associated with new uses, leaving some 80% for eventual disposal.[193]

The following three on-site utilization schemes are being used:

1. Bulk disposal of ash
2. Road and fill area construction
3. Soil additives and fertilizer creating admixtures

We have discussed some of these utilization schemes in chapter 1.

Schivley[193] has listed criteria for selection of on-site disposal areas. They are

A. Site location determinants
   1. Amount of ash that can be placed in a given area
   2. Amount of prefill site preparation necessary
   3. Drainage structures and water control facilities required
   4. Groundwater and hydrologic baseline conditions
B. Site preparation
   1. Site drainage is diverted away from fill area to permit site material to dry and prevent contamination of water after filling operations commence
   2. Vegetation is completely cut and removed from site prior to stripping
   3. Topsoil and cover material is removed and stockpiled for distribution after fill operations are completed
C. Placement of ash
   1. Material should be spread in lifts of 12 to 15 in. maximum thickness
   2. Compaction should be accomplished with a self-propelled vibratory roller or other satisfactory compacting equipment
   3. Water should be added during compaction operations to provide optimum moisture content, according to designed compaction specifications

Disposal and recycling of collected residues on land has been extensively discussed in Reference 54.

The land-disposal techniques considered include landfill, lagooning, and land spreading. According to the report, in situations where the residues are spread onto land that supports either agricultural crops or native vegetation, trace elements in fly ash could adversely affect the productivity of the land or the quality of the vegetation grown. Therefore, a comprehensive understanding of the chemical, physical, and biological changes that occur when residues from coal-fired power plants are mixed with soils is essential to evaluate the effect that these changes will have on the productivity or the land and trace-element composition of the vegetation.

All the data in the literature are consistent with the conclusion that application of fly ash, regardless of its source, increases the pH of soils, irrespective of their type.

Application of fly ash to agricultural soils in amounts greater than a few percent generally has a significant impact on the physical properties of soil, the quantities and effects being influenced by the properties of the specific fly ash and soil involved.

In addition to fly ash from coal-burning plants, municipal wastewater sludge represents a major disposal problem. To alleviate the disposal problems, it was proposed that these two waste products could be combined to create a material which would be suitable for structural fills such as highway embankments.[117] Potential advantages of this technique include:

1. The waste products would be utilized, becoming an asset instead of a liability.
2. Mutual benefits may occur as the result of interactions between the fly ash and sludge. It was expected that combining fly ash with wastewater sludge would result in a decrease in permeability due to moisture-retaining characteristics of the sludge and some fixation of metal ions by silicate complexes in the ash or by hydration reactions occurring in the mixture.
3. The environmental consequences should be minimized compared with other popular ash and sludge disposal practices. For example, there would be no air pollution which can result from sludge incineration and no build-up of heavy metals, which remains a threat from sludge placed on agricultural land.

Tenney and Echelberger[194] have also discussed fly ash as a conditioner of various waste sludges such as those from paper mills, rubber plants, and municipal wastewater. They conducted a study on the effectiveness of fly ash as a dewatering agent for activated sludge. Their results indicated that a specific fly ash concentration of one part fly ash to ten parts sludge, by weight, gave optimum rate of dewatering. This rate corresponded to additions of 50 to 75 g of fly ash per 1 of sludge. Aside from improving the filterability of the sludge-fly ash cake, fly ash lowered the residual moisture content of the cake and thereby increased its potential fuel value. Their study showed a reduction of 20% in the moisture content of the cake.

Tenney and Echelberger[194] concluded that sludge particles are bonded to the surface of fly ash particles by means of chemical and/or electrostatic interactions that give appreciable rigidity to the sludge-fly-ash lattice; therefore, the sludge particles are held securely during vacuum filtration. Such a mechanism increased porosity and allowed for the relatively unrestricted egress of water.

Physical and chemical properties of the fine coal waste in the ground were investigated by Busch et al.[195] In order to obtain more information on coal sludge, especially at various depths, a sampling program was carried out on two typical waste piles. Four holes were drilled and sampled at 5-ft intervals. These samples were carefully extracted from the drill holes and shipped back to the laboratory in a manner that would minimize disturbance. Laboratory tests were performed on these samples to provide data that could be used to evaluate problems involving drainage, liquefaction, strength, and stability. In addition, these samples, plus coarse refuse and red dog from previous sampling, were subjected to a typical coal analysis which could provide information as to the possible use of the waste products as a fuel or as a source for marketable by-products. Because of the apparently substantial percentages of coal in the waste materials, regular coal proximate and ultimate analyses were performed. The ash, volatile matter, and fixed carbon percentages obtained by the proximate analysis define the amount of combustible material in the refuse, while the calorific value defines the rank of the coal. The ultimate analysis describes the elemental composition of the coal portion and the percent ash. The physical properties and field conditions of coal sludge combine to make deposits of this material unstable. The extremely fine size of the material plus the rather well-graded grain sizes make the permeability very slow.

Coal, as mined, contains a great deal of extraneous rock and mineral matter. The inorganic constituents of coals often represent as much as 50% of run-of-the-mine products. These rocks and mineral impurities are expensive to ship, dilute the caloric content of the coal, and produce undesirable gaseous and particulate pollutants when the coal is burned or utilized. Consequently, much of the more highly mineralized coals (about one half of the total mined in the U.S.) is cleaned or processed to remove some of the unwanted mineral and rock materials. The discarded rock, mineral, and coaly matter from coal processing, together with other coal mine refuse, constitute the gob piles and culm banks, which are scattered over thousands of acres in coal-producing regions.

In addition to these well-recognized problems, however, another potential environmental hazard is beginning to gain attention. Coals, and undoubtedly coal wastes, contain a broad array of elements. Many of these trace elements are of considerable concern because of the low tolerances of plants and animals for them. Undoubtedly, many of these trace elements are carried into the environment by weathering, leaching, and burning of refuse. Although the relative amounts of these components per unit of waste is usually small, the total absolute amount of each available in a large waste bank could cause grave consequences in water, soil, or air if they were concentrated by natural processes.[3]

The trace element characterization of coal wastes has been extensively studied by Wewerka et al.[122,196-199] at Los Alamos Scientific Laboratory (LASL).

The overall objectives of the LASL research program were to assess the problem of trace element contamination in coal waste drainages and to identify suitable control technologies; more specifically to:

1. Assess the nature and magnitude of trace elements in the effluents from coal preparation wastes
2. Identify the chemistry of the trace constituents of environmental concern
3. Identify and experimentally verify effective environmental strategies to control the release of hazardous constituents, and
4. Analyze the trade-offs associated with the different control technologies and recommend control measures of necessary research development and demonstration programs

To understand why coals and coal preparation wastes release elements in the amounts they do, Wewerka et al.[196-199] have studied the levels and releases of elements from a low-sulfur Appalachian coal area where the mineral drainage is not highly visible and from high-sulfur Illinois Basin areas where mineral drainage has long been recognized as a severe problem. Both types of wastes are composed primarily of clay minerals, quartz, iron sulfides, and calcite. Low-sulfur Appalachian wastes differ from the high-sulfur Illinois wastes by having < 1% iron sulfide minerals (pyrite and marcasite) as compared to 20 to 30% for the latter. Some 55 elements have been identified and undoubtedly there are more. The most abundant of these (aluminum, iron, and silicon) are from the major minerals. The trace constituents are probably present as minor minerals (even as microparticles), components of residual coal, or substituents in the major minerals. A number of elements that are generally considered to be environmentally sensitive are present in significant quantities (> 30 μg/g of waste).

The high-sulfur (high-pyrite) wastes from the Illinois Basin, when exposed to air and water, produce highly acidic drainages (pH $\sim$ 2 to 4). Even low-sulfur (low-pyrite) Appalachian wastes produce acidic drainages (pH $\sim$ 4), though the total amount of acid is much less. Experiments by Wewerka et al.[196-199] simulating the intermittent rainfall and weathering to which coal waste dumps were subjected have revealed that alternate oxidation of leaching of the pyrite in the waste is a most effective way to regenerate acid leachates continuously. Authors have demonstrated experimentally that intermittently leached coal wastes pose a far greater pollution threat, in both quantity and time, than do those wastes that are always submerged in water or are isolated from air, water, or both in some manner.

Acidic leachates in coal waste dumps are very efficient in dissolving or degrading many of the minerals present and releasing the elements associated with them. For example, aqueous leachates from high-sulfur Illinois Basin coal wastes contained nine elements (aluminum, cadmium, cobalt, copper, fluorine, iron, manganese, nickel, and zinc) in environmentally hazardous concentrations.

It is difficult to assess the overall seriousness of this type of discharge because suitable data are generally lacking on the behavior and effects of the various trace elements in the environment. The Environmental Protection Agency (EPA), however, in 1975 published a set of proposed effluent limitations that apply to discharges from coal-refuse materials. These guidelines are listed in Table 19. Although only a few elements and effluent characteristics are covered by the proposed regulation, these data, nonetheless, serve as a useful point of reference.

IV. RADIOACTIVITY

During the last few years, radioactivity released by coal power stations into the environment has received great attention from the public, as well as from governments and their agencies. Some of the reports have received very much attention and have been published in journals and newspapers.

One of these reports which has been much discussed by the general public is derived from the study conducted by the U.S. Environmental Protection Agency (EPA). The report, "Radiological Impact Caused by Emissions of Radionuclides into Air in the United States", was prepared as a result of recent legislation that requires radionuclides to be included among other sources of air pollution. Preliminary findings suggest that there are greater risks to the public of developing cancer from radionuclides emitted by coal-fired power plants than by normally operating nuclear plants.

Radionuclides — including isotopes of uranium, thorium, tritium, argon, noble gases, iodine, radon, and polonium — are released into the atmosphere from operating the various facilities. These radionuclides are dispersed into populated areas where exposure occurs by breathing or swallowing the materials. When coal is burned, the mineral content is converted to ash and slag. These wastes contain most of the radionuclides originally present, but a fraction of the ash is released into the atmosphere. The quantity released depends on the particulate control system, furnace design, mineral content of the coal, and existing emission control standards.

Table 19
U.S. EPA EFFLUENT LIMITATIONS FOR COAL REFUSE

Effluent characteristic	Maximum for any 1 day (mg/l)	Maximum av. daily value for 30 consecutive days (mg/l)
Iron, total	7.0	3.5
Iron, dissolved	0.60	0.30
Aluminum, total	4.0	2.0
Manganese, total	4.0	2.0
Nickel, total	0.40	0.20
Zinc, total	0.40	0.20
pH	within the range 6.0—9.0	

According to the EPA figures for existing coal-fired plants, the lifetime risk to individuals nearest to the facilities ranges from 60×10^{-6} to 700×10^{-6}. For boiling-water and pressurized-water reactors, the risks are 20×10^{-6}, respectively.

Translated into health effects, the EPA estimates that from 0.0004 to 1.5 fatal cancers can be expected to develop per year of operation for each existing coal-fired power station. There are 250 such stations in the U.S. and the wide variation in range is due to differences in siting and power generating capacity. The accepted number of fatal cancers for each of the 69 nuclear power generating plants is 0.001/year of operation.[200]

The EPA report makes reference to a total of 250 existing and 145 new coal-fired plants, 25 boiling-water reactors (BWR), and 44 pressurized-water reactors (PWR) in the U.S. On a direct comparison at suburban sites between coal and nuclear plants, BWR facilities each can be expected to produce 0.0013 fatal cancers per year and PWR facilities, 0.0009 fatal cancers per year. Existing coal-fired plants, on the other hand, each can be expected to produce 0.10 fatal cancers per year and new coal plants, 0.017 fatal cancers per year.

Principal radionuclides emitted by coal-fired stations include radon, uranium, and thorium. Those emitted by nuclear plants include the noble gases, tritium, and the halogens.

The radioactivity from coal has obtained space even in *Financial Times* [201] which has bought an article describing a study by W. Camplin of U.K. — National Reduction Protection Board (NRPB). This study was the first of the kind to be made in Britain. The results show that public exposure from radioactive discharges to the atmosphere from coal-fired stations in Britain is about the same as the discharges to atmosphere from the whole of the nuclear fuel cycle as practiced in Britain.

The NRPB study, by Camplin, makes no attempt to compare directly the leakiness of the two technologies. Neither does it consider the whole of the coal-fuel cycle, from coal mine to final disposal of the great bulk (99.5%) of the slightly radioactive ash as building material or landfill. It considers only the 0.5% of ash which evades the filters and is emitted from the chimney stack, together with radioactive gases. The report considers the emissions from a typical 2,000 MW coal-fired power station in Britain. Such a station, operating at about 56% load factor, would burn about 4.7 metric tons of coal per year. The aims of the report are: to isolate the sources of activity in the emissions, to model the dispersion of the radioactivity released, and to estimate the radiological impact on those people most highly exposed.

The gateways studied for radionuclides to reach humans are: from external irradiation from the plume of activity in the air, from inhaling the plume, from external irradiation from "fallout" deposited on the ground, from inhaling activity from "fallout" resuspended in the air by wind, rain, etc., and from ingestion of contaminated foodstuffs.

The most complex of these five pathways is through the foodchain, which involves movement of the activity into the soil, then into the plant, and then into an animal or directly to man. The contribution from this source dominates the estimate of the individual radioactive dose. Specifically, animal liver turns out to be the dominant contributor. So the study considers the case of people who eat livers from cattle grazing in the vicinity of the point of maximum deposition of ash from power station emissions.

The estimated annual committed effective dose equivalent received by a member of a hypothetical critical group in the population, as a result of routine atmospheric discharges, is less than 5% of the dose equivalent limit. Although, conceivably, some people in Britain could receive this much, it is likely that individual doses would be much lower.

Second, the collective effective dose equivalent commitment, truncated to 500 years, to the population of Britain from such a station would be 34.000 man-rems.

Third, the highest components of the individual dose estimate are due to ingestion of food contaminated with lead-210, polonium-210, and protactinium 231. The main contribution to the collective dose estimate comes from inhaling thorium.

With these two studies getting such wide publicity, the general public has been interested in the problem of radioactivity resulting from the use of coal.

Here we shall mention in some detail some other reports indicating the magnitude of this problem. Data on uranium and thorium content are summarized in Chapter 2, and there is no need to repeat these figures here. Let us just mention that the measurements performed indicate that for coal the whole thorium series is contained within the matrix, whereas, for petroleum, radium has been leached selectively from the substrate. In other words, coal contains ^{238}U, ^{235}U, ^{232}Th, and their radioactive daughter products in secular equilibrium.[202] Secular equilibrium is a steady-state condition in which the rate of formation of the radioactive daughter products is just equal to their rate of decay; i.e., activities of radioactive parent and daughter are the same. The amount of radionuclides discharged into the atmosphere will depend mainly on the quality of ash collection equipment.

Coles et al.[203] have investigated the pathways of naturally occurring radionuclides during coal usage in the power stations. The measurements were made on radionuclide content of coal, bottom ash, and fly ash collected by electrostatic precipitators (ESP) in two power stations. The fly ash from ESP was classified into four fractions according to the grain size. The obtained samples were analyzed using gamma spectroscopy. It was found that uranium occurred as uranite and coffinite, while ^{232}Th and ^{40}K were associated with clay material. Most of Th, Ra, and U in coffinite ended in bottom ash, while uranium in uranite evaporates and, together with lead, is bound to fly ash.

The amounts of radioactive substance emitted in the airborne effluents of coal-fired plants have been studied by a number of research groups, including Eisenbud and Petrow,[165] Terrill et al.,[69] Martin et al.,[204] Hull,[127] Lave and Freeburg,[205] McBride et al.,[206] and Styron et al.[207]

Eisenbud and Petrow[165] have shown by the analysis of the fly ash produced by combustion of pulverized Appalachian coal that a 1000-MW coal-burning power plant will discharge into the atmosphere from about 28 mCi to nearly 1 Ci/year of radium-226 and radium-228. An oil burning plant of similar size will discharge about 0.5 mCi of radium per year. Comparison of these data with data on the release of fission products from nuclear-powered generating stations shows that when the physical and biological properties of the various radionuclides are taken into consideration, the conventional fossil-fueled plants discharge relatively greater quantities of radioactive materials into the atmosphere than nuclear-powered plants of comparable size.

A detailed report on radiological impact of airborne effluents of coal-fired and nuclear power plants has been published by McBride et al.[206] Data based on the same report are also presented by Torrey.[3] Here is the summary of their reports.

Radiological impact of naturally occurring radionuclides in airborne effluents of a model coal-fired steam plant of 1000 MW(e) is evaluated assuming a release to the atmosphere of 1% of the ash in the coal burned, and compared with the impact of radioactive materials in the airborne effluents of model light-water reactors of 1000 MW(e). The principal exposure pathway for radioactive materials released from both types of plants is ingestion of contaminated foodstuffs. For nuclear plants, immersion in the airborne effluents is also a significant factor in the dose commitment. Assuming that the coal burned contains 1 ppm uranium and 2 ppm thorium, together with their decay products and using the same impact analysis methods used in evaluating nuclear facilities, the maximum individual dose commitments from the coal plant for the whole body and most organs (except the thyroid) are shown to be greater than those from a pressurized-water reactor (PWR) and, with the exception of the bone and kidney doses, less than those from a boiling-water reactor (BWR). With the exception of the bone dose, the maximum individual dose commitments from the coal plant are less than the numerical design guideline limits listed in 10 CFR 50, Appendix 1, for light-water reactors (LWRs).[9] Population dose commitments from the coal plant are higher than those from either nuclear plant, except for the thyroid dose from the boiling-water reactor. The use of coal containing higher uranium concentrations and/or higher particulate releases ($\gg 1\%$), characteristic of the present coal-fired power industry, could result in dose commitments from a coal plant several orders of magnitude higher than those estimated in this study. The study is limited to a comparison of the radiological impacts of airborne effluents from model coal-fired and nuclear power plants and does not compare the total radiological impacts of a coal vs. a nuclear economy. It is concluded that an evaluation of the radiological impact on the environment should be included in the assessment of both coal-fired and nuclear power plants.[206]

Concentrations of 1 ppm uranium and 2 ppm thorium in the coal were selected to develop a source term for the model 1000-MW(e) coal plant. Under these assumptions, and further assuming that all the thorium is in the ash and that 1% of the ash in the coal is released to the atmosphere, about 23.2 g/MW(e)-year of uranium and 46.4 g/MW(e)-year of thorium and associated nonvolatile radioactive daughter products would be released with the ash. Annual releases from a 1000-MW(e) station with the same operating parameters as for the TVA Thomas A. Allen steam plant at Memphis, Tenn. would be 2.32×10^4 of uranium and 4.64×10^4 g of thorium and associated nonvolatile radioactive daughter products.

A source term based on the release of 1% of the fly ash was calculated (see Table 20) assuming that the radioactive daughters of ^{238}U, ^{235}U, and ^{232}Th in the fly ash are in secular equilibrium with the parent elements and are released in the same proportion as the parent elements except for the radon isotopes. All of the radon initially present in the coal is assumed to be released in the airborne effluent. The 1% ash release assumed is nearly an order of magnitude less than the average ash release for the industry in 1972, but approximates the present EPA regulation for the release of particulates to the atmosphere.

Both the model coal plant and the nuclear power plants were assumed to be located in the Midwest with meteorology characteristic of St. Louis, Mo. The surrounding

Table 20
EST. ANNUAL AIRBORNE RADIOACTIVE MATERIALS
FROM POWER PLANT[3]

	Isotope	Release (Ci/year)
U-238 chain		
	U-238	8×10^{-3}
	Th-234	8×10^{-3}
	Pa-234m	8×10^{-3}
	U-234	8×10^{-3}
	Th-230	8×10^{-3}
	Ra-226	8×10^{-3}
	Po-218	8×10^{-3}
	Pb-214	8×10^{-3}
	Bi-214	8×10^{-3}
	Po-214	8×10^{-3}
	Pb-210	8×10^{-3}
	Bi-210	8×10^{-3}
	Po-210	8×10^{-3}
U-235 chain		
	U-235	3.5×10^{-4}
	Th-231	3.5×10^{-4}
	Pa-231	3.5×10^{-4}
	Ac-227	3.5×10^{-4}
	Th-227	3.5×10^{-4}
	Ra-223	3.5×10^{-4}
	Rn-219	3.5×10^{-4}
	Pb-211	3.5×10^{-4}
	Bi-211	3.5×10^{-4}
	Tl-207	3.5×10^{-4}
Th-232 chain		
	Th-232	5×10^{-3}
	Ra-228	5×10^{-3}
	Ac-228	5×10^{-3}
	Th-228	5×10^{-3}
	Ra-224	5×10^{-3}
	Pb-212	5×10^{-3}
	Bi-212	5×10^{-3}
	Tl-208	1.8×10^{-3}
Radon releases		
	Rn-220	0.4
	Rn-222	0.8

Note: Materials released from a model 1000-MW(e) coal-fired power plant (source term). Assumptions are (1) the coal contains 1 ppm uranium and 2 ppm thorium, (2) ash release is 1%, (3) Rn-220 is produced from Th-232 in the combustion gases at the rate of 1.38×10^{-9} Ci/g of thorium, (4) the annual release of natural uranium is 2.32×10^4 g and of Th-232 is 4.64×10^4 g, and (5) 15 sec are required for the gases to travel from the combustion chamber to the top of the stack.

population was assumed to be 3.5 million people out to 88.5 km from the facility, the average population distribution around three midwestern population centers.

The population density in persons per square kilometer assumed for a radial distance of 8 km from the facilities was 37; from 8 to 40 km, 49; and from 40 to 88.5 km, 170.[3]

Table 21
MAXIMUM INDIVIDUAL DOSE COMMITMENTS FROM
THE AIRBORNE RELEASES OF MODEL 1000 MW POWER
PLANT (MREM/YEAR)[3]

Organ	Coal-fired plant	Boiling-water reactor	Pressurized-water react.	10 CFR 50 Append. 1 guides
Whole body	1.9	4.6	1.8	5
Bone	18.2	5.9	2.7	15
Lungs	1.9	4.0	1.2	15
Thyroid	1.9	36.9	3.8	15
Kidneys	3.4	3.4	1.3	15
Liver	2.4	3.7	1.3	15
Spleen	2.7	3.7	1.1	15

Maximum individual doses and population doses out to 88.5 km were calculated for both nuclear plants and stack heights of 50, 200, and 300 m for the coal plant. Radioactive materials released at the top of the stack of the model coal plant were assumed to rise because of the buoyancy of the hot stack gases. The effective release height is the sum of the physical height of the stack and the buoyant plume rise as calculated through the use of Brigg's equations. Information from the 122-m Allen steam plant stack was used in the plume rise calculations. A 20-m fixed height with no plume rise was used for releases from roof vents of the nuclear plants. These heights are characteristic of existing plants.

Atmospheric dispersion of plumes as they are blown downwind from the plants was estimated using the Gaussian plume equation. Radionuclides released as particulates deposit on ground surfaces through the processes of dry deposition and scavenging. The rate of dry deposition, which involves adsorption, particle interception, diffusion, and chemical-electrostatic effects, was estimated by multiplying the concentration of the radionuclide in air at ground level by the deposition velocity. A value of 1.0 cm/sec was used for the deposition velocity of all particulates. Particle sizes were assumed to be small enough that gravitational settling could be ignored. The AIRDOS computer code[208] was used for the atmospheric dispersion calculations, using annual-average meteorological data in terms of joint frequencies of wind-speed categories, atmospheric stabilities, and wind direction. The computer code estimates annual-average concentrations in air at ground level and ground deposition rates for each radionuclide released from the plants for each of 16 compass directions as a function of distance from the source. Each concentration and deposition rate is an average value across a 22.5° sector. Concentrations in air for each sector and distance from the source are used in AIRDOS to calculate dose via inhalation and immersion in air. Ground surface concentrations are used for the estimation of external radiation exposure. The ground deposits are also assimilated into food, which results in additional doses through ingestion.

Tables 21 and 22 give the maximum individual dose commitments and the population dose commitments calculated to result from the estimated releases of radioactive materials from the model 1000-MW(e) coal-fired and nuclear power plants. The source term for the coal plant assumes a concentration of 1 ppm uranium and 2 ppm thorium in the coal and a release of 1% of the fly ash. The maximum individual doses for both the coal and the nuclear plants are the maximum values at a 500-m perimeter.

The maximum individual doses at the 500-m boundary of the coal plant meet the Appendix I regulations with the exception of the bone dose (Table 21). The maximum

Table 22
POPULATION DOSE COMMITMENTS FROM THE AIRBORNE
RELEASES OF MODEL 1000 MW POWER PLANT[3]

Organ	Coal-fired plant stack height (m)				Boiling-water reactor	Pressurized-water reactor
	50	100	200	300		
Whole body	23	21	19	18	13	13
Bone	249	225	192	180	21	20
Lungs	34	29	23	21	8	9
Thyroid	23	21	19	18	37	12
Kidneys	55	50	43	41	8	9
Liver	32	29	26	25	9	10
Spleen	37	34	31	29	8	8

Note: Men-rem/year, 88.5 km radius.

individual doses for the nuclear plants also meet the Appendix I regulations, with the exception of the thyroid dose from the BWR. An actual nuclear plant would have to conform to the Appendix I regulations,[9] i.e., a maximum of 15 mrem/year for the thyroid dose at the site boundary. A lower dose would result from reducing the amount of iodine released, a site-location with a greater site-boundary distance, more favorable meteorology, or a greater distance to the nearest dairy pasture. The data in Table 21 also show that the maximum individual dose commitments from the model coal plant are less than those from a BWR (except for the bone dose), but are greater than the doses from the PWR (except the thyroid dose).

The maximum individual doses at the perimeter of the coal plant are essentially the same for all stack heights from 50 to 300 m. This is the result of the assumptions (1) that the washout coefficient for small particles is independent of the height of the particles above the ground, i.e., all particles at all heights are washed out to the earth in the same time interval for a given distance from the stack; and (2) that the washout effect is much greater than the sum of various dry deposition effects at distances close to the plant. Dry deposition does not make a significant percentage contribution to dose until the plume has traveled far beyond the plant boundary.

Population dose commitments from the coal plant are greater than those from either nuclear plant (Table 22) with the exception of the thyroid dose from the BWR. The ratio of the population doses for the coal plant to the nuclear plants is higher than the same ratio for the individual doses at the plant boundary (Tables 21 and 22). This results from the rapid decay of the short-lived noble gases released from the nuclear plants as they move from the plant boundaries out to 88.5 km.

Table 23 lists the percentage contributions of radionuclides to the population doses from the coal-fired plant. The radium nuclides, ^{226}Ra and ^{228}Ra, are the major contributors to the whole-body dose and most organ doses. However, ^{210}Po is the major contributor to spleen dose, and ^{210}Po and ^{210}Pb together contribute almost half of the dose of the kidneys. The contribution of ^{222}Rn to the doses is insignificant, even though its release rate is much greater than that of any other nuclide in the source term. The lung is the critical organ for ^{222}Rn, but the ^{222}Rn contribution to the total lung dose is only about one part in a million.

According to Styron et al.,[207] for a realistic assessment of the magnitude of release of radionuclides, special attention needs to be given to lead-210 and polonium-210 since they appear to have a large potential for significant environmental impact and have not received sufficient attention in trace-element studies for power plants. An-

Table 23
PERCENTAGE CONTRIBUTIONS OF RADIONUCLIDES TO POPULATION DOSES[3]

Organ	Contribution of radionuclides[a] (%)							
	^{226}Ra	^{228}Ra	^{228}Th	^{230}Th	^{232}Th	^{210}Po	^{210}Pb	^{227}Ac
Whole body	67	21	0.7	3.5	0.7	2.5	2.1	1.0
Bone	59	14	1.8	12	1.8	0.9	4.9	2.6
Lungs	47	15	10	10	10	2.1	1.7	0.7
Thyroid	68	21	0.7	3.5	0.7	2.5	2.1	1.0
Kidneys	28	8.4	0.6	11	0.6	29	18	0.9
Liver	48	15	0.2	4.2	0.2	16	11	4.7
Spleen	42	13	0.4	2.2	0.4	40	1.2	0.6

Note: From the airborne releases of a 1000-MW(e) coal-fired power plant.

[a] Percentage contributions are for coal containing 1 ppm uranium and 2 ppm thorium. The radionuclides are assumed to be released from a 50 m stack at a midwestern site with a plume rise due to buoyancy of the hot stack emissions. The horizontal columns total less than 100% because radionuclides contributing only to a minor extent to the organ doses are not listed. Release heights greater than 50 m result in slightly higher contributions from the radium nuclides and lower contributions from ^{230}Th.

other potentially important parameter in determining radiation exposure to man centers on disposal and utilization of coal ash and refuse. Lee et al.[209] have suggested that emanation of radon-222 from ash disposal ponds will be the most serious radionuclide problem associated with increased use of coal. A potential hazard can be associated with the use of fly ash in cement and concrete blocks and in roadway construction. The radium-226 in these concrete blocks used for home construction may constitute an important source of radon-222 dose to the public.

The U.S. Environmental Protection Agency (EPA) and Department of Energy (DOE, formerly the Energy Research and Development Administration and the Federal Energy Administration) jointly initiated a project with Monsanto Research Corporation (MRC) at Mound Facility (formerly Mound Laboratory) to assess the fate of radionuclides in coal and their associated health and environmental effects. The purpose (preliminary report by Styron et al.[207]) was to broadly survey pathways of radionuclides in the coal fuel cycle, identifying critical questions and providing direction for subsequent, definitive studies of radionuclides in coal and of the technological enhancement of concentrations of naturally occurring radionuclides through use of coal. The scope of this preliminary phase of the study was limited to only one coal province, and Western coal was selected because of its prominence in the National Energy Plan. It seems almost certain that the U.S. will turn increasingly to the vast coal deposits of the Western states, since Western coal reserves (198 billion t or 180 billion metric tons) represent 72% of the identified U.S. coal resources and this coal has a low sulfur concentration. The specific objectives of the study were to (1) investigate the fate of radionuclides in the coal fuel cycle and, where possible, delineate the scope of the potential environmental and human health problem associated with radioactivity in coal; (2) compile and evaluate data for uranium-238, uranium-234, lead-210, polonium-210, and thorium-230 found in Western coal; (3) study the release, fate, and accumulation of radionuclides from a power plant burning Western coal; and (4) assess the possible need for additional control technology or standards.

It was found that radionuclide concentrations in bottom ash and fly ash were higher than in the feed coal, probably due to a loss of mass during combustion. Measurements of radionuclide concentrations in bottom ash (Table 24) ranged from 2.33 to 2.74 pCi/g for uranium-234 and uranium-238, from 4.98 to 5.11 pCi/g for polonium-210, from 2.52 to 5.63 pCi/g for lead-210, and from 2.65 to 3.95 pCi/g for thorium-230. Measurements of concentrations in fly ash ranged from 4.000 to 4.68 pCi/g for uranium-234 and uranium-238, from 5.25 to 5.37 pCi/g for polonium-210, from 4.82 to 5.76 pCi/g for lead-210, and from 3.66 to 4.24 pCi/g for thorium-230. With an ash content of 11.91%, one would expect an increase in concentration of nonvolatile elements in ash over feed coal by a factor of 8.4. The measured ratios of radionuclide concentrations in bottom ash to feed coal were 5.9 for uranium-234, 5.9 for uranium-238, 4.3 for polonium-210, 6.6 for lead-210, and 6.1 for thorium-230. Similar ratios for the fly ash were 9.5 for uranium-234, 9.9 for uranium-238, 4.5 for polonium-210, 7.7 for lead-210, and 7.3 for thorium-230. Ratios less than 8.4 for bottom ash suggest an enriched fraction of the radionuclides is incorporated into stack effluents.

Measurements of the radionuclide concentrations in effluents at the input to the electrostatic precipitator averaged 9.99 pCi/m^3 for uranium-234, 9.15 pCi/m^3 for uranium-238, 9.95 pCi/m^3 for polonium-210, 13.2 pCi/m^3 for lead-210, and 560 pCi/m^3 for radon-222. Measurements at the output of the electrostatic precipitator averaged 7.48 pCi/m^3 for uranium-234, 7.54 pCi/m^3 for uranium-238, 9.17 pCi/m^3 for polonium-210, 14.6 pCi/m^3 for lead-210, and 290 pCi/m^3 for radon-222. These limited data indicate that concentrations of uranium-234 and uranium-238 are reduced by the electrostatic precipitator; whereas polonium-210 and lead-210 are not affected significantly. The data also indicate a reduction in the concentration of radon-222, although the precipitator should have no effect on the radon-222.

Also of interest is the radionuclide balance for unit #2 of the Neal Station as reported by Styron et al.[207] This is shown in Table 25.

We shall also briefly mention here the calculations of doses and concentrations by dispersion modeling as it was done by Styron et al.[207] Atmospheric dispersion calculations were performed to estimate concentrations of radionuclides in air and soil, as well as doses from inhalation and ingestion pathways resulting from radioactive materials in the airborne effluents from the Neal Station. Although inaccuracies are suspected in the measurement of radionuclide concentration of stack effluents, these data are judged to represent a reasonable approximation of emissions from the plant for this preliminary study.

The computer code AIREM (Ref. from Styron)[207] was used to calculate ground-level air concentrations of radionuclides. This computer code used a Gaussian dispersion plume model. It divides the area around the source into 16 sectors and allows up to 12 radial distances. The relative ground-level air concentration, χ/Q (sec/m^3), is calculated for each sector element by summing the contributions to that sector for six stability classes. The air concentration, χ (Ci/m^3), for each nuclide is calculated by multiplying the χ/Q by the source term, Q (Ci/sec).

The AIREM code allows the calculation of the ingrowth of one radioactive decay product per parent nuclide during plume transit. The ground-level air concentration of polonium-218 due to the emission of radon-222 was calculated using this option. Subsequently, the ratio of the concentrations of radon-222 and polonium-218 was used to calculate an effective transit time. This effective transit time was then used to calculate the ground-level air concentrations of the remaining short-lived decay products of radon-222; specifically, lead-214, bismuth-214, and polonium-214.

The meteorological data used in the dispersion calculations were obtained from the local Climatic Center. The three operating units at the Neal Station were considered individually because of the differences in the heights and operating characteristics of

Table 24
RADIONUCLIDE CONCENTRATIONS IN ASH FROM THE NEAL STATION[207]

Sample identification	Mound sample number	^{234}U (pCi/g)		^{238}U (pCi/g)		^{210}Po (pCi/g)		^{210}Pb (pCi/g)		^{230}Th (pCi/g)	
Bottom ash	BAU-1	2.73	0.07	2.33	0.06					2.65	0.85
	BAU-2	2.72	0.09	2.74	0.09						
	BAPO-1					4.98	0.28	2.82	0.16		
	BAPO-2					5.11	0.23	2.52	0.16		
Fly ash	FAU-1	4.68	0.09	4.45	0.09					4.24	0.23
	FAU-2	4.04	0.11	4.00	0.11					3.66	0.13
	FAPO-1					5.37	0.25	4.82	0.14		
	FAPO-2					5.25	0.27	4.98	0.09		

Table 25
RADIONUCLIDE BALANCE FOR UNIT NO. 2
OF THE NEAL STATION[207]

Component	Radionuclide throughput (pCi/sec)			
	^{234}U	^{238}U	^{210}Po	^{210}Pb
Input coal	17,843	16,793	38,484	23,090
Bottom ash	1,446	1,350	2,685	1,946
Fly ash	15,849	15,376	19,302	18,866
Stack	2,644	2,665	3,241	5,160
Total output	19,939	19,391	25,228	25,972
% Relative imbalance	−12	−15	+34	−12

the three stacks. The effective stack height was calculated for each stack. Plume rise was calculated using the Holland equation:[207]

$$\Delta H = \frac{vd}{u} \left(1.5 + 2.68 \times 10^{-3} \, p \, \frac{T_s - T_a}{T_s} \, d\right) \quad (2)$$

where ΔH = plume rise (m)
v = efflux velocity (m/sec)
d = inside stack diameter (m)
u = wind speed (m/sec)
p = atmospheric pressure (mb)
T_s = stack gas temperature (°K)
T_a = ambient air temperature (°K)

The following parameter values were used: u = 4.92 m/sec, p = 920 mb, T_a = 293 °K.

Unit 1	Unit 2	Unit 3
v = 32.6 m/sec	v = 27.4 m/sec	v = 22.0 m/sec
d = 2.87 m	d = 4.65 m	d = 6.09 m
T_s = 421 °K	T_s = 406 °K	T_s = 397 °K

Values of plume rise were calculated to be 69.4 m, 121.5 m, and 148.0 m for Units 1, 2, and 3, respectively. The physical heights of the stacks are 76.2 m, 91.4 m, and 121.9 m; therefore, the calculated values of effective height are 146 m, 213 m, and 270 m.

The maximum dispersion coefficient was based on an estimate of the annual average height of the mixing layer for the specific location of the Neal Station. It is assumed to be 490 m.

The source terms for uranium-234, uranium-238, polonium-210, and lead-210 were based on stack measurements of Unit 2. The measurements of concentration at the precipitator outlet were averaged and multiplied by the nominal gas flow rate to obtain the source terms for Unit 2. The source terms for the other two units were estimated by multiplying those for Unit 2 by the ratios of the power ratings of the other units to that of Unit 2, i.e., multiplying by 150/330 for Unit 1 and by 540/330 for Unit 3. It is realized that the source term is dependent upon other factors besides power level; however, in light of the uncertainties in the stack measurements, this simple approximation should be adequate for the purposes of these calculations.

A different approach was used for estimating the radon-222 source terms. The radon-222 source term was based on two assumptions: (1) radon-222 is in secular equilibrium with uranium-234 in coal, and (2) all the radon-222 in the coal is sent up the stack as the coal is crushed and burned. Therefore, the radon-222 source term value used for Unit 2 is the same as the uranium-234 input rate value from the radionuclide balance for Unit 2. This approach is believed to be superior to using the stack measurement values, because an unknown, but probably significant, fraction of the measured values is from background radon-222. The radon-222 source terms for the other two units were estimated by applying the power level ratios as described above. The source terms for the short-lived decay products of radon-222 were assumed to be zero.

In Table 26, calculated and measured air concentration values are presented for a few locations. It should be pointed out that the measured values are based on a short sampling duration, whereas the calculated values are estimates of annual averages; therefore, comparisons between the two may not be very meaningful. Both the calculated and measured uranium concentrations in the vicinity of the Neal Station are on the order of background. The calculated concentrations of polonium-210 and lead-210 in air are significantly lower than the measured values.

The measured values for both polonium-210 and lead-210 are on the order of the reported background values. The calculated radon-222 concentrations are small in comparison to typical background levels (approximately 0.0002% of background). The deposition rate of each radionuclide was calculated by multiplying the calculated ground-level air concentration by a deposition velocity, which was assumed to be 0.01 m/sec. Ground concentrations were calculated assuming deposition of the radionuclides at the given deposition rates over a period of 20 years including the ingrowth of decay products after deposition of the parent nuclides. The only mechanism assumed for removal of the radionuclides from the soil was radioactive decay. The ground concentrations were divided by the "effective surface density" of a 15-cm layer of soil, 240 kg/m², to obtain the calculated soil concentrations. The calculated soil concentrations are smaller than the measured soil concentrations by a factor of approximately 500 for uranium-234 and uranium-238.

The authors[207] have concluded the following:

1. Atmospheric dispersion modeling of stack effluents for uranium-234, uranium-238, lead-210, polonium-210, and radon-222 indicated there would be no significant impact from either the inhalation or ingestion pathways.
2. Environmental concentrations of radionuclides predicted by atmospheric dispersion modeling to accumulate in soil are two to three orders of magnitude lower than measured concentrations in soil. This difference in values predicted from data on stack effluents and values actually measured is evidence that the power plant stack is not a significant source of environmental radionuclides. Concentrations of radionuclides measured in the environment adjacent to the power plant are not significantly different from background concentrations measured 40 and 96 km away.
3. Occupational radiation doses to workers inside the Neal Station were not found to be significant (< 0.5 mrem/year).
4. Burning coal at the Neal Station with reduced emission controls could increase the release of radionuclides by a factor as high as 5 to 7.
5. Accumulation of radionuclides in ash ponds may present an environmental prob-

Table 26
CALCULATED AND MEASURED GROUND-LEVEL AIR CONCENTRATIONS (aCi/m³)[207]

Sampling station	Location from power plant	^{234}U Calc.	^{234}U Meas.	^{238}U Calc.	^{238}U Meas.	^{210}Po Calc.	^{210}Po Meas.	^{210}Pb Calc.	^{210}Pb Meas.	^{222}Rn Calc.	^{218}Po Calc.	^{214}Pb Calc.	^{214}Bi Calc.
1	3.1 km, NNW	45	41	46	39	56	983	89	5,410	307	287	56	9
2	1.4 km, N	59	48	59	54	72	1,207	114	24,220	395	315	34	3
3	2.3 km, E	32	29	33	37	40	1,298	63	—	219	202	37	5
4	5.3 km, SW	18	38	18	43	22	1,270	35	12,964	122	121	39	10
5	3.7 km, SE	45	47	45	48	55	1,299	88	—	306	278	49	7
Park	96 km, S	—	54	—	62	—	1,572	—	22,274	—	—	—	—
Max. calc.	2.5 km, N	65	—	65	—	79	—	127	—	438	387	58	7

lem, primarily due to the emanation of radon-222. Use of coal with high concentrations of radionuclides could increase the magnitude of this problem.
6. Use of Western coal in the U.S. being mined today in a modern power plant with effective emission controls (ESP) has led to no significant environmental impact due to radionuclides.

According to Styron et al.,[207] there are several potentially problematic areas related to the radiological impact of expanded coal utilization which warrant further attention. The following list contains a number of areas in which information is needed:

1. Evaluate pathways and partitioning coefficients for uranium-238 and radium-226 during coal cleaning, coal combustion, and coal conversion: radionuclide balance for the processes, including waste products
2. Evaluate potential population doses from radium-226 and radon-222 in coal refuse and flue gas desulfurization sludge disposal sites, as well as use of this waste material in manufacturing various products, e.g., concrete blocks and gypsum-based wallboard
3. Conduct interlaboratory calibration for analysis of radionuclides in coal and coal refuse
4. Determine whether uranium-238 and radium-226 are in secular equilibrium in coal
5. Conduct extensive surveys of mines for uranium-238 and radium-226 in coal of all U.S. coal-producing states

V. ENVIRONMENTAL IMPACTS OF NEW TECHNOLOGIES

The new technologies of coal usage have been very carefully studied in order to assess their environmental impacts. Some of the new technologies, e.g., fluidized-bed combustion, are being developed almost entirely to reduce the environmental impact of coal use in energy production.

In this section we shall mention some of the published reports dealing mainly with the environmental impacts of coal gasification and coal liquefaction. Obviously, to discuss all of the reports on this subject would require much more space; therefore, we shall limit ourselves to some of the characteristic research done in this area.

Baria[210] has presented a survey of trace elements in North Dakota lignite, and effluent streams from combustion and gasification facilities.

Sather et al.[147] have studied potential trace element emissions from the gasification of Illinois coal. Trace element analyses of samples of Illinois #5 and #6 seam coals and the unquenched ashes obtained from gasification of these coals in a Lurgi reactor are reported. The results represent only one run of the gasifier at a single set of operating conditions, and the coal samples are representative of only one sample of the coal used in the test. Analytical determinations were done at two separate laboratories with varied procedures; however, the trace element results from the two laboratories agree with each other in most cases within the precision of the measurements. The data itself was not sufficient for a complete assessment of the environmental effects of Lurgi processing of coal. The report concludes with a discussion of areas where additional background information is needed for the assessment of the environmental impact of trace element emissions from coal gasification.

Forney et al.[211,212] have reported on trace element and major component balances around the Synthane PDU gasifier. A series of gasification tests were run in a Synthane PDU gasifier to determine the distribution of 65 trace elements in samples of all feed and products streams. The solid and liquid samples were analyzed for trace elements by spark source mass spectrometric analysis. Trace elements in the gas samples were determined by neutron activation and flameless atomic absorption procedures. Results indicate that trace elements primarily remain in the chars and dusts emanating from the gasifier. Some elements, such as B, Cl, F, and Se are found in the water and some such as Ar, Pb, and Cd are collected in the tars. Mercury analyses were incomplete, but most of the Hg appears in the tars and water, with little remaining in the char or dust.

Attari et al.[213-215] have also studied the fate of trace element constituents of coal during gasification. The distribution of trace elements from Montana lignite and Illinois #6 bituminous coal during a high Btu gasification process was investigated. Volatilization losses based on the feed analysis for 20 elements are reported for pretreatment and gasification of the bituminous sample, but for gasification only of the nonagglomerating lignite sample. Atomic absorption analysis was used to evaluate trace element composition. The major trace element losses during pretreatment and gasification for the bituminous sample were represented by Hg, Cd, Cl, Pb, N, S, and Bi. Major trace element losses in the gasification of the lignite sample included As, Bi, Cd, Cl, Pb, Hg, N, Se, S, and Te. It was concluded that evolution of trace elements from future coal-gasification operations would be substantially less than if the same feedstock were burned in a coal-fired steam plant. Trace element concentration data for various stages of gasification of a sample of Pittsburgh #8 coal were also presented.

Lloyd and Francis[216,217] have determined concentrations of 18 minor and trace elements in the feed coal and residues (chars) from 5 bench-scale gasification units to establish their behavior during gasification. The feed coal in all tests was a western Kentucky bituminous coal. A static-bed gasifier train was utilized for gasification with steam, nitrogen, hydrogen, oxygen, and carbon dioxide. Elemental analysis was performed by X-ray fluorescence. Of the elements studied, one group was stable and nonvolatile (e.g., Al, Si, K, and Ca). Highly volatile As and Zn were substantially vaporized in all atmospheres tested. A third group, e.g., P, S, Mn, and Ni, was predominantly vaporized in oxygen and hydrogen atmospheres, while remaining in the solid phase under low-pressure gasification with steam and carbon dioxide. This was especially true for P and S and may be indicative of the use of carbon dioxide gasification for high-sulfur coals.

Jahnig and Bertrand[82] have discussed environmental aspects of coal gasification. The authors have tried to assess the impact this new coal conversion industry will have on the environment. In this work, the authors examine such factors as the flow rates of streams entering and leaving the process, the components and effect of sour water, trace elements, and clean-up procedures.

Becker and Murthy[218] have summarized the findings of a review of the effects of process operating parameters on entrainment of impurities in fuel gasification systems. Generally reliable and accurate experimental data on environmental effluents from gasifiers has been very limited; however, some limited experimental data was collected on bench-scale Hygas® and Synthane gasification systems. All input and output streams of the two systems were sampled and then analyzed for various trace element contents. Analysis of this experimental data indicated: Be, Cr, Mn, Ni, P, V, and Zn are concentrated in the residual char from the gasifier; significant losses of Se, Te, Hg, and Pb occur in the feed coal where a separate coal pretreatment is used; trace elements in the raw gas stream appear to be collected in the condensate waste water (this was particularly true for B, Cl, F, and Hg); and high concentrations of As or Hg were found in the condensate tar while the particulate matter showed an enrichment in As, Be, F, P, or Se. Generally it was concluded that further efforts should be undertaken to better characterize coal gas streams with respect to minor and trace element components, and that the potential of processes modification for reducing impurities carry-over should be investigated for specific gasifier applications.

Another aspect of this problem is discussed by Koppenhaal and Manahen.[219] In their work, types of organometallic compounds that could be produced in coal conversion processes were inferred from knowledge of the structure of coal, the biological origin of coal, and the bonding tendencies of known organometallic species. Compounds discussed include metal-porphysin, metal-carbonyls, metallocenes, arene carbonyls, metal alkyls, organohydrides, and metal chelates. A diagram of possible metal-chelate structures that may affect trace metal distributions in coal conversion products, and a tabulation of mean analytical values (NAA determined) for chemical elements in coal are included (with a specific reference to ash-free, solvent-refined Kentucky #9 coal). A thorough investigation of pollution technology concerning potential metallo-organic compounds from coal conversion processes was suggested.

Somerville and Elder[120] have discussed in their paper the possible leaching of the trace elements from disposed gasifier ash. Although it may be possible to mitigate this potential impact to within acceptable limits through the use of disposal techniques, it is difficult, if not impossible, to conclusively demonstrate that the disposed ash and sludges will behave in a given manner once actually disposed of in the mined area. This is true, in spite of the current mathematical models that exist, largely because of widely varying boundary conditions and the very complex chemical systems that may exist in the post-mining environment.

According to Somerville and Elder,[120] because of these difficulties it is probably advisable to attack the question of potential environmental impact utilizing a worst case approach. This approach does not address the question of actual impact, but does allow one to estimate the maximum impact that can reasonably be expected.

This report summarizes the findings on the fate of trace and major constituents during gasification as reported by Somerville, et al.[67,121] and by Attari, et al.[213-215] The conclusion of this work was that the analyses of the laboratory-prepared ashes and its leachates were considerably different than those of the Lurgi-generated ashes and its leachates.

Somerville and Elder[120] have made the following recommendations:

1. Groundwater monitoring wells should be established in and adjacent to the mine and waste disposal areas. The wells should be sampled and samples analyzed for trace and major inorganic elements and organic compounds.
2. Trace element emissions from a gasification facility should not be regulated until their impact is well understood and adequate and inexpensive instrumentation is developed.

To determine the fate of trace elements during gasification and their potential environmental impact, several study areas must be researched. These include:

1. A baseline study of the materials that will be moved and processed by the proposed action
2. A determination of the trace elements and their distribution in the products and by-products of the proposed action
3. A study of the environment in the proposed waste mining area in the configuration that it will have during the following operations
4. A study of the fate of the trace elements in the waste materials that are disposed of in the environment[121]

In the report by Somerville et al.,[121] data are presented which partially describe the results of extensive effort (at University of North Dakota, 1976) to describe the trace element emissions from a proposed gasification facility to be located in Dunn County, N.D. The data presented describe the trace element composition of the overburden, Dunn County lignite (DCL), Mercer County lignite (MCL), DCL ash, and MCL gasification products: oil, tar, gas liquor, and ash.

Present are the results of trace elements analyses for 20 overburden cores from 5 locations shown in the proposed mine area. In addition, four lab-prepared composites were subjected to a water leaching test to provide baseline data for a worst case analysis of water leachable elements in the overburden. Severe leach tests were performed and, in the opinion of the authors, the results approximate the maximum solubilities; i.e., the quantity that can be leached rather than the rate of leaching that will occur in the field.

Spark source mass spectrometry was utilized to determine the majority of trace elements reported. Mercury concentrations were determined using flameless atomic absorption, while fluorine concentrations were determined using the ion-selective electrode.

The overburden leachate studies have been done in this way: a 10-g sample was taken from each of the 5 overburden cores, ground to pass a 60-mesh screen and slurried with 50 mℓ of distilled demineralized water. Each sample was then refluxed for 16 to 24 hr at the boiling point of water. After refluxing, the leachate was analyzed for the trace elements. The trace elements and overall material balance were determined by following these general steps:

1. An elemental analysis of Mercer County lignite was completed.
2. The analysis of the products and by-products from the Sasol test of Mercer County lignite was completed.
3. The elemental and chemical distributions of the products and by-products around the number 13 gasifier at the Sasol plant were determined using the data from the above steps.
4. An elemental and chemical analysis of Dunn County lignite was determined.
5. The percentage elemental distribution found in Step 3 (Sasol-Mercer County lignite test) was combined with analysis data of Dunn County lignite (Step 4) to yield the estimated trace element distribution for a gasification plant using Dunn County lignite.

Some of the results are shown in Table 27 and Table 28. Some data on coal composition are taken from report by Sondreal et al.[220]

The points made by Luthy et al.[221] are as follows: commercial-scale coal gasification facilities will generate large quantities of solid waste material, e.g., western coal commercial concept design for a 250 million standard cubic feet per day (MM SCFD) Hygas® process will consume 20,920 t/day of coal and produce 1560 t (dry basis) of solid waste (char); a similar-sized CO_2-acceptor process will consume 22,770 t of coal and yield 3105 t/day (dry basis) of char, ash, and spent acceptor. A 250 MM SCFD Bi-Gas process will produce on the order of 5200 to 12,800 t/day of slag, ash, and char depending upon coal feed stock.

About 2.8 million t of bottom ash and fly ash are expected from the Wesco Coal Gasification (Lurgi) Complex; this material is proposed to be used as minefill. The COED process consumes 25,000 t of dry coal per day and produces on the order of 13,400 t/day of 70 to 80% carbon chars. The disposal of such large quantities of waste material presents unknown environmental problems. The purpose of the study by Lu-

Table 27
ANALYSIS OF TRACE AND MAJOR INORGANIC
ELEMENTS FROM LIGNITIC COAL[121]

Sample element	Analysis (ppm wt)				
	Coal	Gasifier ash	Oil	Tar	Gas liquor
Ag	<0.1	1			
Al	5,666	63,400	150	1,500	2.9
As	8	74	24	7	0.1
Au	<0.1				
B	56	1,680	4	1	0.9
Ba	616	8,270	2	560	0.005
Be	0.27	6		<0.1	
Bi	<0.1		2	3	
Br	0.27	3	<0.2	0.1	0.001
Ca	16,225	181,600	230	2,600	14.6
Cd	<1	0.5	0.5	<0.3	<0.2
Ca	34.6	190		4	0.006
Cl	26.7	67	3	13	1
Co	1.2	13	0.9 0.6	0.001	
Cr	5.3	140	4	5	0.02
Cs	4	9		0.2	
Cu	10.6	27	2	3	0.01
Dy	0.67	8			
Er	<0.1	4			
Eu	0.4	4		0.1	
F	29.3	191	5	51	5
Fe	7,936	78,800	100	1,500	0.2
Ga	5.3	53		0.6	
Gd	0.8	5			
Ge	0.27	2		0.1	
Hf	<0.1				
Hg	0.2	0.055	0.16	2.9	0.17
Ho	0.4	5			
I	0.13	2		0.1	
Ir	<0.1				
K	268	4,600	35	178	0.8
La	16	74		2	0.004
Li	0.67	45	0.1	0.7	0.002
Lu	<0.1	0.5			
Mg	3,877	42,100	19	764	0.6
Mn	70.7	760	0.7	11	0.03
Mo	4	12	6	2	0.04
Na	6,994	58,604	390	700	82.5
Nb	4	37		0.4	
Nd	2.7	18		0.9	
Ni	6.7	25	2	3	0.006
Os	<0.1				
P	236	3,500	2	90	6
Pb	2.7	58	4	14	0.005
Pd	<0.1				
Pr	1.3	8		0.4	
Pt	<0.1				
Rb	6.7	35	0.2	2	0.003
Re	<0.1				
Rh	<0.1				
Ru	<0.1				
S	11,956	12,600	2,200	2,800	122

Table 27 (continued)
ANALYSIS OF TRACE AND MAJOR INORGANIC
ELEMENTS FROM LIGNITIC COAL[121]

Analysis (ppm wt)

Sample element	Coal	Gasifier ash	Oil	Tar	Gas liquor
Sb	0.27	4			
Sc	8	33	2	0.7	0.006
Se	0.4	0.5	0.4	0.2	0.004
Si	9,114	118,100	180	4,200	117
Sm	1.07	7		0.2	
Sn	0.27	4			
Sr	1,729	12,900	0.3	84	0.004
Ta		0.2			
Tb	0.67	3			
Te	0.27	0.3			
Th	4	45		1	
Ti	193	3,420	5	92	0.02
Tl	<0.1	5			
Tm	<0.1	0.5			
U	4	7		0.9	
V	21.3	150	0.2	3	0.001
W	<0.1	2			
Y	13.3	320	1	3	0.004
Yb	<0.1	4			
Zn	6.7	10	2	0.9	0.2
Zr	65.3	520	1	10	0.008

Note: From Indianhead Mine, Mercer County, North Dakota, and in effluent streams, taken during its gasification test at Sasol.

Table 28
DISTRIBUTION OF INORGANIC TRACE AND MAJOR
ELEMENTS FROM LIGNITE COAL[121]

	Charged to gasification (lbs/million lbs dry coal)		Distribution of element (% retained)			
	In coal	Total found	Gasifier ash	Oil	Tar	Gas liquor
Ag	<0.1	0.11	100			
Al	5,666	7,087	98.7	0.02	1.2	0.07
As	8	9	90.8	3.2	4.3	1.7
Au	<0.1					
B	56	186.5	99.3	0.03	0.03	0.7
Ba	616	945	96.7	0.01	3.3	0.01
Be	0.27	0.67	99.1		0.9	
Bi	<0.1	0.19		12.4	87.6	
Br	0.27	0.34	97.2	0.6	1.7	0.5
Ca	16,225	20,210	99.2	0.02	0.72	0.11
Cd	<1	<0.38	14.5	1.6	4.5	79.4
Ca	34.6	21.24	98.8		1.1	0.1
Cl	26.7	9.67	76.6	0.4	7.5	15.5
Co	1.2	1.4	97.1	0.7	2.1	0.1
Cr	5.3	15.9	97.7	0.3	1.8	0.2

Table 28 (continued)
DISTRIBUTION OF INORGANIC TRACE AND MAJOR ELEMENTS FROM LIGNITE COAL[121]

	Charged to gasification (lbs/million lbs dry coal)		Distribution of element (% retained)			
	In coal	Total found	Gasifier ash	Oil	Tar	Gas liquor
Cs	4	1.0	99.0	1.0		
Cu	10.6	3.19	93.5	0.7	5.3	0.5
Dy	0.67	0.88	100			
Er	<0.1	0.44	100			
Eu	0.4	0.45	98.7		1.3	
F	29.3	31.6	66.8	0.2	9.1	23.0
Fe	7,936	8,872	99.0	0.02	0.95	0.01
Ga	5.3	5.9	99.4		0.6	
Gd	0.8	0.55	100			
Ge	0.27	0.23	97.5		2.5	
Hf	<0.1	0.44	100			
Hg	0.2	0.43	1.4	0.5	38.2	59.9
Ho	0.4	0.55	100			
I	0.13	0.23	97.5		2.5	
Ir	0.1					
K	268	520	97.8	0.08	1.9	0.23
La	16	8.3	98.5		1.4	0.1
Li	0.67	5	99.13	0.02	0.79	0.06
Lu	<0.1	0.055	100			
Mg	3,877	4,690	99.1	0.01	0.9	0.02
Mn	70.7	84.5	99.2	0.01	0.7	0.1
Mo	4	1.6	84.5	4.6	7.1	3.8
Na	6,994	6,635	97.5	0.07	0.59	1.9
Nb	4	4.1	99.4		0.6	
Nd	2.7	2.0	97.5		2.5	
Ni	6.7	3.0	93.3	0.7	5.7	0.3
Os	<0.1					
P	236	395	96.5	0.01	1.3	2.2
Pb	2.7	7.2	88.3	0.7	10.9	0.1
Pd	<0.1					
Pr	1.3	0.91	97.6		2.4	
Pt	<0.1					
Rb	6.7	4.0	97.0	0.07	2.8	0.12
Re	<0.1					
Rh	<0.1					
Ru	<0.1					
S	11,956		11.7	0.2	1.3	1.5
Sb	0.27	0.44	100			
Sc	8	3.7	98.0	0.7	1.1	0.2
Se	0.4		13.8	1.3	2.8	1.5
Si	9,114	13,447	96.9	0.01	1.8	1.3
Sm	1.07	0.78	98.6		1.4	
Sn	0.27	0.44	100			
Sr	1,729	1,428	99.7	0.01	0.3	0.01
Ta	<0.1	0.02	100			
Tb	0.67	0.33	100			
Te	0.27		11.1			
Th	4	5	98.9		1.1	
Ti	193	46.4	88.6	0.12	11.2	0.08
Tl	<0.1	0.55	100			
Tm	<0.1	0.055	100			

Table 28 (continued)
DISTRIBUTION OF INORGANIC TRACE AND MAJOR
ELEMENTS FROM LIGNITE COAL[121]

	Charged to gasification (lbs/million lbs dry coal)		Distribution of element (% retained)			
	In Coal	Total found	Gasifier ash	Oil	Tar	Gas liquor
U	4	0.83	93.8		6.2	
V	21.3	16.7	99.0	0.01	1.0	0.01
W	<0.1	0.22	100			
Y	13.3	35.5	99.5	0.03	0.45	0.02
Yb	<0.1	0.44	100			
Zn	6.7	1.5	74.7	1.6	3.4	20.3
Zr	85.3	58	99.0	0.02	0.97	<0.02

Note: From Indianhead Mine, Mercer County, North Dakota, during gasification test of a Lurgi gasifier.

thy et al.[221] is to present a methodology for assessing the environmental problems associated with disposal of coal gasification solid wastes, and to apply this methodology to the case of Hygas® coal gasification process char.

The solid waste evaluated in this study was Hygas® pilot plant western coal char formed at approximately 63% carbon conversion and is not truly representative of the type of material formed from proposed commercial-scale Hygas® coal gasification process. The commercial-scale Hygas® process will produce Illinois coal char formed at 90% carbon conversion. Hence additional studies with Hygas® char samples obtained from latter pilot plant runs where coal type and carbon conversion more closely resemble commercial-scale configuration must be made to compare with the results of this investigation. Nonetheless, the information presented in report by Lithy et al.[221] does provide a methodology for assessing potential problems associated with disposal of a coal conversion process char; and this body of data will serve as a reference point from which to compare results of future tests. As the data base evolves, it will be possible to begin to correlate coal conversion process operating conditions with the physical and chemical characteristics of the solid waste.

Most of the effort in the investigation was devoted to quantifying the extent and rate of release of trace elements from the char when leached with eluants of varying qualities. Careful attention was given to document process operating conditions at the time the char was sampled, and to assure that the char was collected in appropriate fashion. Also considerable effort was made to characterize the physical properties of the char.

Systems for sampling, analysis, and data evaluation to determine the fate of potential pollutants generated during operation of the Hygas® pilot plant and to apply these data to demonstration and commercial plant designs are described in the work by Anastasia et al.[62] Hygas® operates at high temperatures to obtain high reaction rates and at high pressure to increase the equilibrium methane yield. The most reactive coal fraction is hydrogasified to form methane while the less reactive fraction remains in the coal char and is used to generate hydrogen and heat. Of the total methane formed in the process, about 64% to 70% is formed in the gasifier.

The sampling of pilot plant streams was done in the following way: process water streams were sampled every 8 hr and 3 of these samples were composited to represent

24 hr of plant operation. Water samples were analyzed for total dissolved solids, total suspended solids, total organic carbon, total phenols, total sulfur, ammonia, cyanide, thiocyanate, sulfide, chloride, and oil. Solids samples were collected once per day and analyzed for sulfur species and trace elements. The solids streams sampled included the coal feed, the first and second stages of gasification, and the spent char. Samples of oils in the coal feed slurry were also taken once per day to determine organic compositions using a gas chromatograph mass spectrometer.

The trace elements in solids obtained from gasification of subbituminous coal can be split into several categories according to the quantities found in the feed coal as compared to those found in the spent char.

1. Group 1: 90% to 100% recovery in the ash — Be, V, Mn, Zn, Li, Cr, Pb
2. Group 2: 50% to 90% recovery in the ash — Fe, Ba, As, B, F, Tl, Ni, Cu, Mo
3. Group 3: Less than 50% recovery in the ash — Cl, Se, Cd, Hg
4. Elements primarily released from char in the steam-oxygen zone — Fe, As, Se, Tl, Ni, Cu, Cd, Mo
5. Elements primarily released from char in first and second stages of gasification — Ba, B, Cl, F, Hg
6. Elements in chars from first and second stages of gasification with higher concentrations than in the feed coal — Fe, Tl, Ni, Cu, Cd, Pb, Mo, Cr

For the latter group of elements, material balances indicated that 130% existed in gasifier chars compared with the feed coals. Two of these elements, Pb and Cr, showed essentially no disappearance from coal feed to char ash. For three elements (Pb, Mo, and Cd) 208%, 247%, and 262%, respectively, of coal feed input was found in the upper stages of gasification. With the exception of As and Se, all of the elements released primarily in the steam-oxygen zone showed an increased concentration in chars from the upper stages of gasification. The trace element distribution reported here may have resulted from the large temperature differences used in the various zones of the reactor; and in this respect, the Hygas® reactor is not unique, but is comparable to other industrial processes (power plants, steel plants, and other coal gasification reactors) where large temperature gradients exist. Moreover, these trace element data represent a single test of the gasifier, and additional testing is required to determine "average" distributions and also the ultimate fate of the trace elements.[62]

Environmental impact of coal liquefaction has not been studied by so many research groups. Possible environmentally important forms of some trace elements during liquefaction are shown in Table 29.[222-225] Obviously this is a rather incomplete list.

A significant number of trace metals is suspected to remain in the coal liquid. To whatever use these liquids are applied (power generation, petrochemicals, etc.), data regarding the amount of their trace elements are required for a number of reasons. Besides the obvious potential environmental pollution impact, metals are believed to function as poisons of costly catalysts in certain of the proposed coal-liquefaction as well as coal-gasification schemes. Little is known about the quantity and nature of the metal-containing species present in coal liquids.

In the measurements done by Coleman et al.,[89-91] solvent-refined coal (SRC), tetrahydrofuran (THF)-insoluble SRC, THF-soluble SRC, and three of its size-separated fractions have been analyzed for eleven metallic elements via flameless atomic absorption spectroscopy.

The procedure employed was as follows: Approximately 97% of solvent-refined-coal solid product (Pittsburgh #8) dissolves in THF leaving behind 2.8% of a THF-insoluble material. The soluble portion of SRC can be preparatively separated into four fractions

Table 29
FORMS OF SOME TRACE ELEMENTS DURING
LIQUEFACTION[222]

Element	Volatile species	Organic species[a]
As	AsH_3, $AsCl_3$, $AsBr_3$	$RAsH_2$, $RR'AsH$
Sb	SbH_3, $SbCl_3$, $SbBr_3$, $SbOCl$	R_3Sb, $R_4Sb^+X^-$
Hg	Hg_0, $HgBr_2$	R_2Hg, RHg^+X^-
Se	H_2Se, Se^0	R-Se-R'; R-SeO_3H
Fe	$Fe(CO)_5$	$Fe(C_5H_5)_2(CO)_x$
Ni	$Ni(CO)_4$	Ni-asphaltene bonds
Ti	$TiCl_4$	$Ti(C_5H_5)_2$

Note: These forms of possible environmental importance.

[a] R and R' represent organic moieties without regard to composition or structure.

based on molecular size by eluting with THF. Elemental analyses for Mg, Al, K, Cr, Mn, Fe, Co, Ni, Cu have been performed via flameless atomic absorption spectroscopy on mineralized samples of SRC, THF-soluble SRC, THF-insoluble SRC and fractions Numbers 10, 20 and 30. Data on Pb, Cd, and Zn were obtained employing the tantalum boat technique. Insufficient material prevented direct metal determinations on fraction Number 40.

All materials were dried at room temperature ($< 10^{-4}$ Pa) for 12 hr prior to analysis. THF-soluble components were filtered and then evaporated to dryness on a rotary evaporator to remove THF before drying. The organic matrix was destroyed by placing the coal in equal volumes of concentrated H_2SO_4 and 30% H_2O_2 together (H_2SO_5), which appeared to dissolve each coal sample quickly and completely, yielding a pale yellow solution with no observable residue. This ashing technique process has the advantage that it is fast and considerably less hazardous than many other ashing techniques currently being employed with fossil fuels, and does not cause volatilization problems with the metals analyzed. Analyses were performed immediately after ashing on duplicate samples which had been independently weighed and ashed. Each element was monitored as carefully as possible employing ultra-pure reagents, plastic containers, the method of standard additions, and deuterium-arc background correction. Reagent blanks in all cases were found to have concentration less than the detection limits.

The average of two measurements for each element appears in Table 30. Measurable amounts of Mg, Al, K, Mn, Fe, and Zn are found in each sample studied. Because of experimental detection limits, quantitative determination of metal concentration was not possible for Cr, Co, Ni, Cu, and Pb on each specimen.

The SRC sample possesses appreciable quantities of Fe, followed by Al, K, and Mg; whereas the environmentally important Cd and Pb are at relatively low concentrations. Reproducibility for the 2 as-received samples was less than 50% for each, which is good considering the heterogeneity of the sample.

Upon dissolution in THF, most of the metals tend to concentrate in the insoluble fraction but not to the total exclusion of the THF-soluble portion. Mg, Al, K, Cr, Mn, Cd, and Pb are at least 10- to 20-fold more concentrated, and Fe is even more so. On the other hand, Co, Ni, Cu, and Zn appear to be about equally distributed between soluble and insoluble forms, although the detection limits for Co, Ni, and Cu are such that an unequivocal judgment cannot be made.[89-91]

Table 30
TRACE ELEMENT ANALYSIS (PPM) OF SOLVENT-REFINED COAL[89-91]

Sample	Mg	Al	K	Cr	Mn	Fe	Co	Ni	Cu	Zn	Cd	Pb
SRC	93	147	113	5.9	21.6	423	<2	23	<9.5	7.6	<0.07	<0.5
THF-soluble SRC	121	80	151	3.1	22.7	140	<2	38.9	<9.5	29	<0.07	<0.7
THF-insoluble SRC	1601	1659	2338	40.3	280	6100	<2	<20	<9.0	37.7	0.78	0.67
Fraction No. 10	76	90	138	5.1	30.5	97	3.7	32.3	9.2	27.0	<0.07	<0.7
Fraction No. 20	80	55	150	<2	31.8	28.1	<2	36.5	<9.0	14.0	<0.07	<0.7
Fraction No. 30	44	40	22	<8	5.7	19.3	29.1	<40	10.4	9.7	<0.07	<0.6

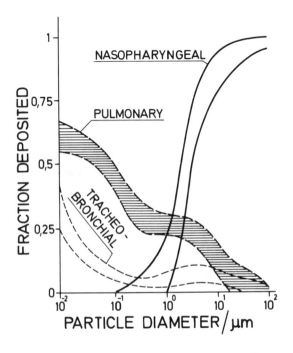

FIGURE 4. Respiratory deposition efficiencies for inhaled particles.[236]

VI. BIOLOGICAL AND HEALTH ASPECTS OF COAL USE

Use of coal as a fuel is expected to triple between 1975 and the year 2000, from approximately 10 quads (1 quad = 10^{15} Btu or 1.05×10^{18} J) to 30 quads. Governmental agencies, environmentalists, and scientists have become increasingly concerned over potential problems associated with large additions of trace elements in the environment from burning fossil fuels.[58,129,226,227] Natusch and Wallace[22] and Linton et al.[33] have reported preferential concentration of the trace elements arsenic, antimony, cadmium, lead, nickel, selenium, thalium, and zinc occurs in the smallest particles emitted from coal-fired power plants; and these particles most easily pass through conventional particulate control devices. These particles are also the ones which will affect all living species the most. Figure 4 illustrates how deposition in the respiratory tract depends on particle size. Billings et al.[23] indicated that about 90% of the mercury in coal burned in a pulverized coal furnace appears as vapor in the flue gas. Enrichment of soil and vegetation in lead has been reported at the Four Corners Plant in New Mexico,[169] and

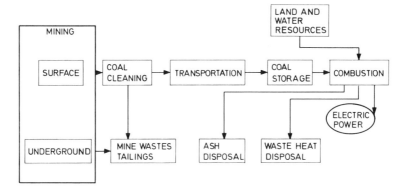

FIGURE 5. Fuel cycle for coal extraction, processing, and combustion. Major health and environmental problems are associated with every stage of the fuel cycle.

enrichment of soil in radium-226 and thorium has been reported in industrialized areas of Poland.[228] Thus, the coal-combustion process releases trace elements to the atmosphere as vapors and particles. These particles have relatively greater concentrations of certain trace elements than the feed coal[13,14] or the collected fly ash.

The problem is even bigger than that. Figure 5 shows the fuel cycle for coal extraction, processing, and combustion. Major health and environmental problems are associated with every stage of the fuel cycle.

The key health and environmental problems from coal extraction, processing, and combustion are:

1. Lack of effective reclamation methods for strip-mined areas: significant land disturbance and land use disruption occur during surface mining operations; restoration of natural habitats, biological productivity, and esthetic character in mined areas under varying climatological, topographical, and soil conditions are essential to the expansion of coal extraction activities.
2. Impairment by surface mining of water availability and quality in semiarid regions: mining processes and especially restoration techniques require large quantities of water (a scarce resource in many areas); additionally, sediment and leachates from the mining operations can pose significant water quality problems.
3. Impairment of agricultural productivity and quality due to coal extraction, processing, and combustion: operations in the coal fuel cycle often occur in the vicinity of agricultural operations; pollutant releases (especially atmospheric) may have deleterious effects on productivity and on the quality of crops and livestock produced.
4. Socioeconomic effects of the coal production/utilization cycle: the population shifts, land-use changes, and economic implications of a major expansion in coal production and utilization may cause serious consequences in terms of social values and welfare at the local, regional, and national levels.
5. Health implications of coal extraction and combustion: occupational health hazards include black lung, silicosis, cancer, explosions, falling rock, etc. Atmospheric pollutants emitted from the coal combustion processes have been implicated as causes or contributory factors in human disease etiologies.

Establishing causal relationships between trace element exposure and specific pathological processes in the environment is hampered by complex, excessive variables which include populations of varying susceptibility, the concurrent existence and interactions of other stresses (pollutants and nonpollutants), and the normal interfering geological background of trace elements, just to name a few. Understanding the environmental health aspects of trace element pollution is a multidimensional problem requiring a multidisciplinary approach. Neither epidemiological nor laboratory data by themselves are sufficient. The meaningful interpretation of the former is frequently plagued by too many variables, and the latter by too few.[136]

Trace metals released during coal conversion are transformed by microorganisms and lead to an overall metal content rise in the microbial biomass. For example, the marine bacterium *Leucothrix* was shown to concentrate metals added to an experimental medium; periwinkles feeding on the bacteria subsequently exhibited an increase in metal content. Trace elements deposited on soil due to ash or tars are likely to undergo microbial degradation. *Thiobacillus ferrooxidans* has been shown, under low pH conditions, to grow on chunks of coal. Microorganisms in lake sediments can transform certain inorganic and organic lead compounds, e.g., lead nitrate and lead chloride, into volatile tetramethyl lead. Mercury, another volatile element present in trace amounts in coal, is highly susceptible to microbial transformations that can lead to more toxic forms such as methylmercury.

Plants affect man; they are a part of his diet, his esthetic environment, and his economic situation; they form the basis of the environmental food chain; and they play a role in the cycling of atmospheric and aquatic pollutants. Plant-pollutant interactions are complex and intermingled; the precise determination of cause and effect is difficult. The interdependence of a variety of factors ultimately determines the effect of pollutants on vegetation: environmental influences (relative humidity, temperature, edaphic conditions, climate, light conditions, ambient levels of other pollutants), and characteristics of the plant itself (species, variety, age, disease, nutritional condition). Although abundant descriptive data regarding the physiological effects of some pollutants are available, the primary sites of action in the plant largely remain to be defined, as do tolerance and detoxification mechanisms.

Such interactions are not limited to pollutant effects on plants. Plants can also contribute to atmospheric pollution by emitting volatile organics containing reactive hydrocarbons in a quantity 6 times greater than the amount produced globally by man's activities, i.e., 175×10^9 kg/year natural vs. 27×10^9 kg/year anthropogenic. Calculations of inputs and outputs of atmospheric trace gases on a global scale indicated that the major source of hydrogen sulfide, ammonia, nitrous oxide, and hydrocarbons was natural emissions, including those from plant foliage. Thus, natural sources produce larger quantities of air pollutants, except for sulfur dioxide, than do anthropogenic sources.

Acid rain, largely the result of sulfur dioxide pollution, may damage plants at distances of up to 1000 km from the pollutant source. Indirect evidence suggests that it has caused a reduction in forest productivity in certain areas, and it has also been shown to cause plant abnormalities and to have a detrimental effect on plant nutrition. Acid rain also can destroy the protective waxy layers of foliage, cause enhancement of disease, and decrease plant productivity. Because plant uptake of heavy metals is increased by lower than normal pH values, a synergistic effect can reasonably be expected where fallout of heavy metals and acid rain appear simultaneously.

Trace elements and heavy metals may be taken up and accumulated by both aquatic and terrestrial plants, with or without injury to the plant itself. Humans have suffered toxic effects from the ingestion of selenium-bearing plants, whereas livestock have incurred molybdenosis and fluorosis from the ingestion of plants containing high concentrations of molybdenum and fluorine.

Trace elements from coal use — including those from catalysts as well as coal itself — pose at least as serious a potential bioenvironmental hazard as do organic effluents. Because some trace elements are resistant to metabolic detoxification and all are nondegradable, they can accumulate in biota. Also, once released, they persist in the environment indefinitely. Some trace metals already exist in the environment at maximum or near-maximum permissible levels.

Of the coal-associated trace elements selected by Comar[229] as particularly important (lead, mercury, arsenic, vanadium, selenium, nickel, and possibly cadmium and fluoride), those capable of long-distance transport (lead, mercury, and cadmium) were isolated by Auerbach[230] as major pollutants capable of bioaccumulation. Catalysts can contribute nickel, cobalt, molybdenum, chromium, vanadium, and zinc.

Because no two elements are alike, the interaction of a trace element with the bioenvironment is rarely predictable. As examples, (1) chromium is most toxic in its most oxidized state (hexavalent), whereas arsenic is most toxic in its most reduced state (as arsine); (2) of the multitude of trace elements in the environment, only mercury, arsenic, and possibly lead are biotransformed by soil microbes into highly toxic, volatile, methylated forms; (3) some elements, such as copper and zinc, that are essential to life are not only toxic at elevated concentrations, but the level of toxicity of one may also be related to the concentration level of the other; and (4) a trace amount of selenium, itself a toxic element, reduces the toxicity of cadmium and mercury. There are many such examples; elements elicit a wide variety of physiological responses in a wide variety of organisms and situations. Nonetheless, efforts are in progress to elucidate the parameters necessary for developing hypotheses concerning the impact of trace elements on ecosystems.

Berry and Wallace[84] presented a literature survey (104 references) discussing the nationwide occurrence of selected trace elements in the environment, their toxicities, ecological pathways, and possible hazards from emissions, specifically as a result of coal combustion. Recommendations were made for additional research needs in problem areas. Several trace element emissions from coal combustion were shown to have a short-term environmental impact on man via their accumulation from the air by the lungs, and their incorporation into his food chain by plant adsorption, e.g., As, Be, Cd, Pb, and Se. Boron was especially toxic to soil, and the fly ash of some coals contained sufficient B concentrations to inhibit plant growth if applied directly to the soil. Further studies are needed regarding elemental concentrations in particulates, the effect of long-term accumulations on agricultural land, emission control by power plants, waste management related to trace element distribution, and consideration of optimal sites for coal power plants based on minimal damage to the environment.

According to the National Academy (U.S.), health effects from occupational exposure to trace metals from steam-electric plant processes have not been documented.

According to Torrey,[3] the stack emissions from coal-burning facilities characteristically have four major types of health impacts: physiological irritation, direct toxicity, carcinogenesis, and physical synergism.

As a defense mechanism, physiological irritation assists the body to reject foreign materials. It is characteristically seen as a local reaction. The challenge from the pollutant has the effect of causing an inflammatory reaction.

Direct toxicity results from the agent interfering with the metabolism of the cell, by either inactivating key enzymes, being metabolized into useless products, or otherwise disrupting normal cell function. In general, substances with toxic effects will also stimulate inflammation, but the response is not always in proportion to the challenge. Inflammation usually occurs at the site of contact, while toxic effects may show up anywhere in the body after absorption.

The pollutant and/or its metabolic by-products may stimulate the development of tumors after some latent period that can range from a few years to several decades. This may occur as the result of an accumulation of gene mutations or chromosome abbreviations due to the biochemical reactions between the genetic material of the cell and the carcinogen.

In the respiratory system in particular, there is a further class of effects that are not directly harmful in and of themselves, but are capable of potentiating the effects described above. The mechanisms for clearing noxious substances from the lungs may be reduced in effectiveness, thereby increasing the residence time of effluents in the lung. This usually results either from a reduction in ciliary action in the bronchial tree; or from a thickening of the protective layer of mucus which interferes with the ciliary action moving foreign particles out of the lung.

Inhalation pathway for dose-to-man assessment of trace element emissions from coal-fired power plants is shown in Figure 6.[17]

The specific trace-metal components of the particulate portion of the SO_x-particulate complex strongly influence the response of experimental animals to the aerosol mixture. Copper, manganese, vanadium, and iron particles plus SO_2 increase in airway resistance in animals several times greater than that produced by SO_2 alone. These metals are known promoting agents of the conversion of sulfur dioxide to sulfuric acid, considered to be the most irritating sulfate form.[231] In experimental heterogeneous droplet systems, manganese and iron are cocatalysts of the conversion of sulfur dioxide to sulfate.[232] Trace-metal synergism with other air pollutants may well be the mechanism of greatest concern to public health, especially considering the long-range transport undergone by gases and small particulates emitted from tall stacks.

Vaughan et al.[50] modeled the behavior and accumulation of trace metals in soil, water, biota, and humans that could be emitted from a hypothetical 1400-MWe coal-fired plant sited in a semiarid western watershed and burning representative western coal. They concluded that the hazard incurred from direct inhalation would be negligible, because respirable (submicrometer) particulates constituted 1/100 or less of the total particle burden, which was an order of magnitude less than present urban air-pollution concentrations. This finding was supported by Baser and Morris,[233] who computed excess deaths attributable to trace metal exposure from a hypothetical 1000-MWe coal-burning plant (Table 31).

The predicted impacts on drinking-water concentrations were several orders of magnitude lower than those proposed as maximum acceptable concentration standards for probability. Cadmium, molybdenum, and tungsten showed 30 to 100% increases in green plants, assuming partial solubility of exogenously supplied metals. The incremental burden for the remaining metals was about 11% of reported content. The size of these changes, which occur over 40 years of plant operation, are small in comparison with those routinely encountered through fertilization of crops. Attention was called to molybdenum, which has a stimulating effect on both anaerobic denitrification and nitrogen fixation reactions. The first reaction could be destructive to agriculture by depleting soils of available nitrogen; the second is a limiting one and critical to agricultural productivity. The study concludes that the ingestion of plants for food appears to be the only consequential pathway of anthropogenic elements to man. In general, excessive bioaccumulation is not expected unless the bioavailability of exogenous fractions is orders of magnitude greater than that of endogenous fractions.

Torrey[3] gives a detailed discussion of health effects associated with specific pollutants. The effects of particulates, nitrogen oxides, and carbon monoxide are discussed separately. Special attention should be paid to long-term effects. They represent the damage accumulated over several years of exposure to the yearly average or back-

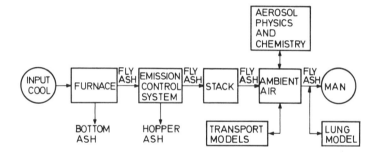

FIGURE 6. Inhalation pathways for dose-to-man assessment of trace element emissions from coal-fired power plants.[17]

Table 31
PROJECTED ANNUAL EXCESS MORTALITY FROM SELECTED TRACE METALS RELATED TO AIR EMISSIONS FROM A 1000-MWe COAL-FIRED PLANT[233]

Trace metal	Range of conc. in coal (ppm)	Excess mortality coefficient[a]	Range of magnitude annual excess mortality (exponents)
Cd	0.01—28	15.1	E-4 to E-1
Cr	4.4—60	9.9	E-3 to E-2
V	10—50	11.2	E-2 to E-2
Fe	1500—7520	15.8	E0 to E0
Mn	6—100	3.1	E-4 to E-2
Ni	4-61	5.3	E-3 to E-2
Zn	5.3—1610	3.7	E-3 to E-1
Hg	0.04—33		E-4 to E-1
Pb	2—106		E-3 to E-1
Be	0.1—1000		E-5 to E-1
As	2—25		E-3 to E-1
Se	1.2—7.7		E-3 to E-2

[a] Unit of coefficient is excess deaths per 10^5 persons per microgram of metal per cubic meter of air.

ground pollution levels. When a new point source of pollution is introduced, there will be a period of several years during which long-term effects, in the form of emphysema, bronchitis, and possibly stomach and pulmonary cancer will begin to develop. Increased mortality from these conditions, however, will not begin to show up for many years after the point source is brought on line. In the meantime, however, increased levels of pollution will show short-term effects — premature illness and death among persons already suffering such debilitating conditions. The magnitude of illnesses will depend in part on the level of pollutant exposure prior to the introduction of the new source. Short-term effects are the direct result of complex interactions between the effects of habituation to pollutants, temperature, and prevalence in the population of easily exacerbated conditions such as emphysema and coronary disease.

Figure 7 shows potentially hazardous elements in coal resonance development as presented by the panel on the trace element geochemistry of coal resources development related to health in a report published by National Academy Press.[54]

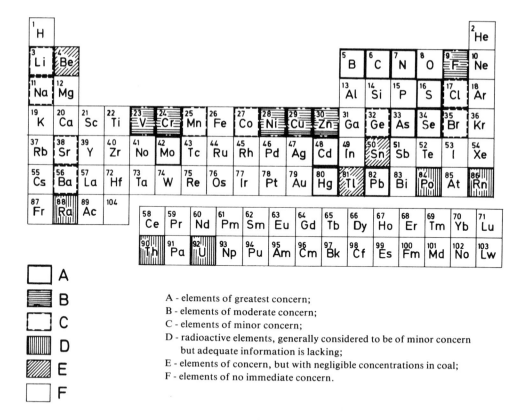

A - elements of greatest concern;
B - elements of moderate concern;
C - elements of minor concern;
D - radioactive elements, generally considered to be of minor concern but adequate information is lacking;
E - elements of concern, but with negligible concentrations in coal;
F - elements of no immediate concern.

FIGURE 7. Potentially hazardous elements in coal resource development.[54]

The panel has classified trace elements that are of concern in coal and residues produced from coal resource development in five categories: those of greatest concern, those of moderate concern, those of minor concern, radioactive elements, and those of concern but with negligible concentrations in coal and coal residues.

Besides the troublesome gaseous compounds of carbon, nitrogen, and sulfur, the elements of greatest concern are arsenic, boron, cadmium, lead, mercury, molybdenum, and selenium. These seven elements commonly occur in coal and in residues from coal cleaning and combustion, at concentrations greater than those encountered on the average in the crust of the earth. The elements carbon, nitrogen, and sulfur and their compounds, although considered to be of greatest concern, are beyond the scope of this report and are not discussed here. Arsenic, cadmium, mercury, and lead are highly toxic to most biological systems when they occur in available form at concentrations above certain critical levels. Levels of molybdenum and selenium derived from airborne deposits resulting from coal development produce concentrations in soil and water that should present no adverse impact on the health of humans or plants. Even if molybdenum and selenium become somewhat enriched in soils, they should not be toxic to plants. However, the disposal of fly ash on soils used to produce forage crops could lead to accumulations of these elements in the forage and render it unsafe for consumption by animals. Adverse impacts of excess boron contamination are limited to possible phytotoxic effects. However, atmospheric deposition of boron onto plants or soil is small, and the probability of encountering levels sufficiently elevated to cause phytotoxicity is remote.

Elements considered by the panel to be of moderate concern include chromium, vanadium, copper, zinc, nickel, and fluorine. The rationale of the panel for grouping these elements in this category is that they are potentially toxic and occur at elevated concentrations in some residues produced during coal combustion, but the probability of adverse impact is considered less than for the elements of greatest concern. Also, vegetation and animal health problems that could occur from excesses of these elements are reversible and can be corrected economically. Copper, nickel, and zinc, in certain site-specific situations, could accumulate over the years in soils to levels that are phytotoxic. Fluorine is a highly volatile element and, in areas where its deposition onto vegetation is great, the forage may not be suitable for consumption by animals.

Those elements considered to be of minor concern include barium, strontium, sodium, manganese, cobalt, antimony, lithium, chlorine, and bromine. Although these elements frequently occur in fly ash at concentrations greater than in the normal geochemical environment, the probability of a specific adverse impact that could not be easily corrected is regarded as remote.

Somerville et al.[120] have discussed the toxicity of trace elements released into the environment from a coal gasification facility.

The criteria they used to determine which trace elements were to be included in the toxicity study were:

1. The concentration of the trace elements in the lignite, lignite ash, and overburden
2. The effluent streams in which the trace element leaves the gasification facility
3. The solubility of the trace element in aqueous solution
4. The plant toxicity level of the trace element

Using the above criteria, 73 elements were screened. The four elements selected for plant toxicity studies were: arsenic, boron, molybdenum, and sodium. Four sets of experiments were run: a baseline study, a gasifier ash impact study, an irrigant study using gasifier ash leachate and gas liquor, and a plant toxicity study.

The results of the gasifier ash impact study indicated that ash in small quantities (< 5%) may be beneficial to plant growth, while excessive additions (20 to 30%) may be detrimental.

The results of the irrigant-inoculated soils study showed that boron is the only element of the ones studied which exceeds toxic limits. A lower germination rate, stunting, and chlorosis among the wheat seedlings were found in the boron toxicity study.

It was observed that the plants growing in sodium-inoculated soils responded better than those growing in untreated soils. In the untreated plants, sodium ranged from 0.01 to 1.57%. In the experiment where sodium was added to the soil, it appeared in the wheat shoots in a range from 0.0176 to 0.0376%. Since the sodium concentration was higher in the ash experiments than in the sodium inoculated experiments, higher levels should have been used in this experiment.

The arsenic level in the analyzed soil samples varied from 2.12 to 20.0 ppm (excluding the arsenic toxicity study). There is no baseline data on the arsenic content in untreated wheat plants. An obvious negative effect of germanium on the wheat seeds and the amount of tissue weights was observed for arsenic inoculations greater than 50 ppm. There was an apparent negative effect on chlorophyll production characterized by the chlorotic appearance of the seedling leaves.

Of the four elements studied in the plant toxicity study, only boron appears in the toxic range. It occurred in levels ranging from 24.8 to 396 ppm (dry) (toxic > 200). All of the other elements appeared to exist in insufficient or at least not excessive quantities.

REFERENCES

1. Zubovic, P., Hatch, J. R., and Medlin, J. H., Assessment of the chemical composition of coal resources, Proc. UN Symp. World Coal Prospects, Katowice, Poland, October 15-23, 1979, 68.
2. Zubovic, P., Hatch, J. R., and Medlin, J. H., Assessment of the chemical composition of coal resources, Proc. UN Symp. World Coal Prospects, Katowice, Poland, October 15-23, 1979, 68. Prospects, Katowice, Poland, October 15-23, 1979, 68.
3. Torrey, S., Ed., Trace contaminants from coal, Noyes Data Corporation, Park Ridge, N.Y., 1978.
4. Loevblad, G., Ed., Trace element concentrations in some coal samples and possible emissions from coal combustion in Sweden, Report IVL-B-358, 1977.
5. Boulding: What is pure coal?, *Environment,* 18 (1), 12, 1976.
6. Bolton, N. E., Carter, J. A., Emery, J. F., Feldman, C., and Fulkerson, W., Hulett, L. D., and Lyon, W. S., Trace element mass balance around a coal-fired steam plant, *Adv. Chem. Ser.,* 141, 175, 1975.
7. Bolton, N. E., Fulkerson, W., Van Hook, R. I., Lyon, W. S., Andren, A. W., Carter, J. A., Emery, J. F., Feldman, C., Hulett, L. D., Dunn, H. W., Sparks, C. J., Jr., Ogle, J. C., and Mills, M. T., Trace element measurements at the coal-fired Allen steam plant, Progress report, February 1973-July 1973.
8. Bolton, N. E., Van Hook, R. I., Fulkerson, W., Emery, J. F., Lyon, W. S., Andren, A. W., and Carter, J. A., Trace element measurements at the coal-fired Allen steam plant, Progress report, June 1971-January 1973, ORNL/NSF/EP 43, Oak Ridge National Laboratory, Tennessee.
9. Code of Federal Regulations, Title 10, Part 50, Appendix I, Revised January 1, 1976, p. 293.
10. Klein, D. H., Andren, A. W., and Bolton, N. E., Trace element discharge from coal combustion for power production, *Water, Air, Soil Pollut.,* 5, 71—77, 1975.
11. Klein, D. H., Andren, A. W., Carter, J. A., Emery, J. F., Feldman, C., Fulkerson, W., Lyon, W. S., Ogle, J. C., Talmi, Y., et al., Pathways of thirty-seven trace elements through coal-fired power plant, *Environ. Sci. Technol.,* 9, 973, 1975.
12. Klein, D. H. and Russell, P., Heavy metals, Fallout around a power plant, *Environ. Sci. Technol.,* 7, 357, 1973.
13. Kaakinen, J. W., Trace element study in a pulverized-coal-fired power plant, Ph.D. dissertation, University of Colorado, 1974, 186.
14. Kaakinen, J. W. and Jorden, R. M., Determination of trace element mass balance for a coal-fired power plant, Proc. 1st Annu. NSF Trace Contam. Conf., 1974, 165.
15. Page, A. L., Elseewi, A. A., and Straughan, I., Physical and chemical properties of fly ash from coal fired power plants with reference to environmental impacts, *Residue Reviews 71,* Springer-Verlag, New York, 1979.
16. Bern, J., Residues from power generation: processing, recycling, and disposal, in *Land Application of Waste Materials,* Soil Conservation Society of America, Ankeny, Iowa, 1976, 226.
17. Ragaini, R. C. and Ondov, J. M., Trace contaminants from coal-fired power plants, From ERDA Energy Res. Abstr. No. 3179, 1976, Report, No. UCRL-76794, 1975, 18.
18. Ragaini, R. C. and Ondov, J. M., Trace contaminants from coal fired power plants, *Int. Conf. Environ. Sensing Assess. Proc.,* Institute of Electrical and Electronics Engineers, New York, 1, 1976, 8.
19. Ragaini, R. C. and Ondov, J. M., Trace element emissions from Western U.S. coal-fired power plants, Preprint for Proc. of 1976 Int. Conf. Modern Trends in Activation Analysis, Munich, Germany, NTIS URCL-77669, 1976, 10.
20. Ragaini, R. C. and Ondov, J. M., Trace element emissions from western U.S. coal-fired power plants, *J. Radioanal. Chem.,* 37, 679, 1977.
21. Ragaini, R. C. and Ondov, J. M., Trace elements from coal-fired power plants, Contrib. to Institute of Electricians and Electronic Engineers, Intern. Conf. Environ. Sensing and Assess., Las Vegas, 1975.
22. Natusch, D. F. and Wallace, J. R., Toxic trace elements: preferential concentration in respirable particles, *Science,* 183, 202, 1974.
23. Billings, C. E., Sacco, A. M., Matson, W. R., Griffin, R. M., Coniglio, W. R., and Hanley, R. A., Mercury balance on a large pulverized coal-fired furnace, *J. Air Pollut. Control Assoc.,* 23, 773, 1973.
24. Gordon, G. E., Davis, D. D., Israel, G. W., Landsberg, H. E., O'Haver, T. C., Staley, S. W., and Zaller, W. H., Study of Emissions from Major Air Pollution Sources and Their Atmospheric Interactions, University of Maryland, prepared for the National Science Foundation for the period Nov. 1, 1972 — Oct. 31, 1974.
25. Toca, F. M., Lead and cadmium distribution in the particulate effluent from a coal fired boiler, Ph.D. dissertation, Health Sciences, The University of Iowa, 163, 1972.

26. Gladney, E. S., Trace element emissions of coal-fired power plants: a study of the Chalk Point Electric Generating Station, Ph.D. thesis, University of Maryland, 1974, 349.
27. Gladney, E. S. and Owens, J. W., Beryllium emissions from a coal-fired power plant, *J. Environ. Sci. Health,* A11, 297, 1976.
28. Oglesby, S., Jr., A Survey of Technical Information Related to Fine-Particle Control, Southern Research Institute, Birmingham, Alabama, Publ. No. EPRI, 259, Electric Power Research Institute, April 1975.
29. **Magee, R. A., Meserola, F. B., and Oldham, R. G.,** Coal Fired Power Plant Trace E. Study, Vol. 1; A Three Station Comparison, Radian Corporation, Austin, Texas, Environmental Protection Agency, Sept. 1975.
30. Davison, R. L., Natusch, D. F. S., Wallace, J. R., and Evans, C. A., Trace elements in fly ash-dependence of concentration on particle size, *Environ. Sci. Technol.,* 8, 1107, 1974.
31. Lee, R. E., Crist, H. L., Riley, A. E., and MacLeod, K. E., Concentration and size of trace metal emissions from a power plant, a steel plant, and a cotton gin, *Environ. Sci. Technol.,* 9, 643, 1975.
32. Gowherd, C., Marcus, M., Guenther, C. M., and Spigarelli, J. L., Hazardous emission characterization of utility boilers, Prep. for Environmental Protection Agency, Task 27, NTIS PB-245 017, National Technical Information Service, Springfield, Va., 1975, 163.
33. Linton, R. W., Surface predominance of trace elements in airborne particles, *Science,* 191, 852, 1976.
34. Smith, W. S. and Gruber, C. W., Atmospheric emission from coal combustion, an inventory guide, Public Health Service Publication No. 999-AP-24, 1966, 112.
35. Winchester, J. W. and Nifong, G. D., Water pollution in Lake Michigan by trace elements from pollution aerosol fallout, *Water, Air, Soil Pollut.,* 1, 50, 1971.
36. ASTM Task Force, Recommended guide for the prediction of the dispersion of airborne effluents (2nd ed.), American Society of Mechanical Engineers, New York, 1973.
37. Slade, D. H., Meteorology and Atomic Energy 1968, U.S. Atomic Energy Commission Report TID-24190, 1968.
38. Turner, R. G., Heavy metal tolerance in plants, p. 339, in Ecological Aspects of the Mineral Nutrients of Plants, Rorison, T. H., Ed., British Ecological Society Symposium, No. 9, 1969.
39. Gifford, F. A., Use of routine meteorological observations for estimating atmospheric dispersion, *Nucl. Saf.* 2(4), 47, 1961.
40. Heffter, J. L., Taylor, A. D., and Ferber, G. J., A regional-continental-scale transport, difussion, and deposition model, National Oceanic Atmospheric Administration, U.S. Department of Commerce, NOAA Tech. Memo ERL-ARL-50, 1975.
41. Nappo, C. J., The simulation of atmospheric transport using observed and estimated wind fields, in Proc. 3rd Symp. on Atmospheric Turbulence, Diffusion, and Air Quality, American Meteorological Society, Boston, 1976.
42. Dickerson, M. H., Mass-consisted wind field analysis for the San Francisco Bay Area, UCRL-74265, Lawrence Livermore Laboratory, University of California, 1972.
43. Bradford, G. R., Dlair, F. L., and Hunsaker, V., Trace and major element contents of soil saturation extracts, *Soil Sci.,* 112, 225, 1971.
44. Poelstra, P., Frissel, M. J., Vander Klugt, N., and Tap, W., Behaviour of Mercury Compounds in Soils, Accumulation and Evaporation, Comparative Studies of Food and Environmental Contamination, International Atomic Energy Agency, Vienna, 1974, 281.
45. Peyton, T. O. and McIntosh, A. W., Cd, Zn, and Pb in soil from a borrow pit and sediment from a pond, *National Science Foundation Trace Contaminants Conf. Proc.,* Oak Ridge National Laboratory, Tennessee, 1973, 589.
46. Benenati, F. E., Zn, Pb, Cd, and As in Soil, Vegetation, and Water Resources, Ph.D. thesis, University of Oklahoma, *Dis. Abstr.* Int. B, 35(a), 4420, 1975.
47. Anderson, W. L. and Smith, K. E., Dynamics of mercury at coal-fired power plant and adjacent cooling lake, *Environ. Sci. Technol.,* 11, 75, 1977.
48. Jurinak, J. J., Grenney, W. J., Woldridge, G. L., Riley, J. P., and Wagenet, R. J., A model of environmental transport of heavy metals originating from stack derived particulate emission in semi-arid regions. 77-RD-27, Research and Development. Southern California Edison Co., Rosemead, Calif.
49. Lyon, W. S., Ed., *Trace Element Measurements at the Coal-fired Steam Plant,* CRC Press, Boca Raton, Fla., 1977.
50. **Vaughan, B. E., Abel, K. H., Cataldo, D. A., Hales, J. M., Hane, C. E., Rancitelli, L. A., Routson, R. C., Wildung, R. E., and Wolf, E. G.,** Review of potential impact on health and environmental quality from metals entering the environment as a result of coal utilization, Pacific Northwest Laboratories, Battelle Memorial Institute, Richland, Washington, 1975.
51. Horton, J. H. and Dorsett, R. S., Effect of Stack Releases from a Coal-Fired Powerhouse on Minor and Trace Element Contents of Neighboring Soil and Vegetation, Presented at Environmental Chem. and Cycling Processes Symp. Augusta, Ga., 1976.

52. Slates, R. V., Methods for analysis of trace elements in coal, coal fly ash, soil and plant samples, from ERDA Energy Res. Abstr. 1977, Abstr. No. 25780, Report, No. DP-1421, 1976, 21.
53. Roffman, H. K., Kary, R. E., and Hidgins, T., Ecological distribution of trace elements emitted by coal-burning power generating units employing scrubbers and electrostatic precipitators, *Pap. Symp. Coal Util.*, 4, 192, 1977.
54. Panel on the trace element geochemistry of coal resource development related to health, *Trace Element Geochemistry of Coal Resource Development Related to Environmental Quality and Health*, National Academy Press, Washington, D.C., 1980.
55. Ondov, J. M., Ragaini, R. C., and Biermann, A. H., Characterization of trace element emissions from coal-fired power plants, From Energy Res. Abstr. No. 29344, 1978, Report No. UCRL-80412, 1978, 21.
56. Hutchinson, G. E., *A Treatise on Limnoloy, Geography, Physics and Chemistry*, John Wiley & Sons, New York, 1957.
57. Heit, M., Review of current information on some ecological and health related aspects of the release of trace metals into the environment associated with the combustion of coal, from Energy Res. Abstr. No. 60838, 1977.
58. Rall, D. P., Report of the committee on health and ecological effects of increased coal utilization, *Fed. Regist.*, 43, 2229, 1978.
59. Moore, G. T., Emissions from a coal-fired power plant: a material balance study of selected volatile trace elements, from Diss. Abstr. Int. B, 1979.
60. Ondov, J. M., Ragaini, R. C., and Biermann, A. H., Emissions and particle size distributions of minor and trace elements at two western coal-fired power plants equipped with cold size electrostatic precipitators, *Environ. Sci. Technol.*, 13, 946, 1979.
61. Stone, L. K., Emissions from coal conversion processes, *Chem. Eng. Prog.*, 72, 52, 1976.
62. Anastasia, L. J., Bair, W. G., and Olson, D. P., Environmental aspects of the hygas process. Report CONF-780305-10, 1978, 13.
63. Braunstein, H. M. and Copenhaver, E. D., Environmental Health and Control Aspects of Coal Conversion: An Information Overview, Oak Ridge National Lab., Tennessee, ORNL/EIS-94, 1977, 538.
64. Council on Environmental Quality (CEQ), Report of the Interagency Working Group on Health and Environmental Effects of Energy Use, Washington, D.C., 1974.
65. El-Ashry, M. T., Trace elements in Pennsylvania shales in relation to environmental health in northeastern Pennsylvania, *Geol. Soc. Am. Bull.*, 82, 1425, 1971.
66. Fennelly, P. F., Durchae, D. F., Klemm, H., and Hall, R. R., Preliminary environmental assessment of coal fired fluidized-bed combustion, Proc. 4th Inst. Conf. Fluid.-Bed Combust., 1976, 475.
67. Somerville, M. H., Elder, J. L., and Todd, R. G., Trace elements: analysis of their potential impact from a coal gasification facility, *Technol. Use Lignite*, 285, 1978.
68. Styron, C. E., Preliminary assessment of the impact of radionuclides in Western coal and health and environment, in Technology for Energy Conservation, Report MLM-2497/OP, National Technical Information Service, Springfield, Va., 1978.
69. Terrill, J. G., Harward, E. D., and Leggett, I. P., Jr., Environmental aspects of nuclear and conventional power plants, *Ind. Med. Surg.*, 36, 412, 1967.
70. Wangern, L. E. and Williams, M. D., Elemental deposition downwind of a coal-fired power plant, *Water, Air, Soil Pollut.*, 10, 33, 1978.
71. Wixson, G. B. and Bolter, E., Control of trace elements during the increasing development of coal for energy, Annu. UMR-DNR Conf. Energy, (Proc.), 4, 92, 1977.
72. Yavorsky, P. M. and Akhtar, S., Environmental aspects of coal liquefaction, Environ. Prot. Technol. Ser. (See-XPARD6), 325, 1974.
73. Yen, T. F., in *Trace Substances in Environmental Health*, Vol. 6, Hemphill, D. D., Ed., University of Missouri, 1973, 347.
74. Harlan, S., Green, J., and Manhan, S., Coal humates for the removal of water pollutants associated with the use of coal, Prepr. Pap. Natl. Meet., Div. Environ. Chem., Am. Chem. Soc., 14, 282, 1974.
75. Jacobsen, J., Iron pollution of the river Skjern and Ringkobing fjord, Western Jutland. Mobilization, transportation, and sedimentation of weathering products from the abandoned brown-coal pits, *Arbog.-Dan. Geol. Unders.*, 57, 1976.
76. Leo, P. P. and Rossoff, J., Control of waste and water pollution from power plant flue gas cleaning systems: first annual R and D report, The Aerospace Corporation (For EPA Contract 68-02-1010), NTIS PB-259 211, National Technical Information Service, Springfield, Va., 1975, 165.
77. Jahnig, C. E., Evaluation of pollution control in fossil fuel conversion processes, Gasification, Section 5, BI-GAS Process, EPA-650/2-74-009-g, NTIS PB-234 694, 1975, 65.
78. Jahnig, C. E., Evaluation of pollution control in fossil fuel conversion process, Gasification, Section 6, HYGAS Process, EPA-650/2-74-009-h, NTIS PB-247 225, 1975, 53.

79. Jahnig, C. E., Evaluation of pollution control in fossil fuel conversion processes, Gasification: Section 7, U-Gas Process, EPA-650/2-74-009-i, NTIS PB-247 226, 1975, 3.
80. Jahnig, C. E., Evaluation of pollution control in fossil fuel conversion processes, Gasification, Section 8, Winkler Process, EPA-650/2-74-009-j, NTIS PB-249 846, 1975, 42.
81. Jahnig, C. E., Evaluation of pollution control in fossil fuel conversion processes, Liquefaction: Section 3, H-Coal Process, EPA-650/2-74-009-m, NTIS PB-249 847, 1975, 60.
82. Jahnig, C. E. and Bertrand, R. R., Environmental aspects of coal gasification, *Chem. Eng. Prog.*, 72, 67, 1976.
83. Jahnig, C. E. and Magee, E. M., Evaluation of pollution control in fossil fuel conversion processes, Gasification Section I: CO_2 Acceptor Process, EPA-650/2-74-009-d, NTIS PB-241 141, 1974, 60.
84. Berry, W. L. and Wallace, A., Trace elements in the environment — their role and potential toxicity as related to fossil fuels — a preliminary study, U.S. Atomic Energy Commission Contract At (04-1) GEN-12, University of California, UCLA 12-946, January 1974, 66.
85. Bertelson, A. B., Pollution control system, U.S. (260978), Patent No. 4116835, 1978.
86. Birge, W. J., Aquatic toxicology of trace elements of coal and ash, *DOE Symp. Ser.*, 48, 219, 1978.
87. Cavallaro, J. A., Deubrouck, A. W., Gibbon, G. A., Hattman, E. A., and Schultz, H., A washability and analytical evaluation of potential pollution from trace elements in coal, *Anal. Methods Coal Prod.*, 1, 435, 1978.
88. Cushman, R. M., Hildebrand, S. G., Strand, R. H., and Anderson, R. M., Toxicity of 35 trace elements in coal to fresh-water biota: a data base with automated retrieval capabilities, from Energy Res. Abstr. No. 47254, 1977, 50.
89. Coleman, W. M., Szabo, P., Wooten, D. L., Dorn, H. C., and Taylor, L. T., Minor and trace metal analysis of solvent refined coal by flameless atomic absorption, *Fuel*, 56, 195, 1976.
90. Coleman, W. M., Perfetti, P., Dorn, H. C., and Taylor, L. T., Trace element distribution in various solvent refined coal fractions as a function of the feed coal, *Fuel*, 57, 612, 1978.
91. Coleman, W. M., Szabo, P., Wooton, D. L., Dorn, H. C., and Taylor, L. T., Minor and trace metal analysis of a solvent-refined coal by flameless atomic absorption, *Fuel*, 56, 1976.
92. Hildebrand, S. G., Cushman, R. M., and Carter, J. A., The potential toxicity and bioaccumulation in aquatic systems of trace elements present in aqueous coal conversion effluents, *Trace Subst. Environ. Health*, 10, 305, 1976.
93. Yeh, J. T., McCann, C. R., Demeter, J. J., and Bienstock, D., Removal of toxic trace elements from coal combustion effluent gas, PERC/RI-76/5, Pittsburgh Energy Research Center, Pennsylvania, 1976, 19.
94. Lee, S. H. D., Johnson, I., and Fischer, J., Study of the volatility of minor and trace elements in Illinois coal, from Energy Res. Abstr., 1978, 3.
95. Ruch, R. R., Gluskoter, J. H., and Shimp, N. F., Occurrence and distribution of potentially volatile trace elements in coal: a final report, Environ. Geol. Notes, 72, Ill. Geol. Surv., p. 96, 1974.
96. Ruch, R. R., Kuhn, J. K., Dreher, G. B., Thomas, J., Frost, J. K., and Cahill, R. A., Potentially volatile trace elements in coal, in American Chemical Society, 168th Natl. Meet., Philadelphia, 1975.
97. Lyon, W. S., Trace element measurement at the coal-fired steam plant, Blackwell Scientific, Oxford, 1977, 144.
98. Swift, W. M., Vogel, G. J., Panek, A. F., and Jonke, A. A., Trace element mass balances around a bench-seale combustor, Mitre Corporation, McLean, Virginia, 1975, 20.
99. Kulinenko, O. R., Quantitative evaluation of trace element mobilities during coal accumulation (in the Donets and Lvov-Volyn basins), *Geol. Uh.*, 32, 86, 1972.
100. Kulinenko, O. R., Geochemical mobility of elements during coalification, From Ref. Zh. Geol. K., 1968, Abstr. No. 7K37, Prir. Tr. Resur. Levoberezh. Ukr. Ispol'z, 2, 62, 1967.
101. Saprykin, F. Ya. and Kulachkova, A. F., Role of natural organic substances during migration and concentration of trace elements, *Tr. Vses. Nauchno-Issled. Geol. Inst.*, 241, 77, 1975.
102. Straughan, I., Elseewi, A. A., and Page, A. L., Mobilization of selected trace elements in residues from coal combustion with special reference to fly ash, *Trace Subst. Environ. Health*, 12, 389, 1978.
103. Szolnoki, J., Role of bacteria in the mobilization and accumulation of metals, *Geonomia Banyasz.*, 8, 51, 1975.
104. Gasior, S. J., Lett, R. G., Strakey, J. P., and Haynes, W. P., Major, minor and trace element balances for the synthane PDU gasifier, Illinois, *Am. Chem. Soc. Div. Fuel Chem. Prepr.*, 23, 88, 1978.
105. Kaakinen, J. W., Jorden, R. M., Lawasani, M. H., and West, R. E., Trace element behaviour in a coal-fired power plant, *Environ. Sci. Technol.*, 9, 862, 1975.
106. Kononenko, N. I. and Oshchepkova, A. P., Formation of the trace element composition of Kuznetsk coal mine waters, From Ref. Zh. Khim. 1976, Nauch. Tr. Perm. N.-i. Ugol'n In't, 20, 80, 1975.
107. Minenko, O. A. and Epifamtseva, M. V., Content of trace elements in mine waters, *Nauch. Tr., Permsk. Nauchno-Issled. Ugol'n. Inst.*, 16, 130, 1973.

168   Trace Elements in Coal

108. Nazarova, L. N. and Berman, V. Yu., Interrelations between the sulfur content in a coal seam and components of the chemical composition of mine waters in the eastern Donets basin, *Gidrokhim. Mater.*, 50, 83, 1969.
109. Zubovic, P., Geochemistry of trace elements in coal, in Symp. Proc. Environmental Aspects of Fuel Conversion Technology, II. EPA-600/2-76-149, U.S. Environmental Protection Agency, 47, 1976.
110. Medlin, J. H., Coleman, S. L., Wood, G. H., Jr., and Rait, N., Differences in minor and trace element geochemistry of anthracite in the Appalachian basin, *Geol. Soc. Am. Abs. Programs*, 7, 1198, 1975b.
111. Magee, E. M., Evaluation of pollution control in fossil fuel conversion processes, Final Report, EPA-600/2-72-101, NTIS PB-255 842, 1976, 292.
112. Shaw, H. and Magee, E. M., Evaluation of pollution control in fossil fuel conversion processes, Gasification: Section I, Lurgi Process, Exxon Research and Engineering Co., Linden, New Jersey, EPA-650/2-74-009-C, NTIS PB-237 694, 1974, 69.
113. Sheibley, D. W., Trace elements by instrumental neutron activation analysis for pollution monitoring, *Adv. Chem. Ser.*, 141, 98, 1975.
114. Weiss, L. H., Coal utilization: the emissions control alternative, Intersoc. Energy Convers. Eng. Conf. 11th, Proc. State Line, Nev., 1976, SAE Pap. 769049, American Institute of Chemical Engineers, New York, 1976, 309.
115. Whitemore, D. O. and Switek, J., Geochemical controls on trace element concentrations in natural waters of a proposed coal ash landfill site, from Gov. Rep. Announce., Index, 159, 86, 1977.
116. Coleman, S. L., Medlin, J. H., Meissner, C. R., Trent, V. A., Windolph, J. F., and Englund, K. J., Environmental consideration of the distribution of trace elements in selected low sulfur bituminous coal and anthracite beds of the Appalachian basin, *Geol. Soc. Am. Abs. Programs*, 7, 1032, 1975.
117. Helm, R. B., Keefer, G. B., and Sack, W. A., Environmental aspects of compacted mixtures of fly ash and wastewater sludge, in Ash Utilization, Proc. 4th Int. Ash Util. Symp., St. Louis, Mo., MERC/SP-76/4, 1976, 396.
118. Herbes, S. E., Southworth, G. R., and Gehrs, C. W., Organic Contaminants in Aqueous Coal Conversion Effluents, Environmental Consequences and Research Priorities, Proc. 10th Annual Conf. on Trace Substances in Environmental Health, Columbia, Missouri, 1976.
119. Loran, B. I. and O'Hara, J. B., Specific environmental aspects of Fischer-Tropsch coal conversion technology, in Environmental Aspects of Fuel Conversion Technology, Report EPA-600/7-78-063, Environmental Protection Agency, Washington, D.C., 1978, 409.
120. Somerville, M. H. and Elder, J. L., Comparison on trace element analysis of North Dakota lignite laboratory ash with Lurgi gasifier ash and their use in environmental analysis, in Environmental Aspects of Fuel Conversion Technology III, Ayer, G. A. and Massoglia, E. F., Eds., Proc. Symp. Env. Aspects Fuel Conv. Tech. Hollywood, 1977, (Report EPA-600/7-78-063).
121. Somerville, M. H., Elder, J. L., and Moran, S. R., Engineering, geological and hydrological environmental assessment of a 250 MMSCFD Dry Ash Lurgi coal gasification facility, *Energy Commun.*, 3, 317, 1977.
122. Wewerka, E. M., Williams, J. M., and Wanek, P. L., Assessment and control of environmental contamination from trace elements in coal processing wastes, From ERDA Energy Res. Abstr. No. LA-UR-76-86, 1976, 7.
123. Egorov, A. P., Laktinova, N. V., and Novoselova, I. V., Evaluation of the admission of trace elements into the environment in the combustion of coals in thermal power plants, *Khim. Tverd. Topl., Moscow*, 5, 68, 1978.
124. Hall, J. H., Varga, G. M., and Magee, E. M., Trace elements and potential pollutant effects in fossil fuels, Symposium Proceedings, Environmental Aspects of Fuel Conversion Technology, St. Louis, Missouri, EPA-650/2-74-118, Environmental Protection Agency, Washington, D.C., 1974, 35.
125. Hausen, L. D., Phillips, L. R., Mangelson, N. F., and Lee, M. L., Analytical study of the effluents from a high temperature entrained flow gasifier, *Fuel*, 80, 323, 1980.
126. Hulett, L. D., et al., The characterization of solid specimens from environmental pollution studies using x-ray and nuclear physics methods, Analytical Chemistry Division, Oak Ridge National Laboratory, NTIS CONF-760311-4, 1976, 5.
127. Hull, A. P., Radiation in perspective: some comparisons of the environmental risks from nuclear and fossil-fueled power plants, *Nucl. Saf.*, 12, 185, 1971.
128. Joensun, O. I., Fossil fuels as a source of mercury pollution, *Science*, 172, 1027, 1971.
129. Lee, R. E. and Von Lehmden, D. J., Trace metal pollution in the environment, *J. Pollut. Control Assoc.*, 23, 853, 1973.
130. Lyon, W. S., Bate, L. C., and Emery, J. F., Environmental pollution of neutron activation analysis to determine the fate of trace elements from fossil fuel combustion in the ecological cycle, Nuclear Activation Techniques in the Life Sciences 1972, International Atomic Energy Agency, Vienna, 1972.

131. Magee, E. M., Hall, H. J., and Varga, G. M., Jr., Potential pollutants in fossil fuels, from Govt. Rep. Announce, U.S. Nat. Tech. Inform. Serv., PB Rep., 1973, 292.
132. Manahan, S. E., Cleaning coal with coal: coal humic acids for removal of acid, iron, heavy metals, and organic pollutants associated with use of coal as a fuel, from Gov. Rep. Announce, Index, U.S. NTIS PB Rep., 1975, 40.
133. National Science Foundation (NSF/RANN), Trace Contaminants in the Environment, 2nd Annu. Trace Contaminants Conf., LBL-3217, Lawrence Berkeley Laboratory, Berkeley, California, 1974.
134. Page, G. C., Fate of pollutions in industrial gasifiers, U.S. Environmental Protection Agency, Office Res. Dev., Washington, D.C., 1978, 191.
135. Williams, J. M., Wewerka, E. M., Vanderborgh, N. E., Wagner, P., Wanek, P. L., and Olsen, J. D., Environmental pollution by trace elements in coal preparation wastes, *Pap. Symp. Coal Mine Drain. Res.*, 7, 51, 1977.
136. Piperno, E., Trace element emission: Aspects of environmental toxicology, Trace Elements in Fuel, Advances in Chemistry Ser. 141, Babu, S. P., Ed., Meet. 166, Am. Chem. Soc. Div. Fuel Chem., 1973, 192.
137. Pollock, E. N., Trace impurities in coal, *Am. Chem. Soc., Div. Fuel Chem. Prepr.*, 18, 92, 1973.
138. Pollock, E. N., Trace impurities in coal by wet chemical methods, *Adv. Chem. Ser.*, 141, 23, 1975.
139. Van Hook, R. I. and Shults, W. D., Eds., Effects of trace contaminants from coal combustion, Proceedings of a Workshop, Knoxville, Tenn., ERDA 77-64, National Technical Information Service, 1976.
140. Schwitzgebel, K., Meserole, F. B., Oldham, R. G., Magee, R. A., Mesich, F. G., and Thoem, T. L., Trace element discharge from coal-fired power plants, *1st Int. Conf. Heavy Met. Environ., Symp. Proc.*, 2, 533, 1975.
141. Lawasani, M. H., Model of fate of trace elements in coal-fired power plant, M. S. thesis, University of Colorado, 1974, 77.
142. Lawrey, J. D., Trace metal dynamics in decomposing leaf litter in habitats variously influenced by coal strip mining, *Can. J. Bot.*, 56, 953, 1978.
143. Cato, G. A., Field Testing: Trace element and organic emissions from industrial boilers, KVB Engineering Inc., Tustin, Calif., Contract No. 68-02-1074, EPA-600/2-76-086b, Environmental Protection Agency, Washington, D.C., 1976, 145.
144. Cavanaugh, G., Burkin, C. E., Dickerson, J. C., Lebowitz, H. E., Tam, S. S., Smithson, G. R., Nack, H., and Oxley, J. H., Potentially Hazardous Emission from the Extraction and Processing of Coal and Oil, EPA Report EPA-650/2 75-038, 1975, 1.
145. Fancher, J. R., Trace element emission from the combustion of fossil fuels, Environmental Resources Conference on Cycling and Control of Metals, Beattelle, Columbus, Ohio, 1972, 109.
146. Lowblad, G. and Greenfelt, P., Heavy metals and other trace elements in black coal and their emission to the air at coal combustion, A literature Survey, Report IVL-B-345, Swedish Water and Air Pollution, Research Lab, Goteborg, March, 1977, 22.
147. Sather, N. F., Swift, W. M., Jones, J. R., Beckner, J. L., Addington, J. H., and Wilburn, R. L., Potential trace element emissions from the gasification of Illinois coals, From ERDA Energy Res. Abstr. 1976, 18519, Report No. ANL-75-XX-1, 1975, 22.
148. Shen, T. T., Cheng, R. J., Mohnen, V. A., Current, M., and Hudson, J. B., Characterization of differences between oil-fired and coal-fired power plant emissions, Proc. 4th Int. Clean Air Congr., 1977, 386.
149. Turner, F. B. and Strojan, C. L., Coal combustion, trace element emissions, and mineral cycles, Department of Energy Symp. Ser. 000078 45, Environ. Chem. Cycling Processes, 1976, Conf. 760429 34-58.
150. Turner, F. B. and Strojan, C. L., Coal combustion, trace element emissions, and mineral cycles, Department of Energy Symp. Ser., 45, 34, 1978.
151. Zoller, W. H., Gladney, E. S., Gordon, G. E., and Bors, J. J., Emissions of trace elements from coal fired power plants, *Trace Subst. Environ. Health*, 8, 167, 1974.
152. Applied Technology Corporation, Sulfur dioxide free two-stage coal combustion process, Environ. Prot. Technol. Ser., 1972, 53.
153. Firek, A. and Sado, A., Thermal processing of pyrite coal wastes and a proposal for the neutralization of sulfur dioxide in roasting gases, Inst. Nowych. Konwersji Energ. Akad. Gorn. Hutn. Krakow Pol. Zesz. Nauk. Adad. Gorn.-Hutn. im Stanislawa Staszica, Gorn, 90, 173, 1978.
154. Marier, P. and Dibbs, H. P., Catalytic conversion of sulfur dioxide to sulfur trioxide by fly ash and the capture of sulfur dioxide and sulfur trioxide by calcium oxide and magnesium oxide, *Thermochemica Acta*, 8, 155, 1974.
155. Slack, A. V., SO_2 Scrubbing, Why Not Recovery?, *Electrical World*, Nov. 1, 1974.
156. Watanable, S. and Tozaka, K., Pulping by sulfur dioxide-magnesium sulfate recovery of chemicals and production of lignit coal from its waste liquor, *Kami Pa Gikyoshi*, 25, 103, 1971.
157. Struthers, P. H., Chemical weathering of strip mine spoils, *Ohio J. Science*, 64, 125, 1964.

158. Gordon, G. E., Instrumental activation analysis of atmospheric pollutants and pollution source materials, In Proceedings Conf-710668, Identification and Measurement of Environmental Pollutants, B. Westley, Ed. National Research Council of Canada, Ottawa, 1971, 138.
159. Ratcliffe, C. T. and Pop, G., Chemical reduction of sulfur dioxide to free sulfur with lignite and coal. I. Steady state reaction chemistry and interaction of volatile components, *Fuel*, 59, 237, 1980.
160. Norwegian Institute for Air Research, *Air Pollution Health Effects of Electric Power Generation, A Literature Survey*, Kjeller, Norway, 1975.
161. Phillips, M. A., Levels of both airborne beryllium and beryllium in coal at the Hayden Power Plant near Hayden, Colorado, *Environ. Lett.*, 5, 183, 1973.
162. Lamartia, C. R. and Lunt, R. R., *Air Pollution Control, Solid and Liquid Waste Disposal Consequences in Chemical Engineering in a Changing World*, Koestier, W. T., Ed., Elsevier, New York, 1976.
163. Scott, R. L. and Mulvihill, J. W., Ambient air quality assessment for the synthane coal gasification pilot plant, Energy and the Environ., Proc. of the 4th Natl. Conf., Cincinnati, Ohio, 1976, 7.
164. Mazza, M. H., Green, D. A., Paris, M. W., and Newton, G. J., Mineral characterization of fluidized bed combustion aerosol ash Montana Rosebud subbituminous coal, from Energy Res. Abstr. 1978, Report No. MERC/TPR-78/1, 1978, 22.
165. Eisenbud, M. and Petrow, G. H., Radioactivity in the atmospheric effluents of power plants that use fossil fuels, *Science*, 144, 288, 1964.
166. Palmer, T. Y., Combustion sources of atmospheric chlorine, *Nature (London)*, 263, 44, 1976.
167. Barnhisel, R. I. and Massey, H. F., The chemical, mineralogical, and physical properties of Eastern Kentucky acid-forming coal spoil materials, *Soil Sci.*, 108, 367, 1969.
168. Barnhisel, R. I. and Rotromel, A. L., Weathering of clay minerals by simulated acid coal spoil-bunk solutions, *Soil Sci.*, 118, 22, 1974.
169. Cannon, H. L. and Swanson, V. E., Contributions of major and minor elements to soils and vegetation by the coal-fired four corners power plant, San Juan County, New Mexico, U.S. Geological Survey Open-file Report No. 75-170, 1975, 36.
170. Crockett, A. B. and Kinnison, R. R., Mercury distribution in soil around a large coal-fired power plant, NTIS PB-269289, National Technical Information Service, Springfield, Va., 1977.
171. Strojan, C. L. and Turnter, F. B., Trace elements and sulfur in soils and plants near the Mohave Generating Station in Southern Nevada, Jt. Conf. Sens. Environ. Pollut., Conf. Proc. 4th, 1978, 537.
172. Wochok, Z. S., Fail, J. L., and Hosmer, M., Analysis of plant growth in fly ash amended soil, in Ash Utilization, Proc. Fourth Int. Ash Util. Symp., St. Louis, Mo., MERC/SP-76/4, 1976, 642.
173. Ardelean, I., Cucu, M., and Andronache, E., Action of silica, coal and calcium carbonate dusts on bacterial infections processes, *Igiena*, 18, 707, 1969.
174. Boyadzhiev, G., Nikolov, Zdr., Nenov, N., and Stefanov, G., Elemental impurities in coal dust from the Dobrudja basin, *Izv. Geol. Inst., Bulg., Akad. Nauk., Ser. Geokhim., Mineral. Petrogr.*, 20, 31, 1971.
175. Doughty, D. A. and Dwiggins, C. W., Jr., Characterization of the Valance State of iron in coal dust, U.S. Bureau of Mines, Report of Investigations 7726, 1973, 15.
176. Duplyakin, A. G., Hygienic assessment of coal dust treated with solutions of prophylactic liquids (calcium chloride, DB, and a mixture of DB and calcium chloride), *Vopr. Gig. Tr. Profzabol.*, 133, 1973.
177. Duplyakin, A. G., Effect of an impurity of some chemical substances in coal dust on phagocytic reactions, *Zdravookhr. Kaz.*, 12, 49, 1978.
178. Freedman, R. W., Procedures for analysis of respirable dust as related to coal workers' pneumociosis, *Anal. Methods Coal Prod.*, 2, 315, 1978.
179. Freedman, R. W. and Sharkey, A. G., Jr., Recent advances in the analysis of responsible coal dust for free silica, trace elements, and organic constituents, *Ann. N.Y. Acad. Sci.*, 200, 7, 1972.
180. Kautz, K., Kirsch, H., and Laufhuette, D. W., Trace element content in coals and fine dust arising from them, *VGB Kraftwerkstechnik*, 55, 672, 1975.
181. Oohama, K. and Chimura, K., Elimination of coal mine dust, Japan, Patent No. 70-11211, 1970.
182. Schultz, H. D., Vesely, C. J., and Langer, D. W., Electron binding energies for silicon minerals occurrings in a respirable coal dust, *Appl. Spectrosc.*, 28, 374, 1974.
183. Seki, K., Katsuki, O., Seisho, M., Shikada, N., Isei, T., Suzuki, T., and Mori, N., Prevention of propagation of coal dust explosion by the salting method. I. Experiments in a small experimental gallery, *Kyushu Kozan Gakkai-Shi*, 36, 322, 1968.
184. Sharkey, A. G., Jr., Kessler, T., and Friedel, R. A., Trace elements in coal dust by spark-source mass spectrometry, Trace elements in Fuel, Babu, S. P., Ed., Advances in Chemistry Series 141, Am. Chem. Soc., Div. Fuel Chem. 166th Meet., Chicago, Ill., 48, Aug. 30, 1973.
185. Meyers, R. A., Coal desulfurization, Marcel Dekker, New York, 1977.

186. Bucklen, O. D., Cockrell, C. F., Donahue, B. A., Leonard, J. W., McPadden, C. R., Meikle, P. G., Mih, L. C., and Shafer, H. E., Coal associated minerals of the U.S., Part 7: Uses, specifications and processes related to coal-associated minerals, RDR-8(7), Coal Research Bureau, National Technical Information Service, PB-168, 1965, 116.
187. Deurbrouck, A. W., Sulfur Reduction Potential of the Coals of the United States, Report of Investigations 7633, Bureau of Mines, U.S. Department of the Interior, Washington, D.C., 1972.
188. Yeager, K., Coal cleaning, *EPRI Journal,* June 1980.
189. Deurbrouck, A. W. and Jacobsen, P. S., Coal Cleaning State of the Art., Conf. on Coal and the Environment, Louisville, Ky., 1974.
190. Schultz, H., Hattman, E. A., and Booher, W. B., Trace elements in coal, what happens to them? Prepr. Pap. Natl. Div. Environ. Chem., Am. Chem. Soc., 15, 196, 1975.
191. Capes, C. E., McIlhinney, A. E., Russell, D. S., and Sirianni, A. F., Rejection of trace metals from coal during benefication by agglomeration, *Environ. Sci. Technol.,* 8, 35, 1974.
192. Ford, C. T., Care, R. R., and Bosshart, R. E., Preliminary evaluation of the effect of coal cleaning on trace element removal, Trace Element Program, Report 3. Bituminous Coal Research, Inc., Monroeville, Pa., 1976, 115.
193. Schivley, W. W., Ash disposal: Current Thoughts on a Growing Problem, presented at 4th Symp. Coal Util., (National Coal Association, Washington, D.C.) 1977, CONF-7710110, 216.
194. Tenney, M. W. and Echelberger, W. F., Jr., Fly ash utilization in the treatment of polluted waters, U.S. Bureau of Mines Info. Circ. 8488, 237.
195. Busch, R. A., Backer, R. R., Atkins, L. A., and Kealy, C. D., Physical property data on fine coal refuse, U.S. Bureau of Mines, Rep. Invest. 8062, 1975, 40.
196. Wewerka, E. M. and Williams, J. M., Trace element characterization of coal wastes, DOE, EPA-600/7-78-028, 1978, 52.
197. Wewerka, E. M., Williams, J. M., and Vanderborgh, N. E., Contaminants in coals and coal residues, *Fourth National Conference on Energy and the Environment,* Los Alamos Scientific Laboratory, LA-UR 76-2197, 1976, 23.
198. Wewerka, E. M., Williams, J. M., Vanderborgh, N. E., Wagner, P., Wanek, P. L., and Olsen, J. D., Trace element characterization and removal/recovery from coal and coal wastes, National Technical Information Service, Springfield, Va., 1978.
199. Wewerka, E. M., Williams, J. M., Wangen, L. E., Bertino, J. P., Wanek, P. L., Olsen, J. D., Thode, E. F., and Wagner, P., Trace element characterization of coal wastes, DOE LA-7831-PR, EPA-600/7-79, 1979, 144.
200. Agres, T., EPA says coal a hazard, *Ind. Res. Dev.,* March, 1980, 47.
201. Fishlock, D., How power from coal leaks radioactivity, *Financial Times,* 27.XI.1980.
202. Goldstein, N. P., Sun, K. H., and Gonzales, J. L., Radioactivity in fly ash from a coal-burning power plant, *Trans. Am. Nucl. Soc.,* 14, 66, 1971.
203. Coles, D. G., Ragaini, R. C., and Ondov, J. M., The behaviour of natural radionuclides in Western coal-fired power plants, *Am. Chem. Soc., Div. Fuel Chem., Prepr.,* 22, 156, 1977.
204. Martin, J. E., Harward, E. D., and Oakley, D. T., Comparison of radioactivity from fossil-fuel and nuclear power plants, in Environmental Effects of Producing Electric Power, Part I, Appendix 14, Committee Print, Joint Committee on Atomic Energy, 91st Congress of the U.S. 1st Session, Wash., D.C., Nov., 1969, 773.
205. Lave, L. B. and Freeburg, L. C., Health effects of electricity generation from coal, oil, and nuclear fuel, *Nucl. Saf.,* 14, 409, 1973.
206. McBride, J. P., Moore, R. E., Witherspoon J. P., Jr., and Blanco, R. E., The Radiological Impact of the Stack effluents of Coal-Fired Power Plants, NTIS report ORNL, 5315.
207. Styron, C. E., Casella, V. R., Farmer, B. M., Hopkins, L. C., Jenkins, P. H., Phillips, C. A., and Robinson, B., Assessment of the radiological impact of coal utilization, Report MLM-2514, UC-90a, 1979.
208. Moore, R. E., The AIRDOS Computer Code for Estimating Radiation Dose to Man from Airborne Radionuclides in Areas Surrounding Nuclear Facilities, ORNL-5245, April, 1977.
209. Lee, H., Peyton, T. O., Steele, R. V., and White, R. K., Potential Radioactive Pollutants Resulting from Expanded Energy Programs, SRI-EGU-4869, Stanford Research Institute, Menlo Park, Calif., 1977.
210. Baria, D. N., A survey of trace elements in North Dakota lignite and effluent streams from combustion and gasification facilities, The Engineering Experiment Station, University of North Dakota, Grand Forks, May 1975, 64.
211. Forney, A. J., Haynes, W. P., Gasior, S. J., and Kenny, R. F., Effect of additives upon the gasification of coal in the synthene gasifier, *Am. Chem. Soc., Div. Fuel Chem., Prepr.,* 19, 111, 1974.

212. Forney, A. J., Haynes, W. P., Gasior, S. J., Kornusky, R. M., Schmidt, D. E., and Sharkey, A. G., Trace elements and major component balances around the synthane PDU gasifier, U.S. Environ. Prot. Agency, Off. Res. Dev. EPA 000076 EPA-600/2-76-149, Symp. Proc. Environ. Aspects Fuel Convers. Technol., II, 67, 1975.
213. Attari, A., Fate of trace constituents of coal during gasification, from Govt. Rep. Announce, 1973, 73(20), 66, U.S. Nat. Tech. Inform. Serv., PB Rep., No. 223001/9, 1973, 39.
214. Attari, A., Mensinger, M., and Pau, J. C., Fate of trace element constituents of coal during gasification, Part II. Institute of Gas Technology, paper presented at 169th meeting, Am. Chem. Soc., Div. Fuel Chem., Philadelphia, Pa., April 6-11, 1975, 15.
215. Attari, A., Pau, J. C., and Mensinger, M., Fate of trace and minor constituents of coal during gasification, from Gov. Rep. Announce. Index (U.S.) 1977, 77(23), 279, U.S. NTIS, PB Rep., No. PB-270913, 1976, 46.
216. Lloyd, W. G. and Francis, H. E., Determination of sulfur in whole coal by x-ray fluorescence spectrometry, Proc. of ERDA Symp. on x-and Gamma-Ray Sources and Appl., Univ. of Michigan, Ann Arbor, May 19-20, 1976, (CONF-760539).
217. Lloyd, W. G. and Francis, H. E., Fate of minor and trace elements in alternate gasification schemes, Proc. Second Symposium on Coal Utilization, Louisville, Kentucky, 1975, 324.
218. Becker, D. F. and Murthy, B. N., Feasibility of Reducing Fuel Gas Clean-Up Needs: Phase I Survey of the Effects of Gasification Process Conditions on the Entrainment of Impurities in the Fuel Gas, NTIS FE-1236-15, 49, June 20, 1976.
219. Koppenhaal, D. W. and Manahan, S. E., Hazardous chemicals from coal conversion processes, *Environ. Sci. Technol.*, 10, 1104, 1976.
220. Sondreal, E. A., Kube, W. R., and Elder, J. L., Analysis of the Northern Great Plains province lignite and their ash: a study of variability, U.S. Bureau of Mines, Report of Investigations, 7158, 1968, 94.
221. Luthy, R. G. Vassilidu, P., and Carter, M. J., Report FE-2496-28, National Technical Information Service, Springfield, Va., 1978.
222. Filby, R. H., Shah, K. R., Hunt, M. L., Khalil, R. S., and Sauther, C. A., Solvent refined coal (SRC) Process, Trace elements NTIS, Report FE-496-T-17, National Technical Information Service, Springfield, Va., 1978.
223. Filby, R. H., Shah, K. R., and Sautter, C. A., Trace elements in the solvent refined coal process, U.S. Environmental Protection Agency, Off. Res. Dev., Report No. EPA/600/7-78/063, 1978, 266.
224. Filby, R. H., Shah, K. R., and Sautter, C. A., A study of trace element distribution in the solvent refined coal (SRC) process using neutron activation analysis, *J. Radioanal. Chem.*, 27, 693, 1977.
225. Filby, R. H., Shah, K. R., and Yaghmaie, F., The nature of metals in petroleum fuels and coal-derived synfuels, Ash Deposits Corros, Impurities Combust. Gases, 51, 1978.
226. Barber, D. E. and Giorgio, H. R., Gamma-ray activity in bituminous, subbituminous and lignite coals, *Health Phys.*, 32, 83, 1977.
227. Vine, J. D., Geology of uranium in coaly carbonaceous rocks, Uranium in Carbonaceous Rocks, U.S. Geol. Surv. Professional Paper, 365-D, 1962, 113.
228. Jaworowski, Z. and Grzybowska, D., Natural radionuclides in industrial and rural soils, *Sci. Total Environ.*, 7, 45, 1977.
229. Comar, C. L., Conference proceedings: Workshop on health effects of fossil fuel combustion products, Electric Power Research Institute, Palo Alto, Calif., PB-242, 1975, 418.
230. Auerbach, S. I., Testimony presented to the subcommittee on environment and the atmosphere. U.S. House of Representatives, Hearings on costs and effects of chronic low-level environmental pollution, Nov. 12, 1975.
231. Amdur, M. O., Toxicological guidelines for research on sulfur oxides and particulates, in Proceedings of the Fourth Symposium on Statistics and the Environment, March 3-5, Washington, D.C., 1976.
232. Beilke, S. and Gravenhorst, G., Heterogeneous SO_2 oxidation in the droplet phase. Prepared for presentation at the Int. Symp. Sulfur Atmos., September 7-14, 1977, Dubrovnik, Yugoslavia.
233. Baser, M. E. and Morris, S. C., Assessment of the potential role of trace metal health effects in limiting the use of coal-fired electric power, Informal Report, Brookhaven National Laboratory, Upton, N.Y.,
234. KVB, Inc., Bench-Scale Study of the Dry Removal of SO_2 with Nahcolite and Trona, Final Report, March, 1981, 114 (EPRI CS-1744, RP 982-8).
235. ERDA 77-64, Effects of Trace Contaminants from Coal Combustion, Proc. of workshop August 2-6, 1976, Knoxville, Tenn., sponsored by Div. Biomedical and Environmental Research, Energy Research and Development Administration, Washington, D.C., 1977.
236. Dept. Health, Education and Welfare, Washington, D.C., Air quality criteria for particulate matter, National Air Pollution Control Administration Publication, AP-49, 115, 1969.

Chapter 3

METHODS OF ANALYSIS

I. GENERAL CONSIDERATION

A. Introduction

Analytical methods used in trace element analysis of coal and its ash may be separated into two categories. One category includes methods which cannot be easily used for multielement analysis on an individual sample. Therefore, many of these methods may require large quantities of sample if more than a few elements are to be determined. These methods include atomic absorption spectroscopy, potentiometry, voltammetry, and absorption spectrophotometry, and they require sample preparation for the coal and fly ash matrices. Sample preparation, usually the wet or dry ashing of coal or the dissolution of fly ash through acid treatment or fusion, offers opportunity for sample contamination. However, with the use of appropriate standards and/or the method of standard additions, detection limits can be good. Precision depends largely on the individual analyst's skill. An additional advantage of methods in this category is that the equipment required is comparatively inexpensive. In this chapter we shall discuss to some extent only the optical methods as used in different laboratories.

Methods which are in the second categories include: detection of characteristic X-rays, mass spectroscopy, activation analysis (neutron and charged particles) as well as some others. They all have multielemental capabilities. In this chapter we shall discuss all multielemental techniques used in different laboratories for trace element analysis of coal.

Over the years a number of techniques have evolved for determining trace elements in coal, coke, and coal ash. More recently, the trend has been to develop instrumental techniques, in preference to chemical, calorimetric, or "wet" techniques, because these procedures tend to be faster, less tedious, and generally yield acceptable results when sufficiently checked.

Several comprehensive reviews and bibliographies[1-6] are available in the literature. Perhaps the best guide to the determination of specific elements in coal is the bi-yearly application review on "Solid and Gaseous Fuel."[7] Recent symposia of the American Chemical Society Fuel Division have dealt with the analysis of coal for trace elements.[8] Several organizations have published individual summaries of procedures used in their laboratories.[9-13]

The present trend is toward development of multielemental instrumental procedures, to quantitatively cover as many elements as are applicable. Because any particular analytical discipline is better suited for certain elements than for others for various inherent reasons, a combination of methods is usually necessary to determine all elements of interest. For example, instrumental neutron activation analysis is based on the detection of induced radioactivity. Detection is dependent on several nuclear factors and elemental concentrations relative to other possible interfering elements. For such a complex matrix as coal, coal ash, or residues, about 40 elements are currently detectable with varying degrees of acceptable accuracy and precision. Other analytical disciplines (such as optical emission and atomic absorption spectroscopy) have corresponding principles, limitations, and areas of application. The combined use of analytical disciplines thus allows overlap, enabling an approach to better accuracy. Many laboratories successfully use this approach.

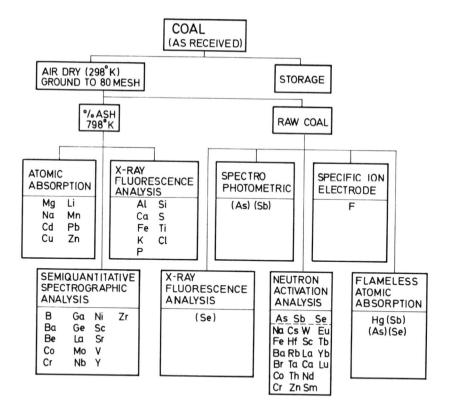

FIGURE 1. Methods used in coal analysis by U.S. Geological Survey.[14]

The preferences for a particular technique are often expressed in the literature. Such preferences are the results of analysts' experiences, equipment available, and many other subjective factors and should not be taken very seriously. The most-used methods for analyzing trace elements in coal are activation analysis, optical emission, atomic absorption, X-ray fluorescence, and mass spectrometry or combinations of these methods. In general, each of these methods is quantitatively applicable to many elements. Mass spectrometry, unless precise and accurate isotopic dilution techniques are employed, is considered to be only semiquantitative. Most colorimetric procedures, although quantitative, are usually specific only for a particular element, hence limiting. Because of the many analytical problems incurred with complex matrixes, polarography has not been extensively developed and applied to coal analysis.

Those methods generally offering lower detection limits are atomic absorption and neutron activation, usually of the order of 1 ppm or better for applicable elements. In the 1 to 10-ppm range and above, X-ray fluorescence and optical emission methods are generally useful for many elements. All of these methods are capable of acceptable precision and accuracy (e.g., ± 10 to 15% of the true value) for the determination of trace elements at concentrations significantly above their detection limits.

Different laboratories have different procedures used in analysis of coal. Let us mention some. For example, the methods used by U.S. Geological Survey[14] are shown in Figure 1. Wewerka et al.[15,16] have considered only neutron activation analysis, atomic absorption spectrophotometry, optical emission spectroscopy, and chemical methods. Preferred methods for elemental analyses of coals and coal wastes as used at Los Alamos Scientific Laboratory are shown in Table 1.

Table 1
PREFERRED LASL METHODS FOR
ELEMENTAL ANALYSES OF COALS AND
COAL WASTES[15]

Element	Method	Element	Method
Li	AA, OES	Br	NAA
Be	AA, OES	Rb	NAA, OES
B	OES	Sr	NAA
F	C	Y	OES
Na	NAA	Zr	OES
Mg	NAA, OES	Mo	OES
Al	NAA, OES	Ag	AA
Si	AA, OES, C	Cd	AA
P	AA, C	Sn	OES
S	C	Sb	NAA
Cl	NAA, C	I	NAA
K	NAA, OES	Cs	NAA
Ca	NAA	Ba	NAA
Sc	NAA, OES	La	NAA, OES
Ti	NAA, AA	Ce	NAA
V	NAA, AA, OES	Sm	NAA
Cr	NAA, AA, OES	Eu	NAA, OES
Mn	NAA, AA, OES	Tb	NAA
Fe	NAA, AA	Yb	NAA, OES
Co	NAA, AA, OES	Lu	NAA
Ni	AA, OES	Hf	NAA
Cu	AA, OES	Ta	NAA
Zn	NAA, AA	W	NAA
Ga	NAA, OES	Hg	NAA, AA
Ge	OES	Pb	AA, OES
As	NAA, AA	Th	NAA
Se	NAA	U	NAA

Note: Methods for analyses were: NAA — Neutron Activation Analysis, AA — Atomic Absorption Spectrophotometry, OES — Optical Emission Spectroscopy, and C — Chemical Methods.

On the other hand, Yaverbaum[17] considered only elements of interest as pollution agents. The most applicable procedures for these elements are shown in Table 2.

Ideally, a method of analyzing trace elements in coal combustion products should (a) determine a large number of elements of interest simultaneously, (b) require relatively little sample preparation, (c) be capable of automation, (d) produce an output compatible with computerized data processing, and (e) be rapid.[17]

Of the methods listed in Table 2, emission spectroscopy is capable of measuring with suitable sensitivity all elements except fluorine and phosphorus. However, if this method is to meet the criteria of rapid output and computerized data processing, a rather sophisticated (and expensive) instrument equipped with direct-reading detectors would be necessary. Calibration of the instrument using materials of composition very

Table 2
APPLICABILITY OF SELECTED ANALYTICAL METHODS FOR TRACE ELEMENTS IN COAL COMBUSTION[17]

Priority	Element or ion	XRS	ES	SIE	AA/FP	FL	CL
1	Fluoride			X			
1	Lead	X	X		X		
1	Mercury	X	X		X		
1	Beryllium		X		X	X	
2	Cadmium	X	X		X		
2	Arsenic	X	X		X		
2	Nickel	X	X		X		
3	Copper	X	X		X		
3	Zinc	X	X		X		
3	Barium	X	X		X		
3	Tin	X	X		X		
3	Phosphorus						X
3	Lithium		X		X		
3	Vanadium	X	X		X		
3	Manganese	X	X		X		
3	Chromium	X	X		X		
3	Selenium	X	X		X		

Note: Analytical methods used were: XRS — X-ray Spectrometry, ES — Emission Spectroscopy, SIE — Specific Ion Electrode Method, AA/FP — Atomic Absorption Spectrometry/Flame Photometry, FL — Fluorimetry, and CL — Colorimetric Method.

close to those being analyzed (a time-consuming effort) also would be necessary. The resulting system would be highly specific to coal, ash, fly ash, etc.; would be very rapid, and would have an accuracy of perhaps 10 to 100%, varying for different elements.

X-ray spectrometry is another method capable of simultaneously measuring most of the elements specified. It is capable of measuring all elements of atomic number greater than about 15 (the atomic number of phosphorus), but it is more sensitive for elements with higher atomic numbers. The X-ray spectrometer has the merit of internal standardization which neutralizes matrix effects and makes for relatively high accuracy. For a production system, the X-ray spectrometer should be coupled to a computer.

The energy-dispersive X-ray spectrometer is particularly suited for coupling to a computer because analysis of the fluorescent X-ray spectrum from the sample is done by a solid-state detector and a pulse-height analyzer that provides digital data for computer processing.

Atomic absorption spectrometry is a method with high sensitivity, particularly since new atomization procedures have been developed; for example, graphite furnaces and plasma units. The method is suited for accurate analysis of special elements because it does not provide information for more than two elements simultaneously. It is most applicable to metals because the sensitivity of the method declines with increasing nonmetallic character of the element. The method is also less sensitive to elements that form refractory oxides (and hence atomize with greater difficulty).

The specific ion electrode method is preferred for measuring trace quantities of fluoride in aqueous solution. Generally, it is used with a pyrohydrolysis technique for liberating the fluoride as HF and collecting it in aqueous solution.[17]

Fluorimetry is a very powerful method of analyzing very small quantities of certain elements. For example, it is the method of choice for uranium and beryllium. It is claimed that concentrations of beryllium as low as 0.1 ppm can be determined in air dust without chemical separations.

A variety of ultrasensitive colorimetric methods are available for many elements that are suited for trace element analysis of coal and coal combustion residues. Only phosphorus is listed here under the "Colorimetric" heading because other methods are more appropriate for the other elements. The colorimetric method is capable of measuring phosphorus in microgram quantities.[17]

The lists of references in Chapter 2 and this chapter include some of the significant literature available for the determination of specific trace elements in coal and coal ash.

We shall describe some of the methods used. No attempt is made here to classify any method as superior to another because each, when properly developed and used, has definite inherent advantages.

B. Sample Preparation

Sample preparation techniques are specific to the method of analysis used. Here we shall describe only some of the general aspects, whereas the details will be mentioned in the reactions dealing with the individual method of analysis.

Let us here mention two reports which may be of interest. First, Montgomery[42] describes in his paper the activities of the American Society for Testing and Materials Committee D-5 on coal and coke, which is responsible for development of specifications for coal and for coke produced from coal; the development of methods of sampling, analysis, and testing. Second, Swanson and Huffman[13] describe guidelines for sample collecting and analytical methods used in the U.S. Geological Survey. According to these authors, the exact type and number of samples of coal and associated rock to be collected are left to the best judgment of the geologist. Samples should be of unweathered coal or rock and representative of the bed or beds sampled; it is recommended that 2 channel samples, separated by 10 to 100 yards and weighing 4 to 5 lb each, be collected of each 5 ft of vertical section. Care must be taken to avoid any sample contamination, and to record the exact locality, thickness, and stratigraphic information for each sample. Analytical methods are described for the determination of major, minor, and trace elements in coal. The methods used to determine these elements include atomic absorption spectroscopy, X-ray fluorescence spectroscopy, optical emission spectroscopy, spectrophotometry, selective-ion electrode, and neutron activation analysis.

Sample dissolution is required for analysis by atomic absorption spectroscopy (AAS)). No single technique is equally satisfactory for dissolution of all samples. Coal requires dissolution techniques capable of decomposing large quantities of organic material. Coal slag and fly ash require techniques capable of dissolving acid-resistant minerals and glasses.

The problems encountered in sample dissolution are discussed in some detail by Slates.[35]

The Bernas[43] procedure is often used for dissolution of slag, fly ash, and soils with very low organic content. Silicates are decomposed in 30 to 40 min at 110°C without volatilization losses in a specially designed vessel lined with polytetrafluoroethylene. The decomposition medium of hydrofluoric acid, boric acid, and aqua regia provides a salt-free matrix that offers negligible interference during atomic absorption analysis. The Bernas procedure permits volume measurements in calibrated glass containers, but does not introduce metallic cations to the solution. This dissolution procedure has

Table 3
CONCENTRATION OF ELEMENTS AS A FUNCTION OF ASHING TEMPERATURE[46]

Element	Uncovered platinum crucibles					Covered platinum crucibles				
	300°C	400°C	500°C	600°C	700°C	300°C	400°C	500°C	600°C	700°C
Sn	<1.9	4.2	2.1	<1.6	<1.6	<2.4	2.9	3.6	2.5	2.6
B	30	32	41	45	41	39	38	42	43	52
Cu	16	14	15	17	25	15	15	15	18	28
Co	15	15	13	14	13	14	14	14	13	13
Ni	39	38	36	37	35	39	37	37	36	35
Be	2.6	2.5	2.6	2.5	2.3	2.8	2.5	2.5	2.5	2.6
Cr	19	17	15	17	17	15	16	17	17	18
V	32	24	19	18	18	31	24	21	18	17
Mo	11	7.8	5.4	5.5	4.7	11	8.4	6.2	5.6	4.0
Ge	5.4	5.4	5.7	5.3	5.6	3.4	5.9	5.5	5.6	5.5

been satisfactorily applied[43-45] to the atomic absorption analysis. The Bernas dissolution procedure may be applicable to the atomic absorption analysis of arsenic, selenium, cadmium, and mercury.[85] Pollock[12] used hydrofluoric acid dissolution for both graphite furnace and hydride volatilization analyses for arsenic.

Very often some kind of ashing of sample is performed. One important aspect to be considered is that volatilization losses of certain elements occur during certain pretreatment and ashing procedures. This can be a limiting factor on what type of sample and analytical procedure may be applied. For example, instrumental neutron activation analysis and X-ray emission spectroscopy can be applied to any matrix and require no sample pretreatment. In contrast, analysis by atomic absorption or optical emission usually requires an "ash" sample, sometimes needing extensive pretreatment. For some elements the ash data cannot be related to the whole coal.

Some of these problems were studied by many authors; however, additional effort is needed. Very often, high-temperature ashing of sample is done. Therefore, it is of interest to describe in some detail work by Dreher and Schleicher,[46] who studied the variation of trace element concentration with ashing condition. In their work, in order to determine whether or not the concentrations of any of the elements of interest were altered with increasing ashing temperature, samples were ashed in uncovered, used porcelain crucibles at 300, 400, 500, 600, and 700°C until the carbonaceous matter was no longer apparent. Two coals were studied and trace element analyses were carried out on each of the ten resulting ash samples. None of the data exhibit losses or gains of trace element concentration in samples ashed between 300 and 700°C, other than a gain in boron concentration and a possible loss of lead. A loss of lead with increasing temperature is expected. For both coals, the boron concentrations increased with ashing temperature from 300 to 600°C, then dropped slightly between 600 and 700°C.

Another pair of coals was chosen to determine whether or not a similar effect took place when coal was ashed without alkaline material. The coals were ashed in an electric furnace, in covered and uncovered platinum crucibles at 300, 400, 500, 600, and 700°C for 20 hr each at a heating rate of 500°C/hr. The results, listed in Table 3, indicated that the boron concentrations again increased when the coal was ashed in uncovered crucibles. The increase in boron concentration is unexplained.

Slates[35] has also investigated sample preparation techniques. Data obtained from the study of two coal samples prepared by high-temperature ashing (HTA) and acid

Table 4
COMPARISON OF SAMPLE PREPARATION FOR ATOMIC ABSORPTION BY
HIGH TEMPERATURE ASHING (HTA) AND ACID BOMB COMBUSTION
(ABC)[35]

Prep. method	ppm in Coal					
	Pb	Zn	Ni	Mn	Cu	Cr
Sample No.1						
HTA	29.1 ± 11.7	5.6 ± 2.8	9.2 ± 1.7	128.7 ± 2.9	8.5 ± 4.1	20.1 ± 4.6
ABC	<MDL	<MDL	26.0 ± 9.8	199.1 ± 10.7	8.7 ± 1.2	23.0 ± 5.5
Sample No. 2						
HTA	11.0 ± 1.6	3.8 ± 0.5	13.5 ± 6.8	137.5 ± 2.6	7.9 ± 3.2	22.4 ± 0.4
ABC	<MDL	<MDL	36.2 ± 9.2	220.5 ± 10.1	8.2 ± 6.1	51.2 ± 9.5

Note: MDL is minimum detection limit.

bomb combustion (ABC) are presented in Table 4. These concentrations were determined through atomic absorption spectrometry.

Low-temperature ashing is done in an oxygen-plasma asher according to the following procedure. Powdered and blended samples of coal material are dried for 24 hr at 60°C after having been placed into ceramic boats. Then the dried samples are ashed at 75°C until all of the organic matter has been removed (constant weight). This usually requires 72 to 96 hr to complete. During ashing the samples are stirred frequently to expose new material to the oxygen plasma. Since the low-temperature ashing is often done, let us describe one of many reports from the literature. Hamrin[47] prepared low-temperature ash (LTA) in a low-pressure, low-temperature, gaseous oxygen plasma using a six-chamber apparatus. Prior to ashing, approximately 40 g of each coal was heated at 378 K for 24 hr to remove moisture. Dried, weighed coal was then placed in 6 90-mm Pyrex® ashing dishes. The dishes were shaken at the beginning and end of the day to expose unashed coal to the plasma. Ashing was usually completed in 72 to 96 hr, and samples were then weighed and ashed for another 24 to 48 hr to insure complete ashing.

The asher was operated at a low power level to keep the temperature down. Approximately 24 W per chamber was used with an oxygen flow of 6670 mm³/sec (400 mℓ/min) at a pressure of 133 Pa (1 torr). Under these conditions the estimated ashing temperature was less than 323 K. Simultaneously, high-temperature ash (HTA) was determined by ASTM procedure. It was noticed that ratio of LTA/HTA yields is greater than one. First, the pyrites are oxidized in high-temperature ashing which amounts to a 66.5% weight loss. Secondly, clay minerals which account for more than half of the mineral matter of some coals contain OH groups and interstitial water which are lost in standard ashing procedures. Finally, carbonates decompose and CO_2 is emitted during ashing. For calcite, this represents a 44% weight loss. It is assumed that none of these processes occurs to any appreciable extent during low-temperature ashing and LTA represents mineral matter as it occurs in nature to the best approximation currently available.[47]

C. Standards

The need for standards in trace element analysis of coal and coal ash in the calibration of apparatus and the evaluation of techniques employed was recognized long ago.

Many research groups have been developing standards for their own use. An early effort in preparation of standards is described in the paper by Peterson and Zink.[48,49] In their work, a synthetic coal ash base for use in preparing standards and approximating the average composition of natural ash was prepared as follows: 29.48 g of $FeSO_4 \cdot (NH_4)_2SO_4 \cdot 6H_2O$ and 10.93 g of $CaSO_4 \cdot 2H_2O$ were weighed into separate Vycor® dishes and placed in a cold muffle furnace. The temperature was raised to 400°C and maintained for 1 hr. The dishes were then removed from the furnace, allowed to cool, and the contents were ground, using plastic containers, plastic balls, and the mixer mill. The following were weighed and placed directly into a plastic bottle: SiO_2 (16.50 g), Al_2O_3 (9.0 g), MgO (0.84 g), TiO_2 (0.65 g), Na_2SO_4 (1.07 g), and K_2SO_4 (0.86 g).

The iron and calcium salts were added to the plastic bottle containing the other compounds, and the bottle was shaken on the mixer mill for 30 min. The mixture was then removed and divided into two approximately equal portions, and each part was transferred to a platinum dish. The dishes were placed in a cold muffle furnace, and the temperature was raised to 400°C and maintained for 6 hr. The mixtures were then removed from the furnace, allowed to cool, again combined in a plastic bottle, and blended for 1 hr, using the mixer mill. Nine sets of synthetic standards were prepared to cover the spectrographically detectable elements. No more than eight elements were included in any one standard series. The first standard in each series was prepared in a base of high-purity quartz.

The head standards were prepared to contain 1.0% each of the analytical elements in a final mixture of 5.000 g. Weights of each compound equivalent to 50 mg of the element were calculated.

The head standard (200 mg) as diluted with 1.80 g of synthetic coal ash base to produce a standard containing 0.100% each of the 8 elements. Mixing was done in a plastic capsule using 2 1/8-in. diameter plastic balls. The capsules were shaken on the Wig-L-Bug® mixer for 55 sec.

Successive dilutions were prepared starting with the 0.100% standard to obtain a series covering the concentration range from 0.100 to 0.0001%.

Gluskoter[10] has also prepared a set of synthetic standards by using the average concentrations of Si, Al, Ca, Fe, K, Mg, and Na, and by using average percent for high-temperature ash (the average values were taken from previously analyzed coals). The concentration values were calculated to their oxide or carbonate equivalents on the ash basis, the type of calculation depending upon the expected combination of each element in high-temperature coal ash. These concentrations were then normalized to 100%.

In November 1971, the National Bureau of Standards (NBS), U.S., issued Standard Reference Material (SRM) 1630, Mercury in Coal, with a provisionally certified mercury content of 0.13 ppm. Later, the provisional value of 2.1 ppm selenium in SRM 1630 was issued.

In mid-1972, an extensive laboratory intercomparison program was initiated by NBS and Environmental Protection Agency (EPA) to determine the accuracy of the current methods of analysis for trace elements in fuels, with the intent of improving the reliability of such determinations. About 50 laboratories, using a variety of methods, participated in the analysis of a sample of coal, fly ash, fuel oil, and gasoline for 18 elements: As, Cd, Cr, Cu, Hg, Mn, Ni, Pb, Se, Tl, Th, U, V, Zn, F, Be, S, and Fe. The intention was to provide standard reference material for each of four matrices: coal, fly ash, fuel oil, and gasoline, certified for 15 elements.

For a trace element concentration to be certified by NBS, it must be determined by at least 2 independent methods, the results of which must agree within a small experimental error range of ±1% to ±10%, depending on the nature of the sample and the concentration level of the element.

The National Bureau of Standards has produced several such standards. The current coal standard most used for trace-element analysis is the National Bureau of Standards Reference Material (NBS SRM) 1632. An excellent standard for determining trace elements in coal ash is NBS SRM 1633. These standards, initially circulated to many laboratories by the NBS and EPA as an extensive comparative analysis study, were further analyzed by several activation analysis laboratories, and comparative data for 33 elements are currently available.[52] Recently, NBS SRM 1632A, a replacement for 1632, and 1635 (a subbituminous coal) have been certified for various elements.

According to the NBS brochure accompanying the standard reference material, SRM is intended for use in the calibration of apparatus and the evaluation of techniques employed in the trace element analysis of coal and similar materials. The material should be dried without heat to constant weight before use.

The recommended procedures for drying are either vacuum drying at ambient temperature for 24 hr. or freeze-drying in which the drying chamber is kept at room temperature. When not in use, the material should be kept in a tightly sealed bottle and stored in a cool, dark place. Long-term (>1 year) stability of this SRM has not been rigorously established. NBS will continue to monitor this material and any substantive change will be reported to purchasers.

The certified values shown in Table 5 are based on at least a 250-mg sample of the dried material, the minimum amount that should be used for analysis. The table also includes, for information only, the values which are not certified, because they were based on a "non-reference" method, or were not determined by two or more independent methods.

Comprehensive round-robin studies of coal samples, which are continuing long-term studies that involve the consideration of additional elements each year, are sponsored by the American Society for Testing and Materials (ASTM) D05 Committee on Coal and Coke. The International Standards Organization (ISO) TC 27 Committee on Solid Mineral Fuels also has extensive consensus data available for mercury, beryllium, cadmium, cobalt, fluorine, chromium, manganese, molybdenum, nickel, lead, strontium, and zinc. As an example, Table 6 shows a comparison of round-robin study on SRM-1633 and NBS certified values.

Many other comparisons have been done. Gluskoter[10] has published a comparison of values for NBS SRM 1632 with reports from nine different laboratories. Values measured for NBS SRM 1632 at four laboratories are listed in reports by Filby et al.[50,51] A very detailed measurement of element concentrations in NBS SRM 1632 and 1633 has been published by Ondov et al.[52] and Lyon.[53] Their findings are shown in Table 7.

Many methods were used for their analysis for NBS SRM. The most popular was neutron activation because of its ability to detect many of the elements present in coal, even at very low concentrations. Emission spectroscopy, atomic absorption, wet chemistry, and X-ray fluorescence were used also. All these methods will be discussed in detail in this chapter.

Every laboratory doing trace element analysis work of coal and coal ash should test its equipment and method used by using SRM. This is the reason why the future work on SRMs is so important.

II. CLASSICAL CHEMICAL METHODS

For years chemical coal analysis could be divided into two categories — the Proximate and the Ultimate. In the Proximate Analysis one determines and reports "Mois-

Table 5
TRACE-ELEMENT CONCENTRATIONS IN NBS, SRM 1632, 1632a, 1633, AND 1635

Element	1632 (bituminous coal)	1632a (bituminous coal)	1633 (coal fly ash)	1635 (subbituminous coal)
Al (%)		(3.1)		(0.32)
As	5.9 ± 0.6	9.3 ± 1.0	61 ± 6	0.42 ± 0.15
Ca (%)		0.23 ± 0.03		
Cd	0.19 ± 0.03	0.17 ± 0.02	1.45 ± 0.06	0.03 ± 0.01
Ce		(30)		(3.6)
Co	(6)	(6.8)	(38)	(0.65)
Cr	20.2 ± 0.5	34.4 ± 1.5	131 ± 2	2.5 ± 0.3
Cs		(2.4)		
Cu	18 ± 2	16.5 ± 1.0	120	3.6 ± 0.3
Eu		(0.5)		(0.06)
Fe (%)	0.87 ± 0.03	1.11 ± 0.02		0.239 ± 0.005
Ga		(8.5)		(1.05)
Hf		(1.6)		(0.29)
Hg	0.12 ± 0.02	0.13 ± 0.03	0.14 ± 0.01	
K (%)		0.42 ± 0.02		
Mg (%)		(0.1)	1.98	
Mn	40 ± 3	28 ± 2	493 ± 7	21.4 ± 1.5
Na		840 ± 40	3070	(2400)
Ni	15 ± 1	19.4 ± 1.0	98 ± 3	1.74 ± 0.10
Pb	30 ± 9	12.4 ± 0.6	70 ± 4	1.9 ± 0.2
Rb		(31)	(112)	
S (%)		1.62 ± 0.03		0.33 ± 0.03
Sb		(0.6)		(0.14)
Sc		(6.3)		(0.63)
Se	2.9 ± 0.3	2.6 ± 0.7	9.4 ± 0.5	0.9 ± 0.3
Si	(3.2)			
Sr			1380	
Th	(3.0)	1.5 ± 0.1	24	0.62 ± 0.04
Ti	(800)	(1800)		(200)
U	1.4 ± 0.1	1.28 ± 0.02	11.6 ± 0.2	0.24 ± 0.02
V	35 ± 3	44 ± 3	214 ± 8	5.2 ± 0.5
Zn	37 ± 4	28 ± 2	210 ± 20	4.7 ± 0.5

Note: Values are in ppm unless otherwise specified. Values in parentheses are for information only.

ture'', "Ash", "Volatile Matter", and "Fixed Carbon". In the Ultimate coal analysis, one actually determines the concentration of hydrogen, carbon, nitrogen, and sulfur; reports in weight percent, and estimates a quantity called "Oxygen by Difference" by subtracting the sum of the above elements plus the "Ash" from 100.

Newer methods have been developed and used for the determination of some quantities. For example, oxygen could be determined by neutron activation (e.g., Volborth et al.,[54]), sulfur by X-ray emission spectroscopy, etc.

In this section we shall mention only some of the classical chemical methods used in coal analysis. To cover all aspects of this approach would be outside the scope of this text. In routine coal quality analyses, calorific value and sulfur content are commonly determined. In a paper by Neavel and Keller,[55] a procedure is described whereby the sulfur content of a coal sample can be estimated with reasonable accuracy from the titration of the calorimeter bomb washing. The procedure requires only an initial estimation of several correction factors and, for subsequent samples, their routine use,

Table 6
COMPARISON OF NBS
CERTIFIED VALUES AND
ROUND ROBIN DATA[a]

Element	NBS value	Round robin
Na		0.32 ± 0.04%
Mg		1.8 ± 0.4%
Al		12.7 ± 0.5%
Si		21 ± 2%
P		
Ca		4.7 ± 0.6%
K		1.61 ± 0.15%
Sc		27
Ti		0.74 ± 0.03%
V	214 ± 8	235 ± 15
Cr	131 ± 2	127 ± 6
Mn	493 ± 7	496 ± 19
Fe		6.2 ± 0.3%
Co		41.5 ± 1.2
Ni	98 ± 3	98 ± 9
Cu	128 ± 5	
Zn	210 ± 20	216 ± 25
As	61 ± 6	58 ± 4
Sr		0.17 ± 0.03%
Ba		0.27 ± 0.5%
U	11.6 ± 0.2	12 ± 0.5

Note: All values are in ppm unless otherwise specified.

[a] NBS values are from National Bureau of Standards, Standard Reference Material, and Round Robin data are from *Anal. Chem.*, 47, 1102, 1975.

along with titration data, in a sample calculation. In the determination of the heat of combustion of coal using an adiabatic calorimeter, the contents of the calorimeter bomb are recovered with a water wash. The washings are then titrated with a caustic solution to determine the amount of nitric acid and sulfuric acid formed in the combustion of the sample, and a correction is applied to the as-determined heat of combustion to compensate for the heat of formation of the acids. ASTM Designation D2015 describes the procedure in detail.

The amount of nitric acid formed during the combustion should be proportional to the amount of air in the bomb and to the amount of nitrogen in the coal. If these two values could be estimated fairly accurately, then it should be possible to subtract the nitric acid contribution from the total acids titrated and calculate the sulfur content of the coal by assuming that the remainder of the acid is sulfuric acid. The estimated sulfur contents of 713 (93%) of 768 samples to which the procedure was applied differed from values determined in the standard sulfur analysis by less than 1% sulfur. The estimated value can be used as a substitute for a separate, standard sulfur analysis, and is especially useful where coal from one source is routinely analyzed. The technique also provides a direct calorimeter correction for heat of formation of sulfuric acid, obviating the necessity for an independent sulfur analysis as specified in standard procedures.[55]

Table 7
CONCENTRATIONS OF ELEMENTS IN NBS SRM 1632 AND 1633 AS REPORTED BY ONDOV ET AL.[52] AND LYON[53]

	NBS coal standard (SRM = 1632)			NBS fly ash standard (SRM = 1633)		
	Lyon[53]	NBS	Ondov et al.[52]	Lyon[53]	NBS	Ondov et al.[52]
Al%	1.90		1.85 ± 0.13	12.5		12.7 ± 0.5
As	5.5	5.9 ± 0.6	6.5 ± 1.4	54	61 ± 6	58 ± 4
Ba	405		352 ± 30	2780		2700 ± 200
Br	14.2			6.0		12 ± 4
Ca%	0.44		0.43 ± 0.05	4.34		4.7 ± 0.6
Cd	0.31	0.19 ± 0.03		1.85	1.45 ± 0.06	
Ce	18.5		19.5 ± 1.0			146 ± 15
Cl	1000		890 ± 125			42 ± 10
Co	5.9	(6)	5.7 ± 0.4	46	(38)	41.5 ± 1.2
Cr	21 ± 2	20.2 ± 0.5	19.7 ± 0.9	138	131 ± 2	127 ± 6
Cs	1.4		1.4 ± 0.1			8.6 ± 1.1
Cu	18	18 ± 2		133	120	
Eu	0.21		0.33 ± 0.04	2.86		2.5 ± 0.4
Fe%	0.84	0.87 ± 0.03	0.84 ± 0.04	6.37		6.2 ± 0.3
Ga	8.5			49		
Hf	0.95		0.96 ± 0.05	10.8		
In	0.07		0.20 ± 0.12			
Hg	0.088 ± 6%	0.12 ± 0.02		0.127 ± 2.2%	0.14 ± 0.01	
K%	0.290		0.28 ± 0.03	1.8		1.61 ± 0.15
La	10.5		10.7 ± 1.2	82		82 ± 2
Mg%	0.248		0.20 ± 0.05		1.98	1.8 ± 0.4
Mn	46 ± 3	40 ± 3	43 ± 4	460	493 ± 7	496 ± 19
Mo	3.4					
Na	390		414 ± 20		3070	3200 ± 400
Ni	14.5	15 ± 1	18 ± 4	109	98 ± 3	98 ± 9
Pb	29.4	30 ± 9		78 ± 4	70 ± 4	75 ± 5
Rb	24		21 ± 2	120	(112)	125 ± 10
Sb	4.45		3.9 ± 1.3	7.8		6.9 ± 0.6
Sc	4.5		3.7 ± 0.3	32		27 ± 1
Se	3.05	2.9 ± 0.3	3.4 ± 0.2		9.4 ± 0.5	10.2 ± 1.4
Si%		(3.2)				21 ± 2
Sm			1.7 ± 0.2	15		12.4 ± 0.9
Sr	144		161 ± 16	1301	1380	1700 ± 300
Ta	0.17		0.24 ± 0.04	1.6		1.8 ± 0.3
Th	3.0	(3.0)	3.2 ± 0.2	26	(24)	24.8 ± 2.2
Ti	930	(800)	1100 ± 100	6420		7400 ± 300
U	1.21	1.4 ± 0.1	1.41 ± 0.01	11.8	11.6 ± 0.2	12.0 ± 0.5
V	40 ± 3	35 ± 3	36 ± 3	240	214 ± 8	235 ± 13
Zn	34	37 ± 4	30 ± 10	208	210 ± 20	216 ± 25
Zr	45					301 ± 20

A review of classical methods for oxygen determination is published by Ode.[28] The procedure used by U.S. Bureau of Mines[56] follows. Coal samples were pyrolyzed in a nitrogen atmosphere, and the gaseous oxygen compounds liberated were converted to carbon monoxide by a platinized-carbon catalyst at 1173 K. After gas cleanup, the carbon monoxide was heated with iodine pentoxide, liberating equivalent amounts of iodine and carbon dioxide. After several reaction steps, the iodine was finally titrated with sodium thiosulfate and the oxygen content calculated has not been adopted as a standard method for oxygen determinations in coal.

Kinson and Belcher[57] have presented results for organic determinations in Australian coals based on a radio frequency heating method. In this method, 10 mg samples of coal or coke are heated to 2223 K in an iron-tin bath in a carbon crucible. Pyrolytic gases are passed through five different beds making up a CO conversion, and oxygen is determined gravimetrically from the CO_2 adsorbed. For mineral matter contents greater than 5%, elaborate corrections for inorganic oxygen were necessary.

In their work, Cavallaro et al.[58] have described methods for the determination of fluorine. The method used for fluorine determination utilizes a fluoride-ion-specific electrode, which is simple and fast to use and is not as subject to interferences as were the methods used in the past—based on bleaching action of fluoride ion on a colored complex.

A 2-g coal sample was mixed with 0.8 g of CaO in a platinum crucible and ashed at 600°C until all carbonaceous matter was oxidized. The residue was fused with 4 g of Na_2CO_3. The fusion cake was leached with phosphoric acid and the fluorine was distilled from a phosphoric acid-sulfuric acid mixture at 135°C. The distillate was made basic to phenolphthalein with 1% sodium carbonate solution and evaporated to about 5 mℓ. The solution was neutralized with 1:1 H_2SO_4 using methyl orange as an indicator. A commercial fluoride-ion-electrode buffer was added (10 mℓ) and the volume was adjusted to 25 mℓ with distilled deionized water. The solution was transferred to a plastic beaker, and the potential measurements were made on an expanded-scale pH meter.[58]

At fluoride concentrations below 10^{-4} M, the electrode response does not follow the Nernst relationship (i.e., the electrode response is not linearly related to the logarithm of the fluoride concentration), and it becomes necessary to add a known quantity of fluoride to the solution in order to bring the concentration responses.

Polarography and fluoride-ion-selective electrodes normally require individual samples for analysis of individual elements. Polarography is not used extensively for trace-element analyses in coal or fly ash, although its sensitivity for several elements (e.g., cadmium) makes it useful for trace analyses after interfering ions are separated. Trace fluoride determinations are commonly made by fluoride-ion-selective electrode, an inexpensive and simple application of potentiometry. A detection limit of 10 ppm for fluoride in whole coal was obtained by Ruch et al.[59-62] Although extensive sample preparation and digestion were required, precision for repeated measurements was good.

In some cases voltammetry is also used for the determination of element concentrations in coal. Voltammetry is an electrochemical method in which application of a negative voltage is used to plate metal ions acid-extracted from the sample onto an electrode. The electrode potential is then varied linearly in an anodic direction, which produces a sharp current peak proportional to the concentration in the sample.[38] Sample preparation for atomic absorption spectroscopy also requires much "wet chemistry"; we shall describe some of the procedures used in the chapter on atomic absorption.

III. OPTICAL METHODS

Optical methods have been in use in chemical laboratories for coal analysis for some time. A large amount of data on elemental composition of coal and coal ashes has been accumulated by the use of emission spectroscopy and atomic absorption spectroscopy. Here we shall describe some of the applications to coal analysis as reported in the literature.

A. Emission Spectroscopy
Emission spectroscopy is a method well-known to many laboratories and is often

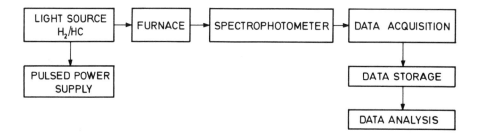

FIGURE 2. The principle of the spectrophotometric system.

used as a direct-reading method or a photographic technique. The principle of the spectrophotometric system is shown in Figure 2. Despite the advantages of optical emission spectroscopy for trace analysis of particulate matter, its use has been relatively limited compared to atomic absorption, neutron activation, and X-ray emission techniques. This is due to misconceptions regarding equipment cost and the complexity of the analytical procedures. A direct reading, optical emission spectrometer is less expensive than the equipment necessary for neutron activation analysis, and its cost is less than, or comparable to, the cost of a modern X-ray spectrometer. Furthermore, since emission spectrometry is a multielement technique, analysis for more than one element is less expensive and more rapid by emission analysis than by atomic absorption analysis.[63]

According to Govindaraju et al.,[64] there are three general types of plasmas: the double-electrode d-c arc plasma jet, the single-electrode capacitatively-coupled microwave plasma (CMP), and the electrodeless, inductively-coupled plasma (ICP). They report that the relatively lower detection limits and generally higher temperature of the ICP make the ICP-OES (inductively coupled plasma-optical emission spectrometer) system relatively free of inter-element effects and capable of rapid multielement analysis of ultra-trace levels. Quantitative spectrographic analysis for trace elements in coal and coal ash have been reported by many workers. We shall describe only some of the approaches applied.

As an example of early work, let us describe the method developed by Peterson and Zink[48,49] and applied to the analysis of ash of U.S. coals. A low-temperature ashing procedure was used to minimize the loss of volatile elements. The analytical procedure provides for the estimation of 68 elements including major, minor, and trace constituents. Standards for the trace elements were prepared in a synthetic base mixture that approximates the average composition of U.S. coal ashes. Excitation was achieved with a d-c arc, spectra were recorded photographically, and interpretation was made by visual comparison of the spectra of unknowns with the spectra of graded standards series. In the analysis of 900 samples, 36 trace elements were found. Here are some details of the method used.

The spectral excitation source used was a d-c arc obtained from a motor generator capable of supplying 15 A at 260 V. A series resistor was used to control the arc current. The 2 spectrographs used were (1) a grating instrument employing 2 gratings, each having a reciprocal linear dispersion of 4 nm/mm in the 1st order, to record spectra from 237.0 to 430.0 nm, and (2) a large Littrow spectrograph with reciprocal linear dispersion of 5 Å/mm at 320.0 nm to record spectra from 344.0 to 900.0 nm. The external optics of each spectrograph were so arranged that radiation falling on the grating or collimator represented sampling of the entire arc column except for the electrode tips which were screened out.

The following reagent-grade chemicals were used in the preparation of the synthetic coal ash base: aluminum oxide (Al_2O_3), calcium sulfate ($CaSO_4 \cdot 2H_2O$), ferrous ammonium sulfate ($FeSO_4 \cdot (NH_4)_2SO_4 \cdot 6H_2O$), magnesium oxide (MgO), sodium sulfate (Na_2SO_2), potassium sulfate (K_2SO_4), silicon dioxide (SiO_2), and titanium dioxide (TiO_2). All compounds were examined spectrographically for trace constituents and were found essentially free of all trace elements sought in coal ash. High-purity compounds of each of the trace elements were used in the preparation of synthetic standards.

Master comparison plates were prepared, each containing spectra of a graded series of standards. A series of chemically analyzed samples was used for the major and minor constituents, and the synthetic standards were used for the trace elements. The lines listed in Table 8 were marked on the appropriate plates. Additional lines were marked as required. Unknown samples were arced, using the identical exposure and excitation conditions as used for the standard. At least one control standard was exposed on each plate to verify the exposure level. The concentration of each element found in unknown samples was estimated by visual comparison with spectra of the standards.

The elements detected in coal ash samples by Peterson and Zink,[48,49] analytical lines concentration ranges found, and approximate limits of detectability are shown in Table 8.

Of the two optical emission spectroscopic methods described, Dreher and Schleicher[46] have used a direct-reading method to analyze Sn, B, Pb, Cu, Co, Ni, Be, Cr, V, Mo, and Ge, and a photographic technique for B, Mn, Cr, Pb, Be, V, Ag, Cu, Zn, Zr, Co, and Ni. Of these elements, 16 were determined in the high-temperature ash of the coals, and 8 of them were cross-checked by both methods. Conventional d.c. arc excitation is used by both methods.

The preparation of high-temperature ash for the analysis is described by Dreher and Schleicher[46] in this way: 2 g of finely ground coal (−60 mesh or smaller) are separated from the larger stock by riffling; weighed into a tared, used silica crucible; and dried in an oven at 110°C for approximately 3 hr. (It has been showed that the walls of new silica crucibles retain some trace element to a greater extent during dry ashing than crucibles that have been used for several ashings). The samples are cooled and weighed for moisture determination. The weighed, moisture-free samples are transferred to a cold muffle furnace, heated to 500°C in 1 hr, and ashed at that temperature until no carbonaceous matter appears. During ashing, the samples are mixed with a platinum wire approximately every hour. The samples are cooled and weighed, ground in a mullite mortar, then dried at 110°C, and stored in a desicator over $Mg(ClO_4)_2$. The authors have prepared their own standards for both direct-reading and spectrographic measurements.

A set of synthetic standards was prepared by mixing commercial SiO_2 base standards containing 1000 ppm of 49 elements (Spex Industries, Inc. 1006) with corresponding 1000 ppm Al_2O_3 base standards (Spex Industries, Inc. 1007) in the ratio of 3:1 (SiO_2 to Al_2O_3). This mixture was then diluted with a 3:1 mixture of Specpure® SiO_2 and Al_2O_3 to concentrations of 1000, 500, 250, 100, 50, and 10 ppm. To 9 parts of each of these mixtures was added 1 part Fe_2O_3, producing final trace element concentrations of 900, 450, 225, 90, 45, and 9 ppm. These standards were, in turn, mixed in a 1:1 ratio with SP-2X graphite powder containing 0.111% indium as internal standard.

The samples (high-temperature ash) and standards, both diluted with an equal weight of SP-2X graphite powder, were then ignited spectrographically.

The results obtained by Dreher and Schleicher[46] are shown in Table 9 and Table 10. Table 9 shows the analytical wavelengths, the concentration ranges encountered in the

Table 8
RESULTS OF SPECTROGRAPHIC METHOD AS APPLIED BY
PETERSON AND ZINK TO COAL ASH ANALYSIS[48,49]

Element	Analytical lines (nm)	Approx. lower limit of detection (ppm of ash)	Conc. range detected (ppm of ash)
Aluminum	Al I-256.799	—	
	Al I-257.510		Major constituent
Antimony	SB I-259.806	20	MDL to 50
Arsenic	AS I-278.020	50	
	AS I-286.045		MDL to 1,000
Barium	Ba I-307.159	20	
	Ba II-455.404		20 to 10,000
Beryllium	Be II-313.107	1	
	Be II-313.042		1 to 100
Bismuth	Bi I-306.772	1	MDL to 10
Boron	B I-249.678	2	
	B I-249.773		10 to 300
Cadmium	Cd I-326.106	5	ND to 0.001
Calcium	Ca II-317.933	—	Major constituent
Cerium	Ce II-422.260	200	
	Ce II-320.171		MDL to 500
Cesium	Cs I-852.100	50	MDL to 100
Chromium	Cr II-286.093	1	
	Cr II-284.325		
	Cr I-425.435		3 to 1000
Cobalt	Co I-345.351	20	
	Co I-340.512		MDL to 1000
Copper	Cu I-324.754	1	
	Cu I-282.437		1 to 500
Gallium	Ga I-294.364	2	2 to 200
Germanium	Ge I-303.906	3	MDL to 300
Indium	In I-410.177	10	
	In I-451.132		MDL to 30
Iron	Fe I-299.639	—	
	Fe I-301.148		Major constituent
Lanthanum	La II-324.513	100	MDL to 500
Lead	Pb I-283.307	1	1 to 2000
Lithium	Li I-670.784	1	
	Li I-323.261		1 to 2000
Magnesium	Mg II-279.553	—	
	Mg II-280.270		Minor constituent
Manganese	Mn I-279.482	1	
	Mn I-403.075		30 to 2000
Molybdenum	Mo I-317.035	1	MDL to 500
Neodymium	ND II-401.225	100	MDL to 500
Nickel	Ni I-341.476	1	
	Ni I-349.296		
	Ni I-310.155		1 to 2000
Niobium	Nb II-309.418	10	MDL to 200
Phosphorus	P I-253.565	1000	
	P I-255.328		MDL to 20000
Potassium	K I-404.720	—	
	K I-404.414		
	K I-766.491		
	K I-769.898		Minor constituent
Rubidium	Rb I-794.760	10	
	Rb I-780.023		MDL to 2000
Scandium	Sc II-424.683	20	
	Sc I-327.363		MDL to 50

Table 8 (continued)
RESULTS OF SPECTROGRAPHIC METHOD AS APPLIED BY PETERSON AND ZINK TO COAL ASH ANALYSIS

Element	Analytical lines (nm)	Approx. lower limit of detection (ppm of ash)	Conc. range detected (ppm of ash)
Silicon	Si I-298.765	—	Major constituent
Silver	Ag I-328.068	1	0.1 to 10
Sodium	Na I-330.232	—	
	Na I-330.299		Minor constituent
Strontium	Sr I-460.733	10	10 to 5000
Thallium	Tl I-276.787	5	MDL to 200
Tin	Sn I-317.505	1	
	Sn I-283.999		1 to 200
Titanium	Ti II-308.802	—	Minor constituent
Tungsten	W I-294.698	100	MDL to 200
Vanadium	V I-318.398	1	
	V I-318.540		1 to 2,000
Ytterbium	Yb II-328.937	1	
	Yb I-398.798		1 to 20
Yttrium	Y II-321.669	10	
	Y II-324.228		10 to 2,000
Zinc	Zn I-334.502	50	MDL to 5,000
Zirconium	Zr II-343.823	50	50 to 2,000

Table 9
RESULTS OF THE SPECTROMETRIC METHOD AS APPLIED BY DREHER AND SCHLEICHER[46]

		Coal			Ash
	Analytical wavelength (nm)	Concentration range (ppm)	Average relative SD (%)	Relative SD range (%)	Detection limit in electrode charge (ppm)
B	249.68	4.6—>224	4.1	0.5—13	0.6
Pb	283.307	2.8—~249	8.0	2.2—15	5
Cu	327.396	4.5—69	5.4	1.3—20	2
Co	345.350	1.0—42	4.1	0.5—19	2
Ni	341.476	3.8—105	4.0	0.9—11	1.5
Be	234.861	0.5—5.9	3.1	0.0—12	0.5
Cr	425.435	4.4—33	3.4	0.9—10	1.5
V	318.540	8.4—~108	6.8	2.4—12	2
Sn	303.412	1.0—51	18.0	4.7—57	6
Mo	317.035	<0.3—30	5.0	1.7—12	0.2
Ge	265.118	<0.7—35	12.0	3.0—29	0.3

coals, average relative standard deviations, the relative standard deviation ranges, and the detection limits in the sample charge determined for the spectrometric method. In most cases, these statistics are based on 4 determinations for each of 98 samples. The generally poor precision for the tin data is probably attributable to the fact that tin concentrations were only slightly above or below the detection limit (~6 ppm in the sample charge).

Table 10 shows the analytical wavelengths chosen, the concentration ranges calculated to the whole coal, average relative standard deviations, and detection limits in ash determined by the spectrographic method.

Table 10
RESULTS OF THE SPECTROGRAPHIC METHOD AS APPLIED BY DREHER AND SCHLEICHER[46]

	Analytical wavelength (nm)	Coal		Ash	
		Conc. range (ppm)	Average relative SD (%)	Upper determination limit (ppm)	Detection limit (ppm)
B	249.77	11—325	12	∼3000	85
Pb	283.307	3.7—370	9	∼3700	30
Cu	327.396	4—65	18	∼500	50
Co	340.51	1.8—49	25	∼1300	10
Ni	341.476	1.3—60	24	∼900	20
Be	313.04	0.20—4.5	14	∼40	2.1
Cr	267.72	<3.8—50	23	∼1000	45
V	318.40	11—133	30	1200	60
Ge	265.118	1.2—117	16	430	30
Ag	328.07	0.16—3.7	—	∼26	∼1
Zn	334.50	7—>200	—	∼4000	100
Zr	339.20	8.4—>263	—	∼3000	50
Mn	257.61	5.2—>400	19	∼5300	100

In the photographic procedure, the lack of suitable internal standard for exposure correction, the attempt to record and determine all elements on one generalized exposure, and the very high concentration of the trace elements in the ash (for some samples as much as 33 times the amount reported in the coal) caused a poor relative standard deviation. However, of the 13 elements determined, only Co, Ni, Cr, and V were less precise than ±20%, a level which the authors feel is suitable for a photographic method.

Emission spectrochemical analysis also has been used often by Gluskoter.[10] In his work, the mixture used for loading the spectrometer electrodes consisted of 40 mg of sample or standard, 10 mg of spectroscopically pure $Ba(NO_3)_2$, and 150 mg of SP-2X graphite powder. These were mixed together on a Wig-L-Bug® shaker for 60 sec in a 2.54 cm in length by 1.27 cm in diameter plastic vial containing two plastic balls 32 cm in diameter. This mixture was then weighed in the appropriate amounts for loading into electrodes. The spectroscopic parameters used by Gluskoter[10] are listed in Table 11.

The direct-reading spectrometer procedures used by Gluskoter.[10] are described in the following way: time-intensity curves were run by the use of standards to determine the proper exposure time for the desired spectral lines. After the exposure time was determined, more standards were arced to establish a calibration curve for each element desired and to apply the proper electronic corrections to each element readout module. The data received from the instrument were relative intensities, standardized by using a spectral line resulting from variable, but known, concentrations of iron. Usually, four electrodes were arced for each sample.

The coordinates of each point used in an element calibration curve were treated by least squares regressions to determine the coefficients of the first or second degree equation that best described the particular calibration curve. By the use of the relative intensity data for unknown samples and the calibration curve coefficients, the concentration of each desired element in the electrode sample was calculated by the use of a computer program.[10]

The results on concentration ranges measured in coal for different elements by op-

Table 11
SPECTROSCOPIC PARAMETERS USED IN WORK BY GLUSKOTER[10]

Instrument	Jarell-Ash 3.4 m Ebert spectrograph	Jarell Ash 0.75 m direct reading spectrometer	Jarell Ash 0.75 m direct reading spectrometer
Arc current (D.C.)	10A	15A	7.5A
Arc gap	4 mm	6 mm	6 mm
Exposure time	80 sec	65 sec	30—40 sec
Atmosphere	80% argon, 20% oxygen	80% argon, 20% oxygen	80% argon, 20% oxygen
Sample electrode	National L-3903 under-cut	National L-3979 thin-wall crater	National L-4006 necked crater 3/16 in. diameter
Counter electrode	National SP-1009	National L-4036 (ASTM C-1)	National L-4036 (ASTM C-1)
Electrode charge	20 mg	15 mg	10 mg
Entrance slit width	10 μm	10 μm	10 μm
Photographic plate and developer	SA-1, D-19		
Step sector	6 step, 2:1 ratio		
Internal standard		Fe, variable internal standard	
Exit slit width		50 μm	50 μm

tical emission methods (direct-reading, DR, and photographic detection, PD) as obtained by Gluskoter[10] are shown in Table 12.

B. Atomic Absorption Spectroscopy

In atomic absorption spectroscopy, (AAS) the sample in solution is atomized by a flame or other energy source, where it produces atomic vapor of the element being analyzed. Monochromatic light which has the same wavelength as that of the required element is then passed through the sample vapor. The atoms present in the ground state of the vapor absorb radiation from the monochromatic light source in proportion to their concentration present in the sample.[65] See Figure 3 for the schematic of atomic absorption apparatus.

Types of interference encountered in using atomic absorption spectroscopy for coal or ash samples include inter-element or chemical interferences, matrix effects (which stem from the large concentrations of acids and solids in solution), and molecular absorptions (which predominately occur from species such as SrO or $Ca(OH)_2$ and result in a positive error in the absorption measurement).[66]

Davison et al.[67] determined Pb, Tl, Ni, As, Cd, and Be in fly ash by flame atomic absorption and found results in good agreement with those from spark-source mass spectrometry, except for Tl. Selenium was converted to H_2Se and then analyzed. Using standard addition calibrations, a precision of ±10% for all analyses could be obtained.

Very often some wet chemistry has to be used for the sample preparation for conventional atomic absorption spectroscopy. For example it is possible to determine Li, Be, V, Cr, Mn, Co, Ni, Cu, Zn, Ag, Cd, and Pb after dry ashing and acid dissolutions. Arsenic and antimony can be determined as their hydrides by AAS after low-temperature ashing. Germanium, tin, bismuth, and tellurium can be determined as their hydrides by AAS after high-temperature ashing. Selenium can be determined as its hydride by AAS after a special combustion procedure or after oxygen bomb combustion.[68,69]

Here is the description of the procedure used by Simon and Huffman[14] in the anal-

Table 12
EXPERIMENTAL PARAMETERS AND RESULTS FOR OPTICAL EMISSION DIRECT READING (DR) AND PHOTOGRAPHIC DETECTION (PD) AS OBTAINED BY GLUSKOTER[10]

	Wavelength (nm)	Method	Conc. range whole coal (ppm)	Average relative SD (%)	Detection limit in ash (ppm)
Ag	328.07	PD	0.01—2.4		
B	249.68	DR	5—264	5.4	1
Be	234.6	DR	0.1—5.5	7.0	0.2
	313.107	PD	0.15—3.4		
Cd	228.80	DR	<0.1—29	16.7	0.7
Co	345.35	DR	0.9—18	5.65	0.3
	345.35	PD	0.4—14		
Cr	425.43	DR	2—82	7.9	0.5
	284.325	PD	1.6—50		
Cu	327.40	DR	2—69	6.9	0.5
	327.40	PD	3.0—111		
Ge	265.12	DR	<0.1—17	15.5	0.8
	303.91	PD	<0.35—18		
Mn	260.57	PD	1.4—346		
Mo	317.03	DR	0.1—25	10.1	0.3
	317.03	PD	0.11—32		
Ni	341.48	DR	2—52	4.7	0.6
	341.48	PD	1.3—52		
Pb	405.78	DR	<1—188	14.3	10
	283.31	PD	1.0—64		
Sr	460.73	DR	11—270	11.0	1
Tl	377.57	DR	0.1—1.3	16.0	0.3
V	318.40	DR	5—80	7.0	3.3
	318.54	PD	3.8—142		
Zn	213.86	DR	<1—191	12.3	1
	334.50	PD	<0.8—592		
Zr	339.20	DR	9—67	8.1	3.3
	339.20	PD	14—103		

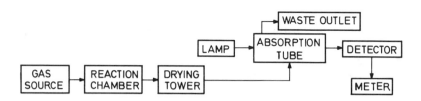

FIGURE 3. The schematic presentation of atomic absorption apparatus.

ysis of coal ash by atomic absorption spectroscopy. Coal ash (0.500 to 1.000 g) is weighed into a 100-mℓ platinum dish; 10 mℓ of demineralized water, 10 mℓ HNO$_3$, and 10 mℓ of HF are added to the dish. The dish is covered and allowed to stand overnight; then 7 mℓ of HClO$_4$ are added to the dish. The dish is placed on a steam bath for 1 hr and then placed on a hot plate to fume off the acids to near dryness. The 25 mℓ of water and 5 mℓ of HCl are added to the dish. The dish is covered, placed on a steam bath, and digested for 30 min. The solution is transferred to a 100 mℓ volumetric flask and diluted to volume with water. Aliquots or dilutions of this sample

Table 13
DETECTION LIMITS FOR ELEMENTS IN
COAL ASH[14]

Element	Sample weight (g)	Lower limit of detection ($\mu g/g$)
Na	0.5	100
Mg	0.5	100
Cu	0.5	10
Li	0.5	5
Mn	0.5	25
Zn	0.5	10
Pb	1.0	25
Cd	1.0	1

are then aspirated into the air-acetylene flame of an atomic absorption spectrometer to determine the elements listed in Table 13. The sample aliquot used for the determination of Mg is made to contain 1% La. The instrument settings used are those recommended by the manufacturer. Background correction is used for the determination of Cd and Pb. Cd and Pb may also be determined by using a heated graphite atomizer. Separations (solvent extraction) are required, and detection limits that are one to two orders of magnitude lower than those given in Table 13 result.

For many elements, e.g., cadmium and lead, solvent extraction offers a means for increasing the sensitivity of the determinations by isolation and concentration of the metals of interest. A procedure developed for detection of cadmium and lead by Cavallaro et al.[58] utilizes the extraction of the iodide complexes of lead and cadmium into methylisobutylketone (MIBK).

A 10-g sample of coal (or a 5-g sample of the sink 1.60 fraction) was weighed into a Vycor® or platinum dish and placed in a cold muffle furnace. The temperature was raised to 500°C in 1 hr, and ashing continued at that temperature for about 16 hr. There is no significant loss at this temperature for the trace metals investigated. The resulting ash was digested in concentrated HCl and filtered, and the insoluble residue was again ashed at 500°C. After treatment with HF and H_2SO_4 to volatile silica, the sample was again leached with HCl and filtered. The residue after ignition was fused in potassium carbonate, the fusion cake was dissolved in dilute HCl, the resulting solution was added to the combined filtrates, and the solution was evaporated to near dryness. The residue was dissolved in HCl, transferred to a volumetric flask, and diluted to volume. Aliquots were taken and standard additions of lead and cadmium were made. Ascorbic acid, potassium iodide, and MIBK were added and the lead and cadmium iodides were extracted into the MIBK. The ketone layer was aspirated into the flame of an atomic absorption spectrophotometer. Methyl isobutyl ketone was used to establish a base line. The amount of analyte present was calculated by means of a linear least squares regression procedure.[58]

The procedure used by Cavallaro et al.[58] for the preparation of coal samples for the determination of Cr, Cu, Mn, and Ni was similar to that used for the preparation of coal samples for Cd and Pb determinations. A 2-g sample of coal (or a 1-g sample of the sink 1.60 fraction) was ashed at 500°C in a platinum crucible, and the ash was then treated with H_2SO_4 and HF to volatilize the silica. After evaporation, the residue was leached with concentrated HCl and filtered. If considerable residue remained, it was again treated with HF and leached with HCl. Finally the insoluble portion was fused at 950°C with $LiBO_2$. The fusion cake, after cooling, was dissolved in 3 N HCl, and the solution was combined with the filtrates. The solution was transferred to a

100-mℓ volumetric flask and diluted to volume, 3 25-mℓ aliquots were taken, and additions of standards were made to each aliquot.

Lithium metaborate has been reported to be a good fluxing agent and has also been used in conjunction with atomic absorption analysis in silicate analysis. A paper by Muter and Nice[70] describes a lithium tetraborate-atomic absorption analytical technique which is being used to analyze coal ash. Lyon et al.[71] have described mercury determination coal and coal ashes by flameless atomic absorption in addition to Zn, Cu, and Pb.

The use of atomic absorption spectrometry for iron determination in coals is described by Gladfelter and Dickerhoof.[72] Analyses of iron content of the hydrochloric and nitric acid extracts in the ASTM procedure for sulfur forms in coals can be accurately determined by atomic absorption spectrometry. Sulfate ion and sulfuric acid were found not to interfere. The results obtained by titration are compared showing that both methods are reliable and may be substituted as there are no significant differences in results between these two methods.

The ASTM method D 2492 (ASTM, Part 26 D 2492, 1974, 507) is a useful analytical scheme for determining the forms of sulfur present in natural coals. It can also be applied to hydrodesulfurized coals.[21] At two different instances in the ASTM procedure, titrations for dissolved iron are required. It has been suggested that atomic absorption techniques may be substituted for these titrations. In the paper by Gladfetler and Dickerhoof,[72] the use of atomic absorption analyses as a replacement for the titrations is described.

The procedure was as follows: a Perkin-Elmer model 303 atomic absorption spectrometer was used in this investigation and 100-mℓ aliquots of either the hydrochloric acid or the nitric acid extractions were placed in 1-ℓ flasks. Exactly 100 mℓ of either 1:7 nitric acid or 2:3 hydrochloric acid (whichever was not used in the extraction procedure) was also added. The samples were then diluted to the proper volume and analyzed on the instrument. The extra acid, hydrochloric or nitric, was added to make the final solution (the matrix) as nearly identical as possible for all samples. The standards were also prepared in this manner. Thus, each sample and all blanks were 1/10 2:3 hydrochloric acid and 1/10 1:7 nitric acid by volume.

In order to determine the possible effects of sulfuric acid and/or the sulfate ion on the atomic absorption analysis of iron, three series of ten solutions each were prepared. Each solution contained 10 ppm iron. The 1st series of 10 was prepared with various sulfuric acid concentrations ranging from 0:00005 M to 1:0 M. The second series of ten had the same sulfuric acid concentrations as the first series but also contained the same hydrochloric and nitric acid concentrations as would be present in all samples analyzed by this procedure. The 3rd series contained the same hydrochloric and nitric acid concentrations as the 2nd series but contained potassium sulfate in concentrations varying from 00001 M to 0:1 M instead of sulfuric acid.

The second series (Fe, H_2SO_4, HCl, and HNO_3) was used to determine whether the matrix correction used in the analysis of standards and unknown samples is effective in removing these interferences. The third series (Fe, K_2SO_4, HCl, and HNO_3) was used to determine whether a difference exists in starting with sulfate ion or sulfuric acid. The solutions that contained potassium sulfate had to be corrected for the small amount of iron contained in this reagent. The values of iron concentrations determined in raw coal were in the region 0.6 to 2.0%.

For a very useful description of the flame atomic absorption analytical procedures, see also papers by Gluskoter and Lindhal[73] and Gluskoter.[10] The procedure is as follows: approximately 0.1 g low-temperature ashed sample, previously dried at 110°C for several hours, was transferred to a 60-mℓ or 125-mℓ linear polyethylene screw-cap bottle. The sample was wetted with 1 mℓ of 1:1 distilled HCl and was dried in a steam bath. The dried sample was then wetted with 0.7 mℓ aqua regia and 0.5 mℓ of Hf was added. The bottle was capped tightly and was placed on a steam bath for approximately 2 hr. After the bottle was removed from the steam bath and was allowed to cool, 10 mℓ of a 50 g/ℓ H_3BO_3 solution was added. The dissolved sample was transferred to a 50 mℓ Pyrex® volumetric flask, was diluted to volume with deionized water, and was returned to the bottle for storage.

As outlined by Pollock,[12] the atomic absorption analyses can be divided into four groups: elements that can be determined by conventional flame analysis (Cd, Co, Cr, Cu, Mn, Ni, Pb, Zn); elements that can be determined by flameless AAS with a graphite furnace (As, Cd, Cr, Pb, Zn); elements that can be determined by evolution of their volatile hydrides (As, Se); and a miscellaneous group that requires special methods (Hg, Se).

The graphite furnace method provides greater sensitivity than conventional flame procedures and greater convenience than hydride volatilization procedures. Cold vapor method has proved highly sensitive and very reliable for mercury analysis. Selenium can be analyzed by the graphite furnace method[74] or by a hydride evolution method.[69]

The major advantage offered by flameless atomization schemes such as the graphite tube atomizer is that the sensitivities and detection limits are often 100 to 1000 times better than with flame atomization for most metals. This is due, to a large extent, to the greatly increased residence time of the atomic vapor in the optical path and also to the total sample being available for absorption. The major disadvantages in this method were that it is more subject to severe interferences and that it is much more time-consuming than flame methods.

Many applications of flameless atomic absorption to coal analysis can be found described in the literature. For example Aruscavage[75] has described determination of arsenic, antimony, and selenium in coal by atomic absorption spectrometry with a graphite tube atomizer. In his work, submicrogram quantities in antimony, arsenic, and selenium in coal samples are determined by an atomic absorption process using an electrically heated graphite atomizer. The samples are decomposed in a mixture of nitric, sulfuric, and perchloric acids and are separated and concentrated in extraction from sulfuric acid-iodide solution into toluene. The results obtained on several intra laboratory reference samples and the National Bureau of Standards coal 1632 are compared with results of other methods. The determination can be made routinely for concentrations as small as 0.1 ppm for arsenic, antimony, and selenium in the coal.

Simon and Huffman[14] have determined mercury concentration in coal using flameless atomic absorption. In their procedure, Hg is determined by the method described by Huffman and others.[76] In this method, 0.200 g of ground coal is digested under oxidizing conditions using the HNO_3-H_2SO_4-$HClO_4$ digestion procedure. Hg in the sample solution is reduced to its elemental state with stannous chloride and then aerated from solution onto a silver screen placed in the vapor train. This silver screen is subsequently heated and the mercury vapor carried by an airstream to an absorption cell, where its concentration is determined by atomic absorption spectrometry. The lower limit of the determination is 0.01 ppm.

Coleman et al.[77,78] analyzed solvent-refined coals (SRC) for 11 metallic elements by flameless atomic absorption spectroscopy (AA). Samples of the SRC, and the tetrahydrofuran (THF)-soluble and insoluble SRC were wet ashed and dried for 12 hr prior to analysis. Matrix effects were compensated for by standard additions and deuterium-arc background corrections. Measurable amounts of Al, Mg, K, Mn, Fe, and Zn were found in all samples analyzed by AA. Analysis of the THF-insoluble and soluble samples found Mg, Al, K, Cr, Mn, Cd, and Pb to be 10 to 20 times more concentrated

Table 14
ANALYTICAL CONDITIONS USED BY GLUSKOTER[10] TO
DETERMINE Cd, Te AND Tl IN COAL

Element	Cd	Te	Tl
Source	HCL	EDL	EDL
Current or power	8 mA	7.2 w	5.8 w
Wavelength (nm)	228.8	214.3	276.8
Slit (nm)	0.7	0.2	0.7
Purge gas/flow (1/min)	Ar/1.2	Ar/1.2 (interrupt)	Ar/1.2
Drying time (sec)	30	30	30
Drying temperature (°C)	150	150	150
Charring time (sec)	20	20	20
Charring temperature (°C)	300	400	300
Atomization time (sec)	8	8	8
Atomization temperature (°C)	2000	500	2300
Background correction	D_2	D_2	D_2
Typical sensitivity (pg/.0044 Abs)	2.5	21	260
Typical detection limit (ppm) (ash, 20 µl sample)	0.05	1.0	5.0

and Fe to have even a greater concentration in the insoluble fraction. Cobalt, Ni, Cu, and Zn appeared to be equally distributed between soluble and insoluble forms.

Let us describe in some detail measurements performed by Gluskoter[10] for the determination of Cd, Te, and Tl in coal samples. The flameless atomizer used in his investigation was a Perkin-Elmer HGA-2000 graphite furnace used in conjunction with a Perkin-Elmer Model 306 atomic absorption spectrophotometer. Absorbance signals were recorded on a strip chart recorder. Corrections for broad band absorption were made with a deuterium arc background corrector. Electrodeless discharge lamps were used for tellurium and thallium determinations, and a hollow cathode lamp was used for cadmium. To compensate for any matrix interferences that might occur in the determinations, the method of standard additions was used for all three elements. The standard additions were made directly into the furnace following the addition of the sample solution. The analytical conditions developed for this study are summarized in Table 14.

Filby et al.[50] have used flameless atomic absorption spectrometry for the determination of Pb, Cd, and Be in coal. Here is the description of the method used for organic solvents. The organic solvent sample, (approximately 8 g), was weighed into 1 25-ml volumetric flask and dissolved in redistilled xylene which contained 0.1 mg/ml dithizone and 3 ml of the xylene-dithizone solution was transferred to a 50 ml Pyrex® screw-cap centrifuge tube. The oil solution was extracted with 3 ml of 40% v/v HNO_3 by shaking for 30 min and then centrifuged for 15 min. The lower aqueous layer was then withdrawn with a capillary transfer pipette. A blank was prepared by extracting 3 ml of the xylene-dithizone solution with 40% v/v HNO_3.

The Pb content of the oil extract was measured by flameless atomic absorption spectroscopy using a Varian Model AA-5 with a Model M-63 carbon rod atomizer (CRA). Table 15 shows the pertinent instrument settings used in the analysis of Pb, Cd, and Be. Argon was used as the sheathing gas and 3 µl of the extract solution were used for the analysis.

Data were recorded on a strip chart recorder, operated at a speed of 5 cm/min. The dual input allowed for simultaneous recording of the absorbance peak at atomization and the integrated form of the peak. Either or both measurements can be used for analytical work. Because absorbance showed a marked dependence on matrix in the case of Pb, an internal standard was necessary (method of standard additions). The

Table 15
INSTRUMENT SETTINGS (VARIAN AA-S WITH
CARBON ROD ATOMIZER M-63) AS USED BY
FILBY ET AL.[50]

Setting	Pb	Cd	Be
Lamp current (mA)	7	4	7
Slit width (μm)	300	300	150
Slit height (mm)	4	4	4
Special band width (nm)	0.99	0.99	0.50
Dry cycle (voltage-time)	7 V-30 sec	7V-30 sec	7V-30 sec
Ash cycle (voltage-time)	8V-20 sec	7V-20 sec	8V-20 sec
Atomize cycle (voltage-time)	9V-3 sec	9V-3 sec	10V-3 sec
Wave length (nm)	217.0	228.8	234.9

procedure used was as follows: 2 mℓ of the 40% v/v - dithizone HNO extract (or a dilution of this extract), the absorbance of which did not exceed 0.2, was pipetted into a 10-mℓ Erlenmeyer flask. The absorbance was measured by minimum of three firings of the CRA. If the agreement was poor (deviation greater than 10%), more measurements were made. Next, 25 $\mu\ell$ (to minimize the volume correction) of a freshly diluted standard solution of Pb (NO$_3$)$_2$ (Pb concentration 2.4 μg/mℓ) sufficient to increase absorbance by about 0.1 was added and the atomization measurements repeated. A second addition of 25 $\mu\ell$ of Pb standard was made to give a final maximum absorbance of not greater than 0.4. Corrections were made for the volume changes due to removal of solution for AAS and addition of Pb-containing solvent. The Pb content of the original extract was determined by a linear extrapolation of the three points using a procedure designed to reveal any significant deviation of the data from linearity.[50]

The 40% HNO$_3$-dithizone extracts of oil samples prepared for Pb measurement were also for the measurement of Cd. Absorbing substances (including dithizone products) cannot be completely removed at the low ashing temperature required to avoid the loss of the volatile element, Cd. Thus, background measurements are important in this work. These measurements were performed using a hydrogen continuum lamp. Although measureable quantities of Cd were found in the organic solvents, the use of an internal standard of CdSO$_4$ solution helped to assure that Cd, if present in the extract, would be detected by AAS. Calculated detection limits of Cd are based on actual measurements of the standards in the aqueous extracts.

Using the internal standard method (Standard BeSO$_4$ solution), Be was also determined in the previously prepared 40% HNO$_3$-dithizone extract of the oil samples. No detectable amount of Be was found in any of the aqueous extracts analyzed. As for Cd and Pb, the external standard method was used to calculate the detection limits for Be. The detection limit for beryllium is approximately 5 ppb for organic solvents and light oils.[50]

IV. ELECTRON MICROSCOPY

In this section we shall cover methods in which the coal sample is bombarded by electron beam. The first possibility is to obtain information on sample from transmis-

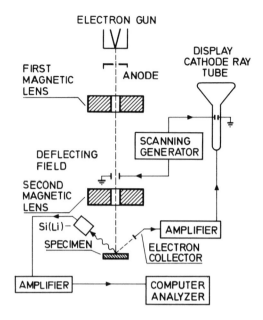

FIGURE 4. The basics of scanning electron microscope.

sion electron microscope (TEM) and scanning electron microscope (SEM). Since the sample emits characteristic X-rays in the process of being bombarded with electrons this can be used to obtain information on element concentration. Electron microprobe has been used for years; its serious disadvantage is low sensitivity because of high bremsstrahlung radiation. However, recently powerful instruments have been developed which combine SEM and energy dispersive system. We shall describe some of the published work in these areas.

Scanning electron microscope techniques have a broad capability for permitting the user to do detailed microstructural analyses of coal at the submicron level. These methods allow one to extend optical microscopical analyses that represent bulk properties. Use of SEM techniques together with energy loss spectroscopy can prove useful in identifying (1) organic coal constituents (macerals), (2) porosity (surface area), (3) mineral matter (potential catalysts), and (4) domains (possible building blocks of macerals). By construction of appropriate cells, one might also have the capability of observing coal microstructures in the electron microscope while changes in the specimen temperature and the pressure of the surrounding atmosphere are being made. *In situ* techniques allow reactions to be monitored as they occur and thus focus attention onto those specific reactions which appear to be the most important during gasification or liquefaction. These may involve the catalytic effect of various minerals, the reaction of the various maceral groups, or any other reactions that occur. SEM also can be used for direct observation of gas/solid reactions. SEM is especially suitable for the study of material response to hydrogenation, dehydration, and gas diffusion. The SEM can be used to examine the surface of fly ash particles for major elemental composition, and by etching away layers from the surface one can also study concentration as a function of depth.

The basic components of SEM are shown in Figure 4. The electron beam from the gun is passed through two electromagnetic lenses which reduce the source image and bring it to focus on the specimen surface. A scan generator in the second (sometimes third if there are three lenses) lens rasters the path of the electrons into a square pattern over the specimen surface. Simultaneously, the beam of a cathode ray tube is rastered across the tube face in exact synchronization. As the microscope beam strikes the specimen surface, secondary electrons (electrons of 5 to 10 eV energy) are emitted: the intensity of the beam striking the cathode ray tube is modulated according to the secondary electron intensity. Thus, in a point-by-point fashion, an image of the specimen is generated on the cathode ray tube face. Other signals such as X-rays, light, and Auger electrons are also produced by the electron beam striking the sample. These also may be collected and used for image generation.

Scanning electron microscopy is done with the specimen at high vacuum. For examining the interiors of particles, the specimen is often ion etched. When the Si(Li) detector is used for detection of produced characteristic, X-rays from the surface constituents can be detected and the spectrum obtained reflects the relative concentration of elements present in the sample.

Because the SEM is a surface technique, preparation of the sample is important. Ash samples that are collected on aluminum backing (e.g., cascade impactor samples) are placed in a vacuum evaporator where a thin ($\sim$50.0 nm) layer of carbon should be deposited over the ash material so that it would be conducting. Without such a coating, charge build-up would be a serious problem.

Several descriptions of such studies done on coal can be found in the scientific literature. Fowkes et al.[79] discussed the details of sample preparation. In their work, ash samples of North Dakota lignite from a laboratory pilot-plant combustor were analyzed for elemental concentrations and distribution relationships using an electron microprobe. The lignite samples used were chosen on the basis of Na content (both high and low Na lignites) and mine location. The lignite samples were then burned in a pilot plant combustor which simulated conditions in a commercial pulverized coal fixed unit. Ash samples were collected from boiler-tube surfaces and a resin binder was added to the sample to permit polishing. The polished samples were then analyzed by an electron microprobe for the following elements: Na, Mg, Ca, Fe, Al, Si, S, and K. Results indicated that Na-S compounds were found in a layer completely surrounding the tube. Sodium concentrations in deposits from high-Na lignites and Fe concentrations in deposits from low-Na lignites were greater near the tube and decreased consistently toward the outer edge of the deposit. The deposits contained higher concentrations of Si and Al than the laboratory-prepared ash: S concentration in the deposit was ⅓ that in the original ash. Overall, 42 discrete groups of elements were identified, each of which may represent more than one chemical compound.

Nandi et al.[80] composed the appearance of polished surfaces of two Canadian coals using the optical microscope and the SEM. In the Moss 3 coal, macerals were distinguishable in the secondary electron image and the backscatter electron image. Coal from the Tantalus Butte Mine, Northwest Territories, Canada, was studied with an SEM equipped with an X-ray electron probe analyzer which scanned the surface for Si, Fe, Ca, and C. Silicon was concentrated in the vitrinite, and Ca, Fe, and C were concentrated in fusinite and semiusinite. See also similar work by Augustyn et al.[31]

The method developed for determining the organic sulfur concentration and the stoichiometry of iron sulfide compounds in coals and chars by Solomon and Manzione[82] employs a scanning electron microprobe which separates the contribution from different sulfur forms by examining the differences in their spatial distribution. The standard ASTM method which calls for the determination of S(O) by subtracting the sulfate and sulfide contribution from the total sulfur is complicated, and often inaccurate. The method developed by Solomon and Manzione[82] is rapid, requires only small samples, and is nondestructive, so results may be repeated. In addition, a knowledge of the sulfide stoichiometry in a char and the original sulfide concentration in the coal yields the sulfide concentration in the char.

The concentration of trace elements in minerals within coal have been studied by Finkelman[83] and Finkelman and Stanton.[84] Finkelman and Stanton[84] indicated that carbon-coated polished blocks of coal, 1 to 2 cm on edge, would be the most suitable type of preparation for this purpose. The SEM system visual resolution of about 10 nm is more than adequate to observe mineral grains and inclusions down to submicrometer sizes. The system can detect X-rays from all elements heavier than neon in concentrations of about 1 wt%. Many of the accessory minerals in the coal are easily detected in the secondary electron images because they appear brighter than the encompassing major mineral species such as the clays and quartz. A 2-μm accessory mineral can easily be distinguished among thousands of clay and quartz grains at magnifications as low as 500 times.[83]

In the measurements by Finkelman,[83] the SEM was operated at an accelerating potential of 20 kV, a condenser current of $\leq$2.0 A, and a working distance of about 10 mm. The polished, carbon-coated mounts were step-scanned at 500 ×, and 200 nonoverlapping areas were examined by both secondary and back-scattered electron modes. Every relatively bright mineral grain >2 μm in diameter in each field of view was analyzed; 20- to 40-sec counts were generally sufficient to qualitatively characterize most particles.

These direct observations on the polished sections were supplemented by using the SEM-energy dispersive system to study mineral grains isolated from the low-temperature ash prepared from the same coal. Among the more interesting phases observed by Finkelman and Stanton[84] in the low-temperature ash were large (as much as 200 μm) grains of a fluorine-bearing amphibole, tourmaline crystals, and particles of magnetite which they postulate to have authigenic, detrital, and extraterrestrial origins. There is a poor correlation between the minerals observed in the polished sections and those derived from the low-temperature ash. This lack of correlation may be primarily due to the fact that each approach concentrates on different grain-size populations: <10 μm in the polished sections and >30 μm in the low-temperature ash. Particles >30 μm are seldom found in the polished blocks, yet these larger grains are among the most obvious constituents of the low-temperature ash which contains an order of magnitude more mineral matter than is found in the polished blocks.

In the measurements by Finkelman,[83] the sample preparation has been done in the following manner: the polished blocks were prepared by mounting oriented pieces of coal in cold-setting plastic resin. Upon setting, the blocks were ground on 600 grit silicon carbide paper, and then successively polished on 5, 0.3, and 0.05 μm aluminum oxide. The assumptions made by the author in order to relate the occurrence of accessory minerals in the coal to the trace elements data were the following:

1. The polished blocks are representative of the chemically analyzed ash.
2. The mineral matter in the polished blocks covers one third of the surface area of the arbominerite (concentrated bonds of mineral matter).
3. The mean diameter for the accessory minerals is 5 μm.
4. Each accessory grain represents about 2500 ppm of the total mineral matter in each field of view.

In view of these assumptions, the calculated elemental concentrations should be considered as semiquantitative. The calculated value for each element is derived by the following formula:

Table 16
COMPARISON OF CALCULATED AND ANALYTICAL
CONCENTRATIONS FOR SELECTED TRACE ELEMENTS IN
WAYNESBURG COAL, U.S.[83]

Element	Primary accessory mineral	No. of grains (N)	Proportion element in mineral (We)	Calculated value (C)	Analytical value
Ti	"Rutile"	139	0.6	1000 ppm	3400 ppm
Rare earths	Various	31	~0.6	230	~200
Cr	Fe-Cr oxide	18	~0.4	90	80
Zr	Zircon	20	0.5	120	110
Zn	"Sphalerite"	8	0.75	70	70
Cu	"Chalcopyrite"	8	0.33	30	47
Ni	Ni-silicate	5	~0.5	30	26
Pb	Various	3	~0.5	20	31
Ba	Feldspars	6	Variable	~100	600
Sr	"Strontionite"	2	0.7	10	240
Y	"Xenotime"	4	0.5	20	18

$$C = \frac{N \cdot We \cdot Wm}{A} \tag{1}$$

C = calculated weight concentration for the element in ppm
N = total number of the primary accessory mineral grains observed
We = weight percent of the element in the accessory mineral
Wm = weight percent of the total mineral matter per unit area represented by each accessory mineral grain (= 2500 ppm)
A = total number of areas scanned (= 200)

The results of this analysis appear in Table 16. Data for the rare earth elements chromium, yttrium, zirconium, zinc, and nickel show an excellent agreement between the analytical values and the values calculated from the abundance of the accessory minerals. This agreement suggests that, in this coal, these elements are predominantly bound in the specific mineral phases indicated in Table 16.

About one third of the titanium can be accounted for as titanium dioxide. Some titanium is bound in the clays (estimated from energy dispersive X-ray analysis to be 500 to 1000 ppm) as well as in other accessory minerals (sphene, ilmenite, pyroxene). However, even when one takes these estimates into account, a large discrepancy remains between the analytical and calculated values for titanium. This discrepancy may be due to the possibility of titanium being organically bound. A similar discrepancy exists for barium and strontium; both are relatively heavy elements and easily detected in the SEM. There are, however, major differences between their analytical and calculated values (Table 16). Both barium and strontium have low ionic potential and thus, in contrast to titanium, both have low organic affinity. These elements are perhaps present in trace amounts (as much as 0.5% in carbonates, feldspars, and clays, or they may be bound in hydrated phases with low mean atomic numbers that would not easily be detected by the SEM in polished blocks of coal. The author has observed minerals of the crandallite group (Ca, Ba, Sr) $Al_3(PO_4)_2 (OH)_5H_2O$ in the low-temperature ash of other coals.

Gluskoter and Lindhal[73] have investigated heavy mineral concentrates, obtained by separating the low-temperature mineral matter residue in bromoform by means of a

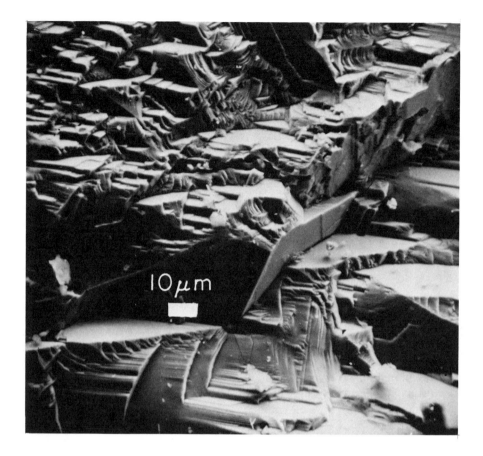

A

FIGURE 5. (A) Scanning electron photomicrograph of sphalente (ZnS) particle. (B) A portion of X-ray energy spectrum (from Gluskoter, H. J. and Lindhal, P. C., *Science,* 181, 264, 1973,[73] with permission).

SEM. Sphalerite (ZrS) was identified in each sample studied, and the identification was confirmed by X-ray diffraction. A scanning electron photomicrograph of a sphalerite particle is shown in Figure 5A; a portion of the X-ray spectrum of sphalerite is shown in Figure 5B. Although this sample had less than one-tenth the Zn content of the three coals with the highest Zn contents, it had the lowest Zn to Cd ratio and therefore gave a clearer indication of the Cd content on the X-ray spectrogram.

The similarity of the Zn to Cd ratios of sphalerite and the Zn to Cd ratios of the whole coal from which the sphalerite was obtained suggests that most, if not all, of the Cd in the coal is within the sphalerite. Because no separate Cd-containing phase was observed, the Cd is thought to substitute for Zn in the sphalerite.[73]

V. X-RAY ANALYSIS

A. X-ray Diffraction

X-ray diffraction analysis of coals and ashes is not a widespread practice. Only a few reports can be found in the literature. As an illustration of the results which one may obtain, we shall discuss in some detail work by Hamrin.[47] In his work, samples

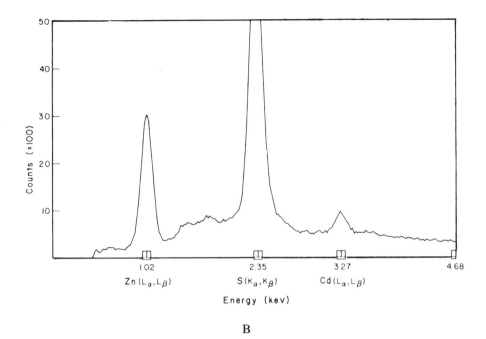

B

of coals and low-temperature ashes (LTA) were sprinkled onto one side of adhesive tape (adhesive on both sides) which was on a glass slide. A spatula was used to spread the sample on the tape. The X-ray results are shown for coals in Figure 6, and it is apparent that even when mineral matter constitutes only 6.93% of the coal identification of illite, kaolinite, quartz, and pyrite was possible. Peaks for the Kentucky #11 coal (top diagram on Figure 6) are much greater, which is to be expected since the mineral matter constitutes 29.2% for this coal (−80 mesh sample). Quartz and pyrites were detected in all coal samples; however kaolinite was not found in the lignite and illite was not detected in the Pittsburgh Seam and Clearfield Mine samples. Szomolnokite, coquimbite, and calcite were also detected. Illite appears in the LTA from the Pittsburgh Seam sample with two weak peaks, but the major peak at 8.8% was not detected in the coal sample, as shown by the X-ray in Figure 7. Calcite was also identified in the LTA but not in the coal diffractogram. Table 17 summarizes the results obtained by Hamrin[47] for X-ray diffraction analysis of coal powders and LTA. The instrument settings were as follows: X-rays Cu K_{ℓ}, Ni filter, 50 kV, 15 mA; 200 cps., time constant 4.0 s, scanning speed 2°/min.

It may be of use here to report major peaks for coal minerals given by

$$2\Theta \text{ (degree)} = 2 \sin^{-1}\left(\frac{7.709}{d(nm)}\right) \tag{2}$$

The major peaks as reported in ASTM file are shown in Table 18.[47]

B. X-Ray Emission Spectroscopy

The use of X-ray emission spectroscopy as an analytical tool has been discussed in many scientific papers, reports, and books, including some by the author of this book (see, for example, References 85-87). The method is widely accepted and is a routine in many laboratories.

X-ray emission spectrometry is a particularly attractive method for analyzing coal

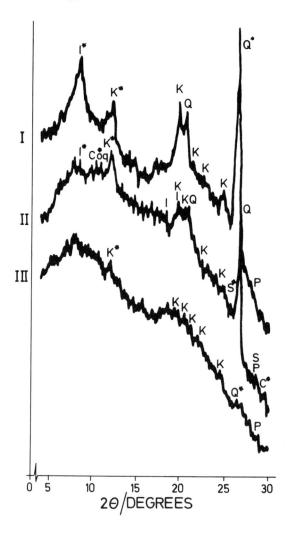

FIGURE 6. X-ray diffractogram of powdered coal samples: I. Kentucky #11; II. Illinois #6; III. Pittsburgh Seam. (After Hamrin[47]).

and coal combustion residues because (a) in principle, little sample preparation is required, (b) a number of elements are determined simultaneously, (c) application is to all elements of atomic number greater than 15, and (d) by the use of internal standards, quantitative information of relatively good accuracy can be acquired.

It is customary to divide X-ray emission spectroscopy according to the means of sample excitation. Characteristic X-rays in the sample material can be excited either by radiation from an X-ray tube, radioactive source, or a particle accelerator. See Figure 8 for the schematic presentation of the principles of X-ray emission spectroscopy. Each method has its own problems and advantages and we shall discuss all three modes as applied to coal analysis.

1. Tube Excited X-ray Emission

Tube excited systems have been used in both wavelength and energy dispersive mode. The wavelength dispersive mode was popular before the development of Si(Li) detectors with good energy resolution. For the sake of completeness we should describe at least one wavelength system used in the trace element analysis of coal.

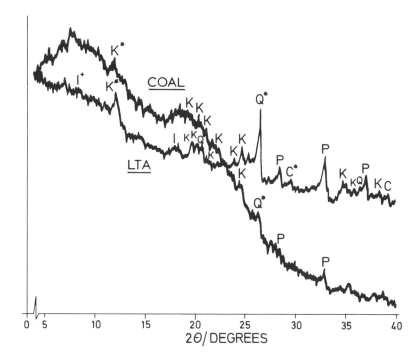

FIGURE 7. X-ray diffractogram of Pittsburgh Seam coal and its ash (after Hamrin[47]).

Kuhn et al.[88-90] have used a wavelength dispersive system to determine the concentration of 21 minor and trace elements in coal. Here is the sample preparation procedure used by these authors. Coal was first ground with a binder and then pressed into a disk. Pellets 2.8 cm in diameter were formed at 40,000 psi (275.790 kPa). For each coal, a low-temperature ash, a high-temperature ash, and the whole coal itself were prepared by the procedure indicated. The spectrometer parameters used by Kuhn et al.[89] are shown in Table 19. The settings are for K_α X-rays for all elements except Pb when L_β line was used. Kuhn et al. have concluded that trace elements determined in this way are limited to those occurring in coals at concentrations of at least a few parts per million. Elements such as selenium, mercury, and antimony, which are generally present in coal at levels below 1 ppm, cannot be determined by this method.

A suite of 24 samples can be analyzed for 21 elements in 3 days by manual instrumentation. Kuhn et al.[89] have done a very careful study of the effect of coal particle size on analytical precision of trace elements.

Nine coals, representing a range of trace element concentrations, were carefully ground to pass screens of various mesh sizes (Table 20). Duplicate 2-g coal samples for each mesh size were weighed and then ground 3 min in a No. 6 Wig-L-Bug® to further reduce particle size. The final grinding eliminated, as nearly as possible, any variation in the pressed coal disks, which were subsequently prepared for analysis. More than 1000 individual determinations were made in this study.

Table 20 gives the combined means of the differences between duplicate trace element determinations for each coal particle size analyzed. Both the means of the absolute differences (in ppm) and the means of the relative differences (absolute difference expressed as a percentage of the concentration) are given. The results show a progressive improvement in precision with decreasing coal particle size.

Table 17
RESULTS OF X-RAY DIFFRACTION ANALYSIS OF SOME COALS AND THEIR ASHES AS REPORTED BY HAMRIN[47]

		Illite	Kaolinite	Quartz	Pyrite	Szomolnokite $FeSO_4 \cdot H_2O$	Coquimbite $Fe_2(SO_4)_3 \cdot 9H_2O$	Calcite	Bassanite $CaSO_4 \cdot H_2O$
Bruceton mine	Coal	VW	S	M	W	ND	ND	ND	ND
	LTA	VW	S	VS	W	ND	ND	ND	M
Pittsburgh seam, W.Va.	Coal	ND	M	W	W	ND	ND	ND	ND
	LTA	NW	S	VS	S	ND	ND	W	ND
Beulah lignite, N.D.	Coal	W	ND	W	VW	ND	ND	ND	ND
	LTA	VW	M	VS	W	ND	ND	?	S
Illinois 6	Coal	VW	M	VS	VW	VW	VW	VW	ND
	LTA	W	S	VS	M	W	S	M	ND
Clearfield mine	Coal	ND	W	M	M	ND	ND	ND	ND
	LTA	W	VS	VS	VS	M	ND	ND	ND
Ky. homestead	Coal	W	W	S	M	W	ND	?	ND
	LTA	W	M	VS	S	S	M	ND	ND
Ky. #11	Coal	S	M	VS	W	ND	ND	ND	ND
	LTA	S	S	VS	M	ND	ND	ND	ND
Elkhorn #1	LTA	S	M	VS	M	M	VW	ND	VW
Ky. #9	LTA	M	S	VS	S	ND	ND	S	ND

Note: Relative diffraction intensity is shown as: VS - very strong, S - strong, M - intermediate, W - weak, VW - very weak or trace, and ND - none detected.

Table 18
MAJOR PEAKS (2θ) FOR COAL MINERALS[142]

Coal minerals	Peaks 1st	Peaks 2nd	Peaks 3rd	ASTM ref. No.
Kaolinite	12.3	62.3	24.9	14—164
Illite	8.8	19.8	26.8	9—343
Montmorillonite	5.9	19.7	17.7	13—135
	6.5	19.9	26.7	13—259
Calcite, $CaCO_3$	29.4	39.3	43.1	5—586
Quartz, SiO_2	26.7	20.9	50.1	5—490
Pyrite, FeS_2	56.5	33.1	37.2	6—710
Szomolnokite, $FeSO_4 \cdot H_2O$	25.9	28.6	35.6	21—925
Coquimbite, $Fe_2(SO_4)_3 \cdot 9H_2O$	10.7	32.4	16.3	6—040
Gypsum, $CaSO_4 \cdot 2H_2O$	11.7	29.2	20.8	6—046
Bassanite, $CaSO_4 \cdot 1/2H_2O$	29.8	14.7	31.9	14—453
Anhydrite, $CaSO_4$	25.5	31.4	38.7	6—226

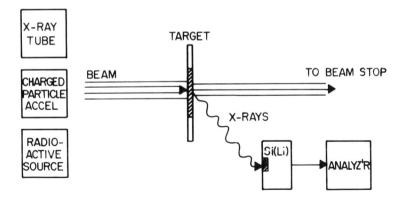

FIGURE 8. Schematic presentation of the principles of X-ray emission spectroscopy.

These data indicate that, for most purposes, acceptable precision can be obtained when −200-mesh coal samples are used. Further improvement is achieved by grinding the samples to −325 mesh, but this is unnecessary except for analyses that are to be used as standard values or for other special purposes.[89]

This work was continued and further reports in the literature are by Ruch et al.[59-62] and Gluskoter.[10] In the report by Gluskoter,[10] determinations were made on whole coal for As, Br, Pb, Zn, Cu, Ni, P, Cl, S, V, Mg, Ca, Fe, Ti, Al, and Si.

A 3-kW chromium X-ray tube was used, and the procedures were as described by Ruch et al.[59-62] The only change in the procedure has been the use of a new diffracting crystal with better sensitivity for the elements determined. A T1AP crystal replaced the EDDT crystal for the determination of elements in the periodic table from Na throught Si. The observed relative standard deviations for elements determined by this technique ranged from 0.35 to 8.4%.

Table 19
SPECTROMETER PARAMETERS USED BY KUHN ET AL.[89] FOR THE ANALYSIS OF COAL SAMPLES

Element	2Θ Angle	Background 2Θ	Crystal	X-ray tube	PHA volts Base	PHA volts Window
Si	108.01	111.01	EDDT	Cr	7	17
Al	142.42	145.95	EDDT	Cr	5	17
Ti	86.12	89.12	LiF	Cr[a]	5	18
Fe	57.51	60.51	LiF	Cr[a]	5	25
Ca	44.85	47.95	EDDT	Cr	14	30
K	50.32	53.90	EDDT	Cr	14	21
Mg	136.69	139.69	ADP	Cr	4	8
V	76.93	80.93	LiF	Cr[a]	5	16
S	75.24	78.38	EDDT	Cr	12	18
Cl	64.94	67.94	EDDT	Cr	11	19
P	110.99	113.99	Ge	Cr	9	15
Ni	48.66	50.36	LiF	Cr[a]	10	27
Cu	45.02	49.67	LiF	Cr[a]	11	28
Zn	41.79	44.25	LiF	Cr[a]	10	22
Pb($L_{\beta 1}$)	28.24	31.24	LiF	Cr[a]	22	28
Br	29.97	35.12	LiF	Cr[a]	25	23
As	34.00	37.00	LiF	Cr[a]	24	23
Co	52.79	53.79	LiF	W	13	16
Mn	62.97	63.97	LiF	W	8	12
Mo	20.33	19.83, 20.83	LiF	W	36	40
Cr	69.35	68.53	LiF	W	7	15

[a] A tungsten tube is used on these elements when already in place.

Table 20
MEAN ERROR FOR ALL ELEMENTS AT VARIOUS COAL PARTICLE SIZES[89]

Mesh size (M)	ppm	Error of mean element conc. (%)
−60	±3.05	8.47
−100	±2.11	6.38
−200	±1.26	4.28
−325	±1.12	2.62
−400	±1.02	1.56
<400	±0.93	1.40

Many reports can be found in the literature about the use of energy dispersive X-ray emission spectroscopy (X-ray fluorescence) for coal analysis. The basic system contains: X-ray power supply, variety of X-ray tubes, Si(Li) detector, fast electronics, and a unit for spectra analysis (multichannel analyzer and/or computer). There are several commercially available systems on the market. Here we shall describe an EG & G ORTEC Energy Dispersive X-Ray Fluorescence Analyzer (TEFA®) and its applications to coal analysis. Figure 9 shows all the components of the system.

The sample preparation for the analysis by TEFA (ORTEC) system includes following steps.[91]

FIGURE 9. Photograph of TEFA system (courtesy of ORTEC, Oak Ridge, Tenn., with permission).

Each sample was received as a homogenized material with a particle size of 100% -100 mesh. Prior to any further sample preparation, each material was dried for 24 hr at 105°C and allowed to cool in a vacuum desiccator.

Following the drying and cooling, each sample was prepared by grinding 4 g of sample, 1 g of boric acid (binder), and 100 mg of sodium stearate (grinding aid) for 6 min in a tungsten carbide rotary swing mill. Following grinding, each sample was pelletized with a boric acid backing for mechanical support in a 1.25 in. die at 15 t/in.^{2}

NBS standard coal samples were exposed to X-ray radiation and emitted characteristic X-ray spectra measured. Logarithmic plots of the X-ray spectra of the elements contained in NBS standard coal as excited under different excitation conditions are illustrated in Figures 10 and 11. Figure 10 is the 0 to 20 keV portion of the X-ray spectrum as excited with Mo filtered radiation from a Mo anode and illustrates the simultaneous measurement of elements from Al to Sr. Greater sensitivities can be obtained for elements with X-rays less than 10 keV with Cu filtered radiation from a W anode operated at 20 keV as shown in Figure 11. By a proper choice of anode material, filter material, and operating voltage the detection sensitivity can be optimized for the elements of interest. The minimum detectable concentration (MDC) of whole coal and coal ash at an analysis time of 100 sec is illustrated in Figure 12.[81] Examination of this figure reveals the trace elements of coal and coal ash can be determined in the approximate range of 1 to 5 ppm at a counting time of 100 sec. Figure 12 can also help analysts to choose the optimum instrumental parameter. The optimum instrumental parameters as determined by Wheeler and Jacobus[91] are shown in Table 21.

Particle size-intensity relationship has been well documented in the literature.[92-96] Wheeler and Jacobus[91] have established a minimum grinding time in a tungsten carbide rotary swing mill for coal and coal ashes. This should be done carefully in order to avoid long grinding time, since after some point no further reduction in particle size any longer influences the fluorescence intensity.

A similar type of study was done for determination of the optimum pelletizing pressure. Examination of data reported by Wheeler and Jacobus[91] reveals that for coal and coal ashes the minimum pelletizing pressure in order to ensure stable intensities as a function of pressure is 15 t/in.^{2}

The X-ray spectra obtained by utilizing the optimum instrumental parameters listed in Table 21 are shown in Figure 13 and Figure 14.

210 *Trace Elements in Coal*

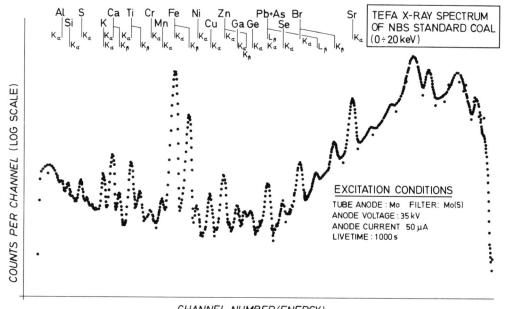

FIGURE 10. TEFA X-ray spectrum of NBS standard coal (0 to 20 keV region). Excitation conditions are tube anode; Mo, filter: Mo; anode voltage: 35 kV; anode current: 50 µA; lifetime: 1000 sec (courtesy of ORTEC, Oak Ridge, Tenn., with permission).

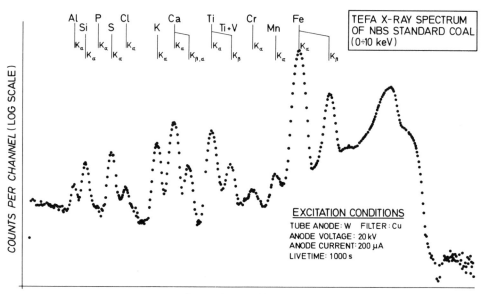

FIGURE 11. TEFA X-ray spectrum of NBS standard coal (0 to 10 keV region). Excitation conditions are tube anode: W; filter: Cu; anode voltage: 20 kV; anode current: 200 µA; lifetime: 1000 sec (courtesy of ORTEC, Oak Ridge, Tenn., with permission).

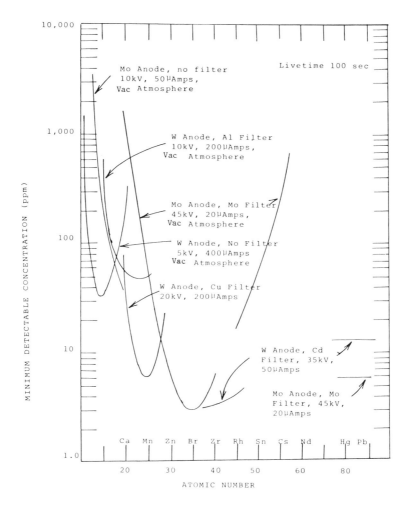

FIGURE 12. Minimum detectable concentrations in coal using TEFA system (courtesy of ORTEC, Oak Ridge, Tenn., with permission).

Table 21
OPTIMAL INSTRUMENTAL PARAMETERS FOR AN EG&G ORTEC XRF SYSTEM[91]

Elements determined	Si, P, S, K, Ca, Ti, Fe	Na, Mg, Al, Cl	Ba, Sr	V, Cr, Mn	Co, Ni, Cu, Zn, Pb, Se, Br, As, Th, Rb, U	Ag, Cd
Material	Coal-ash	Coal-ash	Coal	Coal	Coal	Coal
Anode	Rh	W	Rh	W	Rh	W
Filter	None	None	Sn	Cu	In	Mo
Anode voltage	10 kV	6 kV	50 kV	25 kV	50 kV	50 kV
Anode current	200 µA	200 µA	200 µA	140 µA	175 µA	200 µA
Energy scale	0—10keV	0—10keV	0—40keV	0—10keV	0—20keV	0—40keV
Atmosphere	Vac	Vac	Vac	Vac	Vac	Vac
Counting time	200 sec	200 sec	200 sec	200 sec	200 sec	200 sec

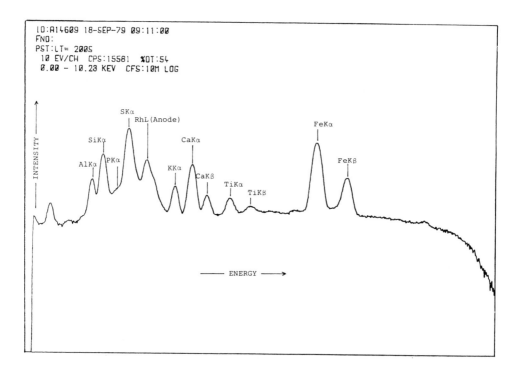

FIGURE 13.   X-ray spectrum (0 to 10.23 keV region) from coal sample (TEFA® — ORTEC).

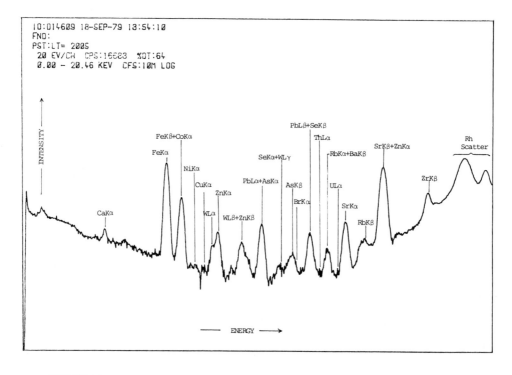

FIGURE 14.   X-ray spectrum (0 to 20.46 keV region) from coal sample (TEFA® — ORTEC).

Figure 13 illustrates a logarithmic plot of the 0 to 10.2 keV X-ray energy region which was excited with direct bremsstrahlung radiation from a rhodium anode which maximizes the sensitivity for the elements of silicon through iron; and, thereby, provides a simultaneous analysis of these elements.

Figure 14 shows a 0 to 20.46 keV X-ray energy region with peaks in the spectra indicating the presence of elements in sample up to Zr.

In order to obtain quantitative values for the concentration of elements present in the coal and coal ashes, the spectral data obtained by TEFA system are analyzed utilizing an exponential software routine[97,98] which provides a linear least squares fit to the interelement corrected intensities. The program corrects the observed X-ray intensities for absorption and/or enhancement effects due to the presence of other elements in the system. The concentration of the ith element is given by the following equation:

$$C_i = a + bI_i [\exp(-M_{ji} I_j)] \tag{3}$$

where: M_{ji} = interaction coefficient for element j on element i
   I_j = intensity of the jth (interferring) element

The interaction coefficients are determined by a nonlinear multiple least squares fit of the standards concentration/intensity data. This requires a minimum of n + 6 standards where n is the number of interfering elements. Elemental concentrations were calculated with an iterative process using Equation 3, plus a spectral deconvolution Equation 4, in order to correct for overlapping lines. Spectral deconvolution equation reads:

$$C_i = a + b [I_i K_\alpha + I_j K_\beta - (I_j K_\beta / I_j K_\alpha)(I_j K_{\alpha s})] \tag{4}$$

where: C_i = concentration of element i
   a = X intercept of calibration curve
   b = slope of calibration curve
   $I_i K\alpha$ = intensity of the Kα of element i in the sample
   $I_j K\beta$ = intensity of the Kβ of element j in the sample
   $I_j K\beta / I_j K\alpha$ = intensity ratio of Kβ/Kα on pure element j
   $I_j K\alpha_s$ = intensity of the Kα of element j in the sample

The results of the analysis of coal ash obtained in such a way are summarized in Table 22.[91] Though the elemental compositions are expressed as oxides, this does not imply their presence as such in the whole coal. In conclusion, the major ash-forming constituents of whole coal and ultimately potential ash content can be performed by energy dispersive X-ray fluorescence in approximately 200 sec. Complete quantitative analysis of coal and ash would consume 800 sec. Accuracies approaching 1/10 of 1% absolute can be achieved in the analysis of the major elements, but require interelement correction either for absorption/enhancement effects or spectral overlaps. Two hundred second minimum detectable concentrations range from approximately 1 ppm for elements near arsenic to about 1/10 of 1% for sodium.

The application of tube excited X-ray system for the analysis of coal has been published by many authors including Javerbaum,[17] Spacek,[99] Lloyd and Francis,[100] Cooper et al.,[101-105] De Kalb and Fassel,[106] Prather et al.[107] and many others.

As an example of adjusting the existing roentgen apparatus in the laboratory for the X-ray fluorescence (XRF) work, let us mention a report by Orlić et al.[108] Usual

Table 22
ANALYSIS OF COAL ASH[91]

Sample I.D.	Na$_2$O List	Na$_2$O Calc	MgO List	MgO Calc	Al$_2$O$_3$ List	Al$_2$O$_3$ Calc	SiO$_2$ List	SiO$_2$ Calc	P$_2$O$_5$ List	P$_2$O$_5$ Calc	SO$_3$ List	SO$_3$ Calc	K$_2$O List	K$_2$O Calc	CaO List	CaO Calc	TiO$_2$ List	TiO$_2$ Calc	Fe$_2$O$_3$ List	Fe$_2$O$_3$ Calc	Total List	Total Calc
D13	1.45	1.59	2.01	1.82	21.73	21.90	51.64	51.50	0.22	0.29	8.10	7.89	0.25	0.25	7.80	7.53	1.28	1.27	4.95	4.76	99.43	98.80
D14	2.20	2.34	1.23	1.18	29.04	28.82	49.77	49.50	0.20	0.23	5.10	5.65	0.28	0.30	5.34	5.09	1.06	1.13	4.51	4.49	98.73	98.73
D15	2.43	2.11	1.05	1.00	29.58	29.83	51.94	51.43	0.25	0.23	4.35	5.33	0.23	0.19	4.36	4.29	1.28	1.31	3.69	3.60	99.16	99.32
D16	2.85	3.28	1.72	1.81	14.76	15.15	47.37	47.17	0.28	0.37	11.13	11.10	0.19	0.12	13.96	13.67	1.08	0.97	6.64	6.53	99.98	100.17
D22	3.63	3.66	2.09	1.98	21.32	21.25	48.73	48.82	0.54	0.40	7.63	7.84	0.27	0.27	7.80	7.63	1.16	1.19	6.59	6.42	99.76	99.56
D23	3.77	3.71	2.08	2.14	16.61	16.56	43.76	43.70	0.40	0.46	11.50	11.17	0.17	0.16	14.28	14.52	1.04	1.05	5.84	5.90	99.45	99.37
D24	4.21	4.06	1.49	1.37	19.22	19.33	52.04	52.05	0.36	0.28	7.28	7.37	0.28	0.36	6.24	6.32	1.07	1.21	7.50	7.44	99.69	99.79
D25	2.55	2.63	1.05	1.08	27.80	27.73	50.85	51.09	0.40	0.28	4.95	5.04	0.21	0.24	4.01	4.17	1.41	1.42	6.72	6.64	99.95	100.32
D27	0.63	0.53	1.27	1.36	24.51	25.07	56.07	56.50	0.26	0.22	5.63	4.89	0.44	0.44	4.04	4.13	1.29	1.28	5.96	6.10	100.10	100.32
D28	2.18	2.21	1.39	1.51	17.83	18.05	53.82	53.26	0.28	0.28	8.25	8.37	0.25	0.27	8.29	8.55	1.31	1.23	6.48	6.43	100.08	100.16
D32	0.70	0.70	1.43	1.48	24.19	24.38	54.32	54.71	0.25	0.25	6.63	6.31	0.19	0.20	6.29	6.28	1.18	1.22	4.34	4.45	99.52	99.93
G.B.	(0.49)[b]	0.46	(2.25)	2.31[a]	(17.70)	17.63	(46.07)	46.10	(0.39)	0.42	(11.71)	11.84	(1.58)	1.69[a]	(10.51)	10.60	(0.71)[a]	0.73	(8.49)[a]	8.45	99.90	100.23
Avg. err.	0.19		0.08		0.36		0.37		0.06		0.42		0.03		0.15		0.06		0.10			
Std. dev.	0.28		0.11		0.52		0.55		0.07		0.63		0.05		0.20		0.07		0.14			
Stat. err.	0.10		0.03		0.05		0.04		0.002		0.01		0.002		0.01		0.003		0.01			

[a] Beyond limits of Standards.
[b] Value not used in calibration - analyzed as unknown.

working conditions for Philips roentgen apparatus Model PW 10/30 with Mo anode were 26 kV, 12 mA with X-ray beam passing through Zr, Ti, and Mo filters to reduce its intensity. The Si(Li) X-ray detector used has an energy resolution of 180 eV (at E_x = 6.4 keV).

For accurate analysis it was necessary to have constant photon flux, which was determined by the degree of voltage and current stability. To prove source stability, the authors have inserted very thin gold wire in collimator system and all of the spectra obtained in a testing period of 2 weeks had Au characteristic X-ray lines whose intensity had to be proportional to X-ray tube flux (if the counting rate is relatively low). The results of this experiment showed that the intensity was constant in the range of counting statistics error. From time to time a standard sample was also analyzed in order to check flux and geometry consistency.

Some authors used very thin targets (<5 mg/cm²) in order to avoid complications with matrix effects. Such an effort in sulfur analysis is described by De Kalb and Fassel.[106] The X-ray fluorescence determination of sulfur and other constituents in coal ordinarily requires correction for a variety of interelement effects, which can be troublesome and time-consuming. For mineral specimens with 6 to major constituent elements, a large number of well-characterized samples or synthetic standards are required to establish appropriate correction factors when bulk samples are analyzed. However, when the sample is prepared as a thin-film specimen, interelement corrections can be neglected, and only a few samples which have been analyzed previously for the elements of interest are needed for calibration. The adaptation of such a method of sample preparation to coal samples is described and illustrated in detail. Precision and accuracy data for the determination of sulfur in thin-film coal samples are presented in the paper by De Kalb and Fassel.[106]

Orlić et al.[108] have prepared thin coal samples on Formvar® foils. The Formvar foils have been prepared by dropping a small quantity of the solution (solvent was 1,2-dichlorethane) on the surface of redistilled water, where it spreads to form a film which can be picked up on an aluminum frame. Coal previously ground to powder (with particles smaller than 10 μm) was put on the Formvar® foil and spread forming an approximately 1 mg/cm² thick layer. Another Formvar foil was put over to prevent target wasting.

The intermediate thick targets (∼150 mg/cm²) were also prepared in forms of pellets with diameter of 3 cm and 2 mm thick. The disadvantage with thin targets, i.e., small masses (∼1 mg), is in inhomogeneity of the targets, which resulted in low but significant discrepancies (up to 10%) in relative intensities of some peaks obtained by using different thin targets of the same sample.

With pellets as targets this is not the case. However, such intermediate thick targets required consideration of the X-ray absorption in the matrix. This can be accomplished in two different ways: by mathematical corrections of integrated line intensities or by using a new efficiency curve obtained for this type of targets.

Figure 15 and Figure 16 show typical spectra obtained with Mo-tube from coal and coal ash samples. Elemental concentrations were deduced from the measured intensities of characteristic X-ray lines by the computer program. Efficiency curves for this system (relative to Y) for thin as well as for intermediate thick targets were obtained by interpolation between experimentally determined values for efficiencies of following elements: S, Ca, Cr, Fe, Zn, As, and Sr (Figure 17). For each of the mentioned elements five coal targets doped with different element concentrations were prepared and analyzed. Obtained calibration curves were linear (see Figure 18) and enabled one to determine the experimental efficiencies (cycles per second per percent) for given elements.

216 *Trace Elements in Coal*

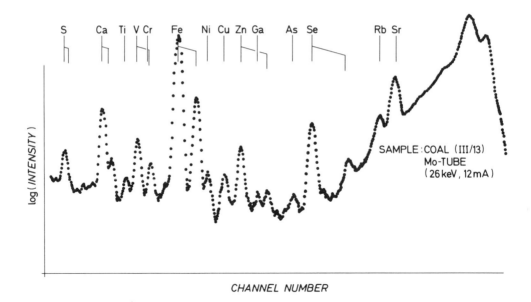

FIGURE 15. Mo-tube excited X-ray spectrum of coal.

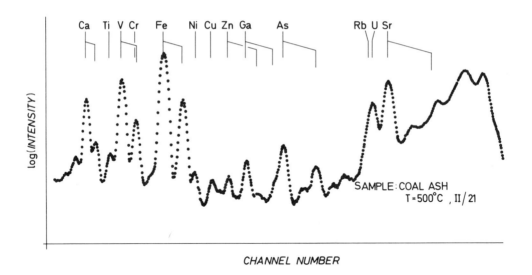

FIGURE 16. Mo-tube excited X-ray spectrum of coal ash.

To avoid influence of various matrices, different coal material was used in preparing the targets for each element. Interpolation between these obtained efficiencies was satisfactory, which means that variations in coal matrices do not influence the results significantly. Such an efficiency curve has an advantage of rapid and relatively accurate determination of concentrations (± 15%) of all elements whose characteristic lines are visible on the spectrum.

However, to make the analysis more accurate the following effects have to be taken into account:

1. Mass absorption coefficient of the sample material (μ) is greatly affected by con-

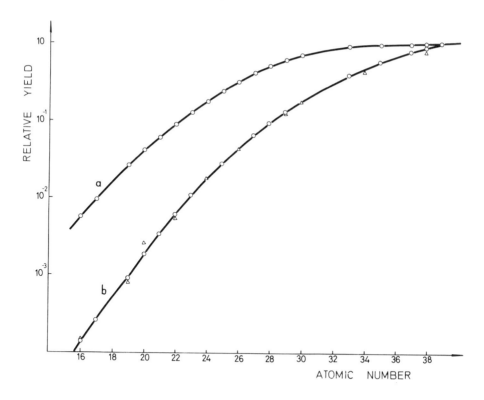

FIGURE 17. K_α peak intensity yield relative to yttrium, both for thin targets (a), and intermediate thick coal targets (b). The triangle symbols present efficiencies obtained by computer program.[108]

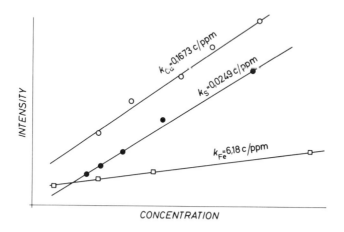

FIGURE 18. Intensity of some K_α lines as a function of element concentration.[108]

centration of heavier elements visible in the spectrum (which vary from sample to sample) and
2. Enhancement effects.

To eliminate absorption effects, a computer program based on the following equation was developed:

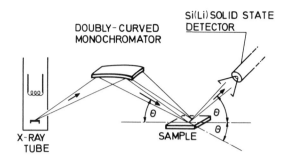

FIGURE 19. Schematics of the system with doubly curved monochromator (after Sparks et al.[111]).

$$T_i = \frac{I_{mi}}{I_{oi}} = \frac{1 - \exp[-m(\mu(E_o) + \mu(E_i))]}{\mu(E_o) + \mu(E_i)} \quad (5)$$

where T_i is the transmission factor relative to iron for i-th element and represents the ratio between measured intensity I_{mi} for i-th element and its unattenuated value I_{oi}; m is mass per unit area of the specimen (g/cm²); and $\mu(E_o)$, $\mu(E_i)$ are mass absorption coefficients of the sample material for the primary radiation and for fluorescent radiation of the i-th element, respectively (cm²/g).

To check the program, an "artificial" coal sample was prepared, measured, and analyzed. The agreement between calculated and "true" values was found to be satisfactory.

The sensitivity of the tube-excited system can be improved by using a monoenergetic X-ray beam to excite the sample. There are several ways of obtaining monoenergetic radiation from an X-ray tube: filters, secondary radiators, and monochromators. The use of a doubly curved monochromator placed between sample and detector is described by Lyon.[109,110] Figure 19 shows schematically the system designed and used by Sparks and co-workers.[111] Samples are usually mixed with a binder (stearic acid) and pressed into pellets. These are then placed in the sample position.

Quantitative results are obtained by identifying each observed peak and measuring its intensity (integrated area) relative to the intensity of a pure element standard. Concentrations in the sample are then calculated by means of a correction equation that uses fundamental constants to take into account absorption effects of the matrix. If the intensity of fluorescence from a given element in a sample is denoted I_s and that from the corresponding pure-element standard by I_p, the concentration is given as follows:

$$C = \frac{I_s}{I_p} \frac{A_s}{A_p} [1 - \exp(-A_s \rho T/\sin\theta)] \quad (6)$$

where: ρT = Product of density of the sample and its thickness,
θ = Angle of incidence for excitation radiation and angle of takeoff for fluorescence radiation,
A_s = Mass absorption coefficient of sample for excitation radiation plus mass absorption coefficient of sample for fluorescence radiation,
A_p = Mass absorption coefficient of pure element standard for excitation radiation plus mass absorption coefficient of standard for fluorescence radiation.

Table 23
VALUES OBTAINED FOR NBS COAL (SRM 1632) AND
FLY ASH (SRM 1633) BY LYON[109] USING
MONOENERGETIC EXCITATION IN XRF

Element	NBS coal SRM 1632	XRF x ± σ	NBS fly ash SRM 1633	XRF X ± σ
V			214 ± 8	
Mn			495 ± 30	
Fe	8600	7050 ± 470		46.460 ± 4886
Ni	15 ± 1	14.5 ± 1.3	98 ± 3	109 ± 17.2
Cu	18 ± 2	18 ± 3.0	(120)	133 ± 7.4
Zn	37 ± 4	34 ± 3.6	210 ± 20	208 ± 11.7
Ga		8.5 ± 1.3		49 ± 6.0
As	(6)	5 ± 2.8	61 ± 3	54 ± 7.6
Se	(3)	5 ± 1.0	9 ± 1	15 ± 1.5
Br		21 ± 2.6		
Pb	30 ± 9	17 ± 6.7	70 ± 2	87 ± 14.5
Rb		24 ± 3.0	(112)	120 ± 8.3
Sr		144 ± 7.4	(1380)	1301 ± 99.3
Y				80 ± 13.2

Sparks et al.[112] derived this equation in a more extended format in which they showed that other constants such as the cross section for exciting fluorescence, fluorescence yield, intensity distribution factor, detector efficiency, and power output from the radiation source are involved. They also showed that constants A_p can be interpolated from literature tables. I_p is measured experimentally for the particular spectrometer being used in the analysis: it can be regarded as a calibration factor. The A_s factor, the mass absorption coefficient sum for the sample, must be either calculated or measured. It can be obtained in a direct manner by measuring the fluorescent intensity from a pure element before and after covering it with the sample. A number of NBS standards were subjected to XRF analysis, and the results are shown in Table 23. Agreement appears satisfactory, with the exception of Fe and Pb in coal.

2. Radioactive Source Excitation

Almost everything that was said for tube-excited systems can be applied to the radioactive source excitation. The only difference is that the excitation of sample is done by one or more radioactive sources instead of X-ray tubes. α, β, γ, and X-ray emitting radioisotopes have been used as sources to excite characteristic X-rays in sample. The most widely used method is by primary or secondary X-rays.

A good source should give a simple line spectrum at an appropriate energy, no high energy β or γ radiation, and have long enough half-life and high enough specific gravity. Some of the most frequently used radioactive sources are ^{55}Fe ^{57}Co, ^{109}Cd, and ^{241}Am.

Probably the most convenient source for the analysis of coal and coal ashes is ^{109}Cd ($t_{1/2}$ = 1.3 years, E = 22.2 keV and 88.0 keV). For lighter elements one should use ^{55}Fe ($t_{1/2}$ = 2.7 years, E = 5.9 keV); while for heavy ones (up to Th and U), the best results are obtained with ^{57}Co ($t_{1/2}$ = 0.7 years, E = 122.1 keV, and 136 keV).

The use of ^{109}Cd is illustrated in Figure 20, in which the comparison of the X-ray spectra of coal and fly ash is shown. Presence of a number of elements can be seen in the spectra accumulated for 100 min. With a source activity of approximately 3.7 × 10^7 Bq, one can get a spectrum with good statistics in about 1000 sec using samples prepared as pellets with diameter of 3 cm and 2 mm thick.

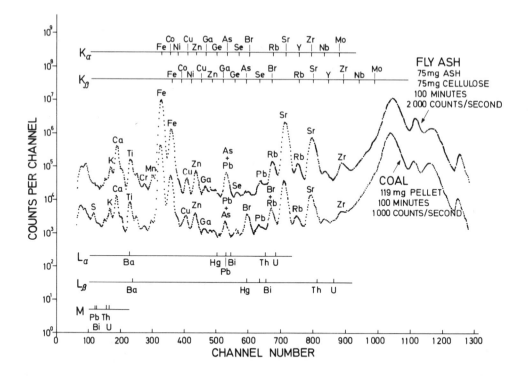

FIGURE 20. Comparison of the X-ray spectra of coal and fly ash as excited with a ^{109}Cd radioactive source. The location of other X-ray lines not observed is also illustrated at the top and bottom of the figure. (From ORTEC Report, 1980[174]).

A very interesting system has been developed by Miklavžič et al.[113] for the study of heavy elements (especially Th and U) in coal and coal ashes. It is based on the detection of heavy element $K_{\alpha 1, \alpha 2}$ lines excited by ^{57}Co source and it is the optimum for uranium concentration detection. The sample mass is deduced from the intensity of Compton scattering, and therefore uranium concentration in the sample can be determined with only measuring ratios of two peak intensities. To illustrate the kind of spectra obtained, Figure 21(a) shows the X-ray spectrum of lignite sample containing 120 ppm of uranium. Figure 21(b) is a blow-up of the region containing peaks of interest for uranium concentration determination. The same type of measurements can be done with fly ash, and the part of the X-ray spectrum containing uranium $K_{\alpha 1, \alpha 2}$ lines looks very much the same. The spectrum of Figure 22 shows X-ray spectrum obtained from an fly ash sample containing 180 ppm of uranium. This system has been used routinely at the Institute "Jozef Stefan", Ljubljana, Yugoslavia, and it was found satisfactory for uranium determination in the geological samples too.

3. Proton Induced X-ray Emission (PIXE) Spectroscopy

Proton-induced X-ray emission (PIXE) spectroscopy has been discussed by the author of this book in many papers and some of his previous books.[85-87] Therefore we shall not go into a great many details here. PIXE has not been extensively used in coal analysis for several reasons, one being its availability to laboratories which have access to proton accelerators (usually Van de Graft accelerators or cyclotrons). The trace element analysis by PIXE (proton induced X-ray emission) requires preparation of thin samples. The usefulness of this method depends also on the method of target preparation. Careful choice of backing materials is very important, since the back-

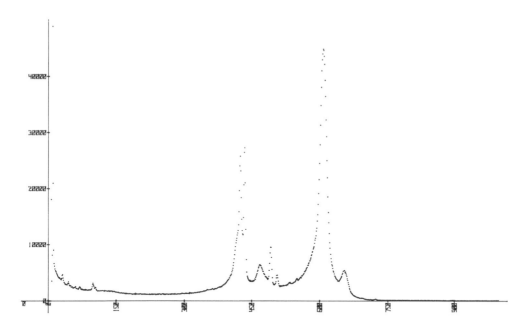

FIGURE 21A. X-ray spectrum of lignite sample containing 120 ppm of uranium excited by ^{57}Co radioactive source.

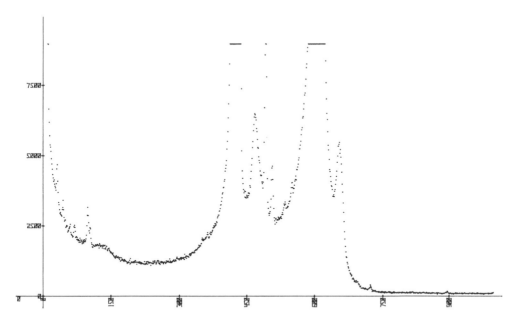

FIGURE 21B. Blow-up of the region containing peaks of interest for uranium concentration determination.

ground highly affects the accuracy of the measurement and the detection limit. The ideal backing is a thin film, composed of elements of low atomic number, and possessing the properties of high mechanical and chemical strength. Measurements were done with different materials (Kapton,® Mylar,® Formvar® and carbon); Formvar® foils were chosen as a backing material by most researchers. Blank Formvar® foils ensure

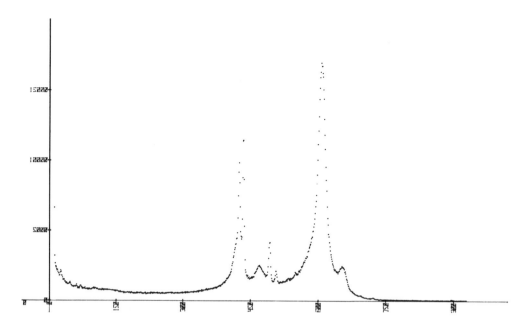

FIGURE 22. X-ray spectrum of fly ash containing 180 ppm of uranium excited by ^{57}Co radioactive source.

low background level intensity and have no additional peaks superimposed on background which means that there are no impurities such as Fe, Zn, Cl, etc. in significant quantities.

For thin targets, major constituent for background is due to bremsstrahlung from secondary electrons produced in sample and backing material, and from electrical charging of the target. To reject target charging, several thin backings were prepared by vacuum evaporation of aluminum on Formvar® foils.[108] Such backings are more fragile than pure Formvar foil and breakage rates, both during specimen preparation and irradiation, are much higher. Hence, this method was abandoned by the authors[108] and they used a specially constructed electron gun which proved to be satisfactory. Another advantage of using thin targets was an expected linear relationship between peak area and amount of corresponding element in sample. For example, such as coal, whose matrix is made mostly of light elements, as C, H, O, N, with heavier elements being present only in small concentrations, matrix absorption of characteristic x-line is negligible for sample thicknesses smaller than 5 mg/cm^2.

Let us mention some of the work done by PIXE on coal analysis.

An early PIXE system developed at Rice University[114] has been used also for coal analysis. The system was based on sample bombardment with 3 MeV protons from the Van de Graaff machine. Using standards of known composition, an efficiency prefabricated curve for the whole system was determined. It is shown in Figure 23. Using only data from these measurements, any unknown spectrum can be analyzed for the ratios of elemental concentrations. Figure 24 shows an X-ray spectrum obtained by irradiation of coal target and some of the concentration ratios deduced (for details see Valković et al.).[114]

In a paper by Simms et al.,[115] analytical techniques are described in which protons from an accelerator are used to provide quantitative determination of most elements from Li to Pu. The sample is prepared as a thin film and bombarded with protons to produce monoenergetic X-rays and γ-rays from all elements in the sample. Energy sensitive Si-Li and Ge-Li semiconductor devices are used to detect X-rays and γ-rays,

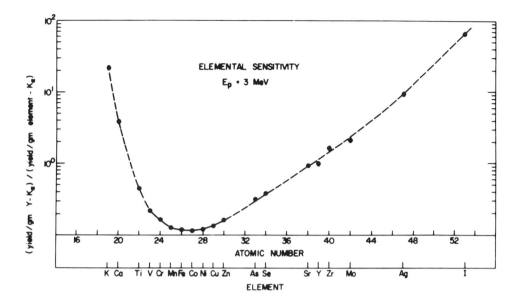

FIGURE 23. Efficiency of PIXE system described by Valkovič et al.[114] for proton energy of 3 MeV.

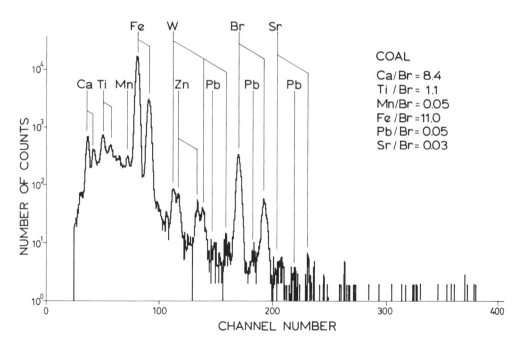

FIGURE 24. X-ray spectrum obtained by 3 MeV proton bombardment of coal target.

respectively. The energy of the radiation identifies the element and the intensity determines the concentration of the element in the sample. A PDP 15/45 on-line computer is used to process data from one sample while data is being accumulated from the next sample. The analysis results are available from the computer at approximately the same time that new data accumulation is finished. Protons have an advantage over X-rays excitation because good sensitivity can be obtained for a much larger group of elements in a single measurement. In addition, the light elements Li, B, F, Na, Mg, Al, and Si can be detected by γ-ray emission following nuclear excitation. Results of elemental analysis of coal (standard reference material) obtained by the technique described are given. It is concluded that proton induced x-ray emission is a powerful technique for multielemental analysis of a wide variety of coal samples. Very light elements can be observed by detecting γ-rays and X-rays in the same experiment.

In the work by Cronch et al.,[116] proton-induced X-ray emission has been used to determine elemental concentrations in solid coal samples. The coal samples were irradiated with 2.5 to 5.5 MeV protons. Concentrations were determined from characteristic X-ray yields taking into account matrix absorption. The precision is shown by replicate analysis and the accuracy by comparison with results obtained by other laboratories using different techniques.

In order to increase sensitivity for lighter elements, the beam energy used in experiment by Orlić et al.[108] was 1.7 MeV. Beam currents were approximately 30 nA or less (beam density is ~100 nA/cm²) to avoid a pile-up effect and sample burning. An intrinsic Ge X-ray detector with a 140 eV resolution (for E = 5.9 keV) was usually positioned at 90° to the incident beam direction outside the vacuum system. X-rays passed through a 100 μm Kapton® window in the chamber and through 1 cm of air before entering the detector through a 25 μm Be window.

The counting rate was kept below 500 cps. The beam current was measured by Faraday cup, far enough to prevent the background X-rays produced in the cup from reaching the X-ray detector. More informaton about this particle-induced X-ray system can be found in the paper by Budnar et al.[117]

Figure 25 and Figure 26 show typical X-ray spectra resulting from the bombardment of coal sample with proton beam. The thickness of the Kapton® adsorber was used to regulate which part of the energy spectra was to be enhanced. Spectrum on Figure 25 was obtained by using 250-μm thick Kapton® adsorber, which very effectively reduced the X-ray from Ca, S, and other light elements. When this absorber was replaced with a 100-μm thick one it was possible to obtain spectra dominated by peaks due to S and Ca characteristic X-rays. Elemental concentrations were deduced from measured intensities of characteristic X-ray lines by the computer program. An efficiency curve for PIXE system was measured for 14 elements from Z = 15 (P) to Z = 39 (Y). Relative yields normalized to iron were determined using thin multielement standards. They were prepared by dropping 10 μℓ of standard solution on Formvar foil. Obtained relative efficiency curve is shown on Figure 27. It represents element concentrations relative to iron. Iron concentrations were determined in an independent measurement by a tube excitation system. Iron could be used for relative measurements because of its relatively high concentrations in all samples (~1%) and its uniform distribution in coal materials studied.

Minimum detection limit (MDL) defined as $N = 3\sqrt{N_B}$, (where N is number of counts in a peak corresponding to a characteristic X-ray line and N_B is the number of counts due to background) was calculated for all the elements from the observed intensities of K_α X-ray lines. This is shown as curve b on Figure 28. The same figure shows (curve a) a minimum detection limit obtained by the same authors from a Mo-tube excited system. It is evident that combination of these two excitation modes can give results needed in coal and coal ash analysis. Comparison of the results obtained by three modes of sample excitation is shown in Table 24.[108] It can be concluded that all three excitation modes produce very similar results. However, there is a remarkable difference in the sensitivities for particular elements. The proton beam (E = 1.5 MeV) excitation mode shows the higher sensitivity for low Z elements (Z < 26). However,

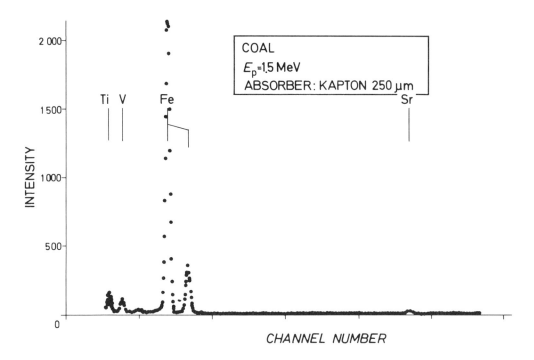

FIGURE 25. X-ray spectrum obtained by bombardment of coal sample with 1.5 MeV protons; 250 μm Kapton absorber was placed between detector and the sample.

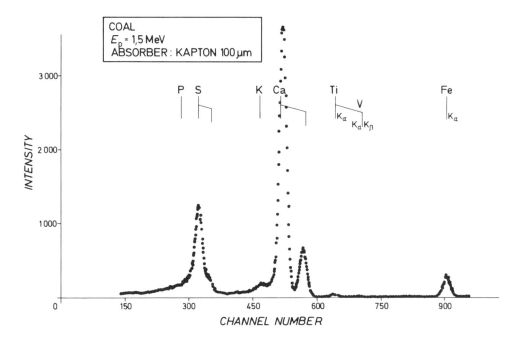

FIGURE 26. X-ray spectrum obtained by bombardment of coal sample with 1.5 MeV protons. A 100 μm thick absorber was placed between detector and sample.

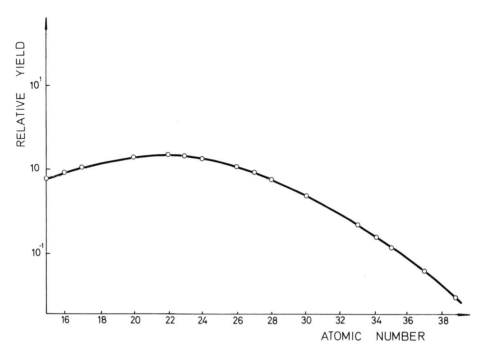

FIGURE 27. Relative yields of K X-rays for different elements when bombarded with 1.5 MeV protons.

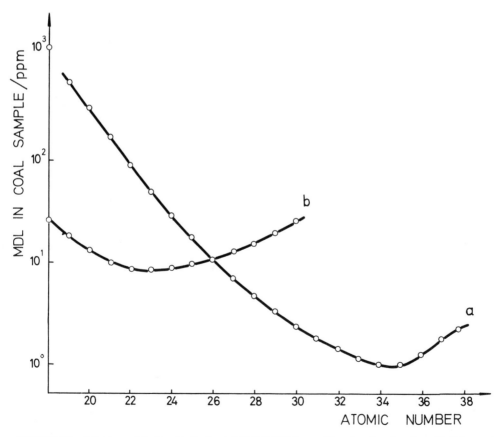

FIGURE 28. Minimum detection limits for 1.5 MeV PIXE (curve b) and system based on Mo-tube excitation (curve a). For details, see text (after Orlić et al.[108]).

Table 24
COMPARISON OF THREE
DIFFERENT MODES OF COAL
SAMPLE EXCITATION[108]

	XRF-Mo tube	PIXE	XRF-^{109}Cd
S	12.4%	13.05%	—
K	—	532 ppm	—
Ca	1.8%	1.82%	—
Ti	361 ppm	380 ppm	—
V	22 ppm	43 ppm	—
Cr	15 ppm	23 ppm	—
Fe	3300 ppm	3300 ppm	—
Ni	23 ppm	—	—
Cu	25 ppm	—	—
Zn	41 ppm	—	—
As	25 ppm	—	—
Se	43 ppm	—	—
Rb	13 ppm	—	—
Mo	—	—	94 ppm
U	55 ppm	—	60 ppm

for elements heavier than iron, X-ray tube excitation is more suitable because of better sensitivity. The ^{109}Cd radioactive source was used only for molybdenum and uranium determination partly because of its low activity. This method was used alternatively for detection of X-ray lines in the energy region where spectra obtained by tube have significant background. However, with the ^{109}Cd source of higher activity, much better results could be obtained for a number of elements.

Chen et al.[118] have demonstrated the feasibility of using the proton microprobe to study trace elements in coal. The authors have presented trace element analyses of the major maceral groups (vitrinite, exinite, and inertinite). In their measurements, blocks of the coal samples were first studied petrographically with the optical microscope, and areas representative of the three maceral groups were designated for analysis by the proton microprobe. After that, the samples were bombarded at a proton energy of $E_p = 3.0$ MeV at beam currents between 50 to 150 pA. The induced X-ray spectra were measured for up to 5400 sec. After test runs with a beam size of 2 μm (vertical) × 4 μm (horizontal), a complete data set with increased statistical precision was obtained with a beam size of 4 μm × 6 μm. As a check on the variability of the elemental concentrations depending on the area of the sample analyzed, data were also accumulated with a beam size of 15 μm × 50 μm. To obtain the elemental concentrations from the X-ray spectra, a procedure similar to one developed by Uemura et al.[119] for thick samples was used. The concentration W_i of element i is given by the equation

$$\frac{P_i}{B} = \frac{2 \cdot 10^3}{A_i} \frac{\sigma xi}{d\Sigma_b/dk} W_i \tag{7}$$

where P = X-ray counts in the peak
       B = background counts per eV under the X-ray peak
       A = molecular weight of the element
       σ_x = X-ray production cross-section (cm²)
       $d\Sigma_b/dk$ = bremsstrahlung background cross-section per X-ray energy interval (cm²/keV) from the target matrix.

Using the formula given by Uemura et al.[119] for $d\Sigma_b/dk$, which is based on a simplified theoretical treatment of the bremsstrahlung production, the above equation gives concentration to within a factor of 2.

To improve the accuracy, an experimental $d\Sigma_b/dk$ at different X-ray energies were determined by bombarding a pure carbon foil of known thickness (0.16 mg/cm²) at $E_p = 3.0$ MeV.

In order to apply the above procedure to thick coal samples, the $(d\Sigma_b/dk)$ experiment was determined at several proton energies below 3.0 MeV (using the 0.16 mg/cm² carbon foil) to simulate the degraded proton energy as the beam traverses a thick coal sample. The $(d\Sigma_b/dk)_{experiment}$ was then determined for each proton energy.[118]

VI. NEUTRON ACTIVATION

Some aspects of the application of neutron activation to trace element analysis of fossil fuels have already been discussed by the author of this book; see for example Valković.[85,86] Neutron activation analysis has been an accepted method for trace element analysis for many years. A variety of neutron sources are used for sample irradiation, the most common being thermal neutron flux from a reactor. In addition, fast neutrons (14.4 MeV) are being used, as well as neutrons from ^{252}Cf-source. For the characteristics of these sources see book by this author.[86]

A. Neutron Activation with Thermal Neutrons

Thermal neutrons (energy 0.023 eV) can be obtained from nuclear reactions in fluxes ranging from 10^{11} to 10^{15} cm^{-2} s^{-1}. When any element of atomic number Z and atomic weight A is placed in a flux of neutrons, there is a finite probability that the element will capture a neutron to yield a new isotope of the same element but with a mass one unit heavier. Thus

$$\left(\frac{A}{Z}\right) + n \rightarrow \left(\frac{A+1}{Z}\right) + \gamma \tag{8}$$

This is usually abbreviated as

$$X(n,\gamma)\Upsilon \tag{9}$$

An example would be

$$^{59}\text{Co} + n \rightarrow {}^{60}\text{Co} + \gamma \tag{10}$$

or

$$^{59}\text{Co}(n,\gamma)\,^{60}\text{Co} \tag{11}$$

If, as is often the case, the newly formed nuclide is radioactive, one may determine its presence by measurement of emitted radiation. In the example, the ^{60}Co decays by beta emission followed by two quanta of energy (gamma rays). The ^{60}Co decay can be represented

$$^{60}\text{Co} \xrightarrow[5.27 \text{ years}]{\beta,\gamma} {}^{60}\text{Ni}$$

^{60}Ni is stable so the decay chain ends; in other instances, however, several additional decay steps may occur.

The production of a radioactive species by neutron (or particle) capture is given by the activation equation:

$$A_o = N\phi\sigma \left[1 - e^{-\left(\frac{0.693\, t_i}{T_{1/2}}\right)}\right] \tag{12}$$

where A_o = disintegration rate of induced radionuclide at end of irradiation
       N = atoms of element irradiated
       ϕ = neutron of particle flux
       σ = cross section of element
       t_i = irradiation time
       $T_{1/2}$ = half-life of induced radionuclide: $\lambda = 0.693/T_{1/2}$ = decay constant.
       $(1-e^{-\lambda t_i})$ is often called S, the saturation factor.

The above equation indicates that the amount of A produced (and hence the sensitivity of any activation analysis) is a linear function of the amount of element to be determined, the flux, and the cross section. The half-life, present in the exponential term, determines the irradiation time: obviously activities with short half-lives may be impractical for use. The sample to be irradiated is placed into the irradiation capsule and transferred to irradiation site and back. Afterwards the gamma ray spectra are recorded and counted.

Three basic methods are used in applying activation analysis: single element comparator, multielement comparator, and multielement absolute. The latter is referred to as instrumental neutron activation analysis (INAA). The single element comparator technique is the simplest and oldest. One merely irradiates a sample and a standard (comparator) containing a known weight of the element sought and then compares the activity induced in each, under standard radioactivity counting conditions. Almost always chemical separations are performed so as to isolate the desired radioelement in standard and sample from interferences. The radiochemical separation technique usually follows these steps:

1. Dissolution of sample and standard
2. Addition of a known weight of carrier (i.e., a nonradioactive species of the radioelement sought), followed by complete mixing and interchange, to both sample and standard
3. Separation, isolation, and determination of yield of carrier element in both sample and standard
4. Counting of sample and standard, correction for yield through Step 3, and comparison of corrected count of sample and standard to determine amount of element in sample

The multielement comparator technique begins like the single element method: irradiation of sample and known standards of the elements to be sought. However, an instrumental, nondestructive counting technique is generally used and individual radioelements are compared in sample and standard.

In the INAA technique, the gamma-ray spectra of all activated nuclei in the sample are usually measured using a Ge(Li) detector coupled to an analyzer/computer. Data obtained from the Ge(Li) detector are stored in the analyzer memory, and from an energy calibration of the detector system, one can obtain the exact energy of each gamma-ray photopeak; this enables positive identification of the radionuclide to be made. The integrated area of each photopeak can then be used to determine the quantitative amount of the radionuclide present.

This can be done using the same procedure to measure standard reference materials and to deduce the calibration curve for the system used. In such a way, in fact, efficiency of the detector for the geometry used is determined.

In most laboratories the peak searching, identification, integration, and calculation of concentrations are all carried out by a mini-computer, using detector efficiencies obtained experimentally and gamma-ray branching ratios from published tables.

Neutron activation of coal and coal ash samples has been done in many laboratories. We should mention here a four-laboratory comparative analysis of NBS reference materials.[120] The NBS coal and fly ash Standard Reference Materials (SRM 1632 and 1633) were analyzed for 37 elements in coal and 41 elements in fly ash, mainly by the use of instrumental neutron activation analysis (INAA), augmented by instrumental photon activation analysis (IPAA), and direct counting of natural γ-ray activity. For most elements measured, there was excellent interlaboratory agreement among the four participating laboratories and with the National Bureau of Standards values.

In cases of most elements for which comparisons can be made, instrumental nuclear methods used in a round-robin study of the standards provided more accurate average concentrations and smaller interlaboratory dispersions of values than the other major techniques used: atomic absorption spectrometry and optical emission spectroscopy.

We shall now describe some of the published work. Sheibley[121,122] described in this way the method used in their laboratory:

The sample aliquots were encapsulated in polyethylene vials for the irradiation period. Two aliquots of the sample plus two standards were irradiated in a pneumatic transfer irradiation system, one set (sample plus standards) for a long time period (12 to 24 hr) and the other for a short time period (5 min). After irradiation the samples were immediately removed from the vials. The short-time sample was counted after decay intervals of 3 min, 30 min, and 24 hr. The long-time sample was counted after a decay interval of 3 weeks; sometimes the decay intervals were 7 to 10 days. The typical neutron flux was 10^{13} neutrons/cm^2/sec. Counting data were processed through computerized data reduction codes; 20 to 30 elements were reported.

The flow chart showing the scheme for irradiation counting and data reduction of various samples is shown in Figure 29.

NBS-EPA Round-Robin Coal and fly ash samples were analyzed and the obtained results are shown in Table 25 and Table 26.

Five aliquots of coal were analyzed to determine the mean value and the standard deviation. The column "Counting Precision Range" refers to the counting error associated with the peak area of the isotope used in the calculation. Results are reported on 43 elements in Table 25. The elements antimony and lanthanum exhibit a larger percent standard deviation than the range of counting precision. These elements might not be homogeneous with respect to the coal matrix. Precision at 1σ ranges from 2% for cerium and iron to 40% for tungsten.

The data on fly ash reported in Table 26 was based on 5 aliquots; 43 elements are also reported here. Precision ranges from less than 5% for aluminum, sodium, strontium, and vanadium to 50 and 55% for magnesium and ytterbium, respectively.

In the measurements by Abel and Rancitelli,[123] coal and fly ash samples were weighed into small bags formed by folding and heat-sealing polycarbonate film. The openings were also heat-sealed after the samples were placed into the polycarbonate

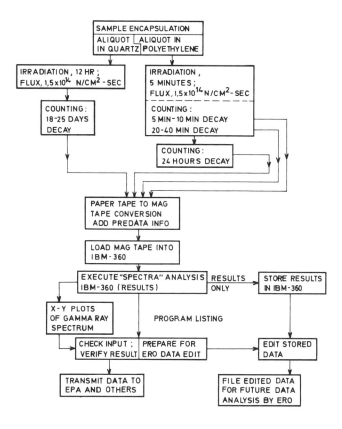

FIGURE 29. Flow chart showing scheme for irradiation, counting, and data reduction of various samples (after Sheibley[121,122]).

bags. The polycarbonate film was used because of its low trace metal content. The heat sealing was then accomplished using a Teflon®-coated "Quik Seal" thermal impulse sealer. The polycarbonate bags were afterwords placed into a clean polyethylene vial (approximately 2.5 cm long × 1.0 cm ID-2/5 dram). The vial containing the coal and fly ash samples was then placed into another clean vial (5.5 cm long × 1.43 cm ID-2 dram) prior to neutron activation.

Standards used for comparators were either well-documented materials or were prepared in the laboratory by pipetting known quantities of elements onto high-purity cellulose material. Standards were weighed into polycarbonate and packaged for irradiation in the same manner as the samples.

The analysis consisted of two irradiation periods and three counting intervals as shown in Table 27. Gamma-ray spectrometry was accomplished by Ge(Li) detector, typically with 15 to 20% efficiency relative to 3 × 3 in. NaI(Tl) detector and <2-KeV resolution at the 1332 KeV ^{60}Co energy. The γ-ray spectra from samples and standards were stored on magnetic tape and the data reduced by a PDP-15 computer. A computer program was developed to identify the radionuclides present from their characteristic γ-ray energies, calculate net peak areas, and convert net peak areas to element weights by direct comparison to known standards, thus calculating the concentration of the elements in the original samples.

In the measurements 50 to 100 mg of coal and 10 mg of fly ash were irradiated. After irradiation, the primary activity is that of ^{28}Al ($t_{1/2}$ = 2.3 min), which decays rapidly compared with most of the other isotopes. Figure 30 illustrates the initial activ-

Table 25
PRECISION ON NBS-EPA ROUND-ROBIN COAL SAMPLE[122]

	Mean (ppm)	±1σ (ppm)	Standard deviation (%)	Counting precision range at 1σ (%)
Al	15,700	1,550	9	0.6—1
As	5.9	0.5	9	10—12
Au	0.146	0.048	33	10—40
Br	20	3	15	9—12
Ba	337	42	12	5—8
Ca	4,070	560	14	8—15
Ce	17.340	0.089	2	1—2
Cl	750	75	10	2
Co	5.48	0.15	3	1—13
Cr	19	0.8	4	3—5
Cs	2.55	0.06	2.3	8—10
Cu	14.1	0.9	6	3—5
Dy	0.85	0.06	7	2
Eu	0.312	0.037	12	0.2—0.3
Fe	7,517	119	2	1
Ga	5.4	0.8	14	11—15
Ge	70	5	7	35—50
Hf	0.92	0.05	6	5—10
Hg	0.95	0.09	10	25—33
I	2.78	0.38	14	12—30
In	0.04	0.01	25	25—30
Ir	2.48	0.27	11	5—12
K	3,500	360	10	3—4
La	11.3	3.3	30	6—12
Lu	0.416	0.017	4	5—8
Mg	980	250	26	12—33
Mn	38.0	2.6	7	0.5
Na	370	33	9	2—3
Nd	6.4	1.5	24	40—55
Rb	19	1.9	10	10—20
Sb	6.4	1.6	24	8—15
Se	3.8	0.51	13	25—33
Sm	1.3	0.19	15	2—5
Sn	125	20	16	10—15
Sr	93	9.2	10	8—11
Ta	0.360	0.028	8	15—20
Tb	0.03	0	0	0.5—7
Th	3.1	0.2	8	2—20
Ti	1,312	150	12	10—20
U	0.980	0.078	8	8—12
V	36	4	11	5—10
W	1.9	0.8	40	30—80
Yb	0.55	0.04	8	15—20

ities and their decay with time. If the sample is counted immediately after irradiation, the ^{28}Al, with its higher energy photons (1779 KeV), interferes in the measurement of most other radionuclides. Conversely, if a decay interval of 20 min is allowed before counting, the ^{28}Al activity has decreased, but the disintegration rates of other radionuclides such as ^{51}Ti, ^{66}Cu, ^{52}V, and ^{27}Mg have also decayed to a point where detection is difficult. Therefore, the counting period was optimized to a 5-min counting interval beginning when the sample has been out of the reactor for 10 min.

Table 26
PRECISION ON NBS-EPA ROUND-ROBIN FLY
ASH SAMPLE[122]

	Mean (ppm)	±1σ (ppm)	SD (%)	Counting precision range at 1σ (%)
Al	109,600	4,020	3.7	0.06-1
As	69.5	7.6	11	8-10
Ba	2,734	167	6	6-8
Br	12.1	1.5	12	35-50
Ca	41,000	3,600	8.8	7-16
Ce	129	10	7	2-10
Cl	185	44	24	10-40
Co	38.6	3.7	9	1-10
Cr	122	12	10	2-4
Cs	13.8	1.4	10	8-12
Cu	142	9	6.2	3-4
Dy	7.6	2.4	31	1-8
Eu	2.42	0.16	6.6	1-2
Fe	52,780	5,600	11	1
Ga	38.3	6.3	16	8-20
Ge	476	166	35	—
Hf	7.62	0.56	7.3	3-20
Hg	3.7	1.1	30	15-30
In	0.156	0.035	23	35-50
Ir	18.6	3.3	18	8-10
K	21,800	2,400	11	3-4
La	77	8	10	3-8
Lu	3.8	0.5	14	5-8
Mg	15,970	8,060	50	15-33
Mn	466	31	6.6	0.3
Na	2,658	129	4.9	1-2
Rb	115	15	13	8-14
Sb	12.08	0.86	7	10-15
Sc	27.5	2.4	9	0.5-4
Se	12.7	1.8	15	20-30
Sm	10.05	0.58	5.8	2-3
Sn	740	210	28	8-20
Sr	869	33	3.8	5-8
Ta	2.74	0.25	9	15-20
Tb	0.22	0.04	16	3-6
Th	25	2	8	2-3
Ti	8,900	752	8.5	10-12
U	8.40	0.56	6.7	8-15
V	230	10.6	4.6	5-7
W	12.7	1.1	8.8	30-50
Yb	6.2	3.4	55	3-25
Zn	700	220	31	10-30
Zr	640	140	22	25-35

The second irradiation involved exposure of 100 mg samples of both coal and fly ash, along with appropriate standards, to a neutron flux of 5×10^{12} neutron cm^{-2}sec^{-1} for 6 to 8 hr followed by 2 counting intervals after appropriate decay periods. Figure 31 illustrates the activities of a few of the intermediate and long-lived radionuclides produced during the 6-hr irradiation and their decay with time after irradiation. This figure illustrates the presence of two groups of radionuclides with significantly differ-

Table 27
IRRADIATION AND COUNTING PROCEDURES FOR COAL AND FLY ASH AS USED BY ABEL AND RANCITELLI[123]

Irradiation Interval/ Neutron Flux (Neutrons/cm²/sec)	Decay interval	Count interval	Instrumentation	Elements determined
		1,000—10,000 min	NaI (Tl) Multi-dimensional-ray spectrometer	K, Th, U
0.5—1 min/1 × 10¹³	10 min	5 min	Ge (Li)	Al, Br, Cl, Cu, Mg, Mn, Na, Ti, V
6—8 hr/5 × 10¹²	4—7 days	10 min	Ge (Li) or anti-coincidence shielded Ge(Li)	As, Au, Ba, Br, Ca, K, La, Lu, Na, Sm, Yb
	30 days	100—1,000 min		Ag, Ce, Co, Cr, Cs, Eu, Fe, Hf, Hg, Ni, Rb, Sb, Sc, Se, Sr, Ta, Tb, Th, Zn

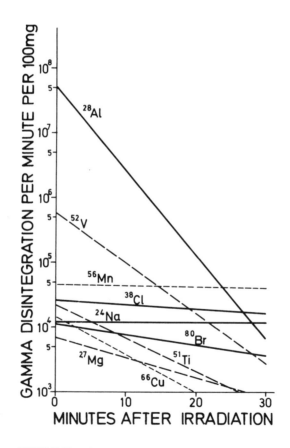

FIGURE 30. Concentrations of short-lived radionuclides produced by neutron activation of coal (after Abel and Rancitelli[123]).

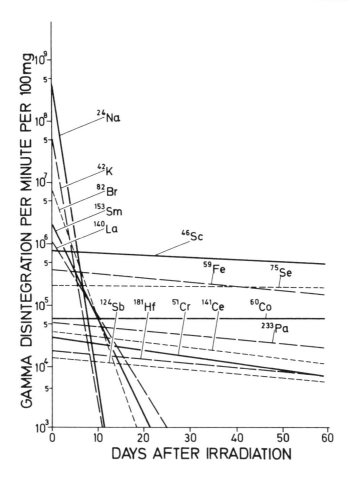

FIGURE 31. Concentrations of long-lived radionuclides produced by neutron activation of coal (after Abel and Rancitelli[123]).

ent half-lives. This is the reason the samples are counted twice with each count optimized to determine one of the two groups of radionuclides. The first counting interval is for 10 min and is begun approximately 5 to 6 days after irradiation. At this time as can be seen in Figure 30, the predominant activity, ^{24}Na ($t_{1/2}$ = 15 hr), present immediately after irradiation, has decayed more than most other radionuclides of interest. The ^{24}Na, which emits a high energy γ-ray, thus no longer significantly interferes in the measurement of low energy γ-rays. Radionuclides which can be measured during the first counting period, as shown in Table 27, include As, Au, Ba, Br, Ca, K, La, Lu, Na, Sm, and Yb. The determination of Na and Br from both Irradiation I and Irradiation II provides an internal check to insure consistent accurate results.

Figure 31 also indicate that a decay period of approximately 25 to 30 days reduces the intermediate-lived radionuclides such as ^{24}Na, ^{140}La, and ^{82}Br to insignificant activity levels without seriously affecting the ability to measure the remaining isotopes of interest. The second count is then conduced for 100 to 1000 min on a Ge(Li) or anticoincidence shielded Ge(Li) spectrometer and provides concentrations for Ag, Ba, Ce, Co, Cr, Cs, Eu, Fe, Hf, Hg, Ni, Rb, Sb, Sc, Se, Sr, Ta, Tb, Th, and Zn.[123]

Radiochemical separations are necessary for many elements when only a NaI detector is available. Even with a Ge(Li) detector, a radiochemical separation increases the sensitivity and accuracy and permits the determination of some elements whose radioactivities are masked by stronger activities in the multielement spectrum of a coal sample. For example, mercury, selenium, gallium, and zinc in most coals are below the limit of detection instrumentally even with the resolution of a Ge(Li) detector, but can be determined after radiochemical separations as is described by Frost et al.[124]

Mercury determination in NBS SRM 1630 coal has been described by Frost et al.[124] in the following way. The irradiated coal sample was placed in a ceramic combustion boat with added mercuric oxide carrier and burned in a slow stream of oxygen in a simple apparatus consisting of a straight quartz combustion boat with added mercuric oxide carrier tube connected to a straight condenser surrounded by a trap of liquid nitrogen. The ash and the tube were then heated to about 809°C to drive all volatile material into the cold trap. The products were dissolved in nitric acid. The activity of a vial containing a 2 N HNO$_3$ solution of the collected material was counted with a Ge(Li) detector, measurements being made at the 0.077 MeV gold X-ray activity arising from the decay of ^{197}Hg.

The only interference was from ^{82}Br, which was more serious if counting was done with a NaI detector. Bromide was removed by adding bromide carrier to the nitric acid solution of the products at 40°C and precipitating with silver nitrate. Mercuric bromide is soluble in warm dilute nitric acid and is quantitatively retained in solution.

The procedure for rubidium and cesium was described by Frost et al.[124] in the following way:

To determine rubidium, the irradiated coal sample and added rubidium carrier are digested and the organic matter destroyed by heating with nitric and sulfuric acids, followed by nitric and perchloric acids. The rubidium is separated from the gross activity of the sample by precipitation with ferric hydroxide. The rubidium is then precipitated as the cobaltinitrite and finally as the chloroplatinate. The activity of the chloroplatinate caused by the 1.8 MeV β^- particle of 18.7 day ^{80}Rb is counted with a Geiger counter.

In the determination of cesium, the coal sample and cesium carrier are digested, and the perchlorate separation and ferric hydroxide scavenging precipitation are made as in the procedure for rubidium. Cesium is then separated from the remaining solution by precipitation of cesium bismuth iodide. The final separation is made by precipitation of cesium chloroplatinate, which is counted with a Geiger counter for the 0.66 MeV β^-decay of ^{134}Cs ($t_{1/2}$ = 2.1 year), or counted for the 0.13 MeV γ-ray associated with the isomeric transition of ^{134m}Cs ($t_{1/2}$ = 2.9 hr).

The procedure for selenium was as follows:[124] The irradiated coal sample with selenium carrier added is wet-ashed with a nitric-perchloric acid mixture under good reflux. Detailed tracer experiments were run to show that trace levels of selenium are not lost from the sample during ashing and equilibration of the carrier with the sample. The selenium (VI) is reduced to selenium (IV) and distilled into 2 M HCl, with two additions of hydrochloric and hydrobromic acid. Elemental selenium is precipitated, dissolved in nitric acid and hydrogen peroxide, and reprecipitated. The ^{75}Se activity of the precipitate was counted with a NaI crystal. The procedures for gallium, arsenic, antimony, bromine, zinc, and cadmium were also described.

Frost et al.[124] also described the method for determination of uranium in the coal: the coal samples are ashed at 600°C, and the ash is dissolved by digestion with nitric and hydrofluoric acids. Traces of hydrofluoric acid are evaporated, and dilute nitric acid solutions of the ash samples are irradiated. The irradiated ash solution, with nitric acid added, is heated to dryness, and the residue is taken up in hydrochloric acid. The solution is treated with hydroxylamine hydrochloride to reduce the neptunium quantitatively to the extractable neptunium (IV). The sample in hydroxylamine hydrochloride-hydrochloric acid solution is passed through a column of thenoyltrifluoroacetone in xylene as stationary phase on borosilicate glass powder as support, and the column is rinsed with hydrochloric acid-hydroxylamine hydrochloride solution. The neptunium is eluted with 6 M HCl, followed by ethyl alcohol and 6 M HCl, and the activity of the solution is counted by measuring the 0.106 MeV photopeak of ^{239}Np.

The application of neutron activation analysis to trace-element analysis of coal was also described by Lyon and Emery,[125] Yaverbaum,[17] Millard,[126] Gluskoter,[10] Weaver,[134] Lyon et al.,[71] and others. Table 28 shows the detection limits and nuclear properties of isotopes used for the analysis of coal by Gluskoter.[10] The author used a sample of approximately 1 g, and activation was done in TRIGA MKII® reactor.

B. Neutron Activation Using ^{252}Cf Source

^{252}Cf is a very versatile neutron source for the examination of both stationary and moving samples of coal and coal ashes. Rhodes[127,128] has showed that it is demonstrated that it is possible to determine essentially all of the major and minor elements in coal necessary for process control by not restricting oneself to a single neutron-gamma technique. Furthermore, the optimum nuclear reaction for each element can then be chosen, so reducing the demands on the gamma-ray detector and electronics. This in turn makes possible the use of the simplest and most practicable instrumentation for reliable plant operation.

Greenwood[129] has summarized the recent development in using the downhole sondes with germanium detectors and ^{252}Cf sources. Measurement of prompt gamma rays from neutron-induced reactions has many advantages when compared to conventional delayed neutron activation analysis. In their paper, Nargolwalla et al.[130] describe a mobile well-logging system and discuss its application to simultaneous logging of Ni, Fe, and Si in lateritic nickel deposits; copper and silicon in porphyry copper deposits; and to delineation of coal seams and the determination of S in the coal. The characterization of coal seams by well logging is a problem in which there is particular interest at the present time. Senftle et al.[131] have reported on the results of an in-field test of a PNAA logging system. In these tests, using a continuous logging mode, the increased ^{1}H(n,γ) signal coupled with the decreased ^{27}Al signal served as sensitive indicators of the presence of a coal seam. Using static sonde positions, elemental analyses for H, C, N, S, Al, Si, Fe, Ti, and Cl were obtained using the PNAA method alone. Assuming an average oxygen content of the coal these elemental analyses allow ash content and heat of combustion to be assessed. For high-ash coal Loska and Gorski[132] determined the relationship

$$(Ash\ \%) = 3.12\ [SiO_2 + Al_2O_3]^{0.672} \qquad (13)$$

and for coal containing less than 15% O, Mott and Spooner[133] suggested the following expression for heat of combustion:

$$Q(cal/g) = 80.3\ [C] + 339.0\ [H] - 39.7\ [O] + 22.5\ [S] \qquad (14)$$

References to other activities related to the development of *in situ* methods using prompt neutron activation analysis for minerals exploration can be found in the paper by Greenwood.[129]

Yeager[135] has described the development of the first on-line instrumentation for the continuous nuclear analysis of coal (CONAC).

CONAC uses a small specimen of radioactive californium (^{252}Cf) to bombard a coal

Table 28
DETECTION LIMITS AND NUCLEAR PROPERTIES OF ISOTOPES USED
FOR THE ANALYSIS OF COAL[10]

	Isotope produced	Half-life	Cross section (barns)	Counting period[a]	Major gamma-rays utilized (keV)	Limit of detection (ppm)	Av. relative SD (%)
Na	^{24}Na	15 hr	0.53	A,B,C	1368	0.5	5
Cl	^{38}Cl	37 min	0.40	A	1642	20	15
K	^{42}K	12.4 hr	1.2	B,C	1525	3	10
Sc	^{46}Sc	83.8 day	13	D	889, 1120	0.01	5
Cr	^{51}Cr	27.8 day	17	D	320	1	10
Mn	^{58}Mn	258 hr	13.3	A,B	846, 1811	0.1	5
Fe	^{59}Fe	45 day	1.1	D	1099, 1292	200	10
Fe	^{54}Mn	291 day	0.4	D	835	1000	15
Co	^{60}Co	5.26 year	37	D	1173, 1333	0.5	5
Ni	^{58}Co	71 day	0.2	D	810	5.0	30
Zn	^{65}Zn	245 day	0.5	D	1115	5.0	30
Zn	^{69}Zn	13.8 hr	0.1	B,C	439	50	25
Ga	^{72}Ga	14.2 hr	5.0	B,C	834, 630	0.5	15
As	^{76}As	26.4 hr	4.5	C	559, 651	0.2	20
Se	^{75}Se	120 day	30	D	136, 264	0.1	15
Br	^{82}Br	35.3 hr	3.0	B,C	554, 777	0.5	20
Rb	^{86}Rb	18.7 day	0.7	D	1079	1.0	20
Sr	^{87m}Sr	2.8 hr	1.3	A,B	388	5.0	10
Mo	^{99}Mo	67 hr	0.15	C	141	5.0	20
Ag	^{110m}Ag	253 day	3.5	D	657, 937	1.0	30
Cd	^{115}Cd	53 hr	0.3	C	528	5.0	50
In	^{116m}In	54 min	160	B	417, 1097	0.01	30
Sb	^{122}Sb	2.7 day	6.5	C	564	0.2	20
Sb	^{124}Sb	60.3 day	2.5	D	1691	0.1	10
I	^{128}I	25 min	6.2	A	443	0.5	25
Cs	^{134}Cs	2.05 year	31	D	797, 569	0.05	15
Ba	^{131}Ba	12 day	8.8	C,D	496, 216	30	10
Ba	^{139}Ba	83 min	0.35	A,B	166	200	20
La	^{140}La	40.2 hr	8.9	C	1596, 487, 329	0.1	5
Ce	^{141}Ce	33 day	0.6	D	145	0.5	15
Sm	^{153}Sm	47 hr	210	C	103	0.05	5
Eu	^{152}Eu	9.3 hr	2800	A,B,C	122, 344, 963	0.10	5
Eu	^{152}Eu	12.5 year	5900	D	1408	0.05	5
Tb	^{160}Tb	72 day	46	D	879, 1178	0.05	10
Dy	^{165}Dy	2.35 hr	700	A,B	95, 361, 633	0.1	10
Yb	^{175}Yb	4.2 day	55	C	396, 282	0.5	25
Yb	^{169}Yb	32 day	5500	D	198, 110	0.1	10
Lu	^{177}Lu	6.7 day	2100	C	208	0.05	15
Hf	^{181}Hf	42.5 day	10	D	481, 153	0.05	15
Ta	^{182}Ta	115 day	21	D	155, 222, 1221	0.01	10
W	^{187}W	23.0 hr	38	B,C	480, 686	0.2	30
Au	^{190}Au	65 hr	99	C	411	0.01	40
In	^{233}Pa	27 day	7.4	D	312	0.2	10
U	^{239}Np	56 hr	2.7	C	277, 228	0.1	20

[a]

Counting period	Irradiation	Flux (n·m^{-2}·sec^{-2})	Decay interval	Count interval
A	15 min	2.0×10^{12}	30 min	500 sec
B	15 min	2.0×10^{12}	hr	2000—5000 sec
C	2 hr	4.1×10^{12}	24 hr	4000—7000 sec
D	2 hr	4.1×10^{12}	day	6—10 hr

stream with neutrons. As the energetic neutrons collide with the atomic nuclei of the coal constituents, they lose energy to these nuclei, especially to those of hydrogen. When the neutrons have lost sufficient energy in this way there is a probability that they will be captured by the nuclei instead of bouncing off them. The magnitude of this probability depends primarily on the physical characteristics of the individual elements. It does not depend critically on either the chemical environment of the nuclei or on the size distribution of the coal sample.

When a neutron is captured by an atomic nucleus, gamma rays are emitted that have a frequency characteristic of the element involved. CONAC uses radiation detectors to count the number of gamma rays at each characteristic frequency. The number for each is proportional to the abundance of the element associated with that frequency. These analyzers represent the culmination of a research and development effort sponsored by Electrical Power Research Institute (EPRI), Science Applications Inc. (SAI), and Kennedy Van Saun (KVS) (a subsidiary of McNally-Pittsburg Manufacturing Corp.) and carried out by SAI and KVS (see also: July-August 1980 EPRI Journal[175]). In 1981 the first of the instruments, a continuous sulfur analyzer, underwent acceptance testing at the Detroit Edison Company 3000 MW Monroe plant. A second instrument, a batch-sample sulfur analyzer, will be installed at the Tennessee Valley Authority Paradise plant. A third instrument, a prototype that performs the equivalent of a complete element and ash analysis of coal on-line, is also scheduled for installation at the TVA Paradise plant later this year. The commissioning of these analyzers represents a significant advance in instrumentation for coal quality control.[135]

Here is the description of the uses for CONAC as described by Yeager.[135] in EPRI Journal:

> Control of coal blending is the first problem to which the on-line nuclear analyzer is being applied. At its Monroe plant, Detroit Edison has elected to meet strict state SO₂ emission requirements by blending more-expensive low-sulfur Eastern coal with less-expensive high-sulfur Eastern coal. Although coal blending avoids the high initial costs of alternative emission control techniques, such as stack gas scrubbing, its long-term cumulative costs are high because low-sulfur coal is increasingly expensive. These costs can be minimized by using only as much of the low-sulfur coal as necessary to meet the emission standards. However, since the variation in the sulfur content of both the high- and low-sulfur coals is large, there is no single optimal blending ratio; instead, this ratio is a dynamic parameter that can change within minutes.
>
> Knowing the range of each coal sulfur content allows a blending ratio to be chosen that will ensure that emission levels are not exceeded, but this means using more of the costly low-sulfur coal than is necessary. To predict what savings would result from optimizing the blending ratio on the basis of feedback from an on-line sulfur analyzer, Detroit Edison used data on plant operating parameters and the sulfur content and heating value ranges of its targeted coals in a mathematical model. The model predicted that an annual savings in fuel costs of over $10 million could be realized by using optimized blending ratios rather than a fixed blending ratio.
>
> SAI and KVS built a simplified CONAC instrument called a sulfur meter. This instrument, which has been running on-line at Monroe since March 1981, gives the necessary feedback signal on sulfur content to enable the continual optimization of the coal blending ratio. Eventually it will control the speed of two rotary plows that will continuously load the main plant feed belt from piles of high- and low-sulfur coal. (This system is scheduled to come on-line in 1982).
>
> Currently at Monroe, the coals are loaded by stacker reclaimers and blended according to a fixed ratio per fill (about 4 hr). Coal entering the plant is sampled by an automatic sampler, and the rejects of the secondary samples (about 3 t/hr) are analyzed by the sulfur meter. Results on the sulfur content of the coal in each fill are then used by plant operations and fuel supply personnel to set subsequent fill blending ratios to meet 24-hr emission requirements.
>
> Figure 32 shows output from the sulfur meter at the Monroe plant over a 12-hr period. During this time, coals of various sulfur levels were fed into the plant silos. For the first 5 hr, 100% low-sulfur coal and blends, of high- and low-sulfur coals were used. For most of the next 6 hr, 100% high-sulfur coal was analyzed. At about the 6th hr the plant coal feed was stopped and the sulfur meter measured the same coal for approximately an hour; the small deviations in the measured sulfur content during this time indicate the precision of the instrument (= 0.04 wt% sulfur). After the stationary coal measurements were made, the coal feed was restarted. Subsequent deviations in the measured values indicate variations in the sulfur content of the high-sulfur coal.

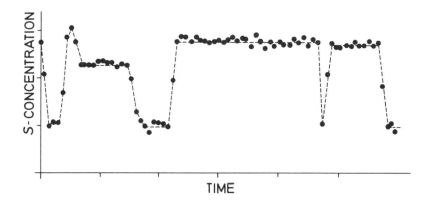

FIGURE 32. On-line sulfur meter readings of high- and low-sulfur coal feeds into the power plant (after Yeager[135]).

Two additional CONAC instruments — a batch sample sulfur analyzer and a total elemental analyzer — will be installed at the TVA Paradise coal preparation and power plant. The Paradise power plant, which consists of 2 704-MW units and 1 1150-MW unit, is required to meet an SO_2 emission standard of 3.1 lb/ 10^6 Btu (3-hr average). The strategy for complying with this limitation calls for coal with a sulfur content equivalent to 5.7 lb/10^6 Btu of SO_2 to be used in all 3 units. Units 1 and 2, the smaller units, also have scrubbers to further reduce SO_2 emissions to 0.9 lb/10^6 Btu. To produce coal of the required quality, a new coal preparation plant was begun in 1978.

The preparation plant, constructed by Roberts & Schaefer Co. will wash up to 2000 t/h of 3 in. × 0 coal from western Kentucky Nos. 9, 11, and 12 seams for use by the Paradise generating units. It consists of four independent modules that use heavy-media separation for 3-in. × 28-mesh coal and hydrocyclones in combination with froth flotation for coal below 28 mesh. Clean coal from all four modules is conveyed to another building for temporary storage and sampling.

Both the batch-sample analyzer, known as a rapid sulfur meter (RSM), and the total element analyzer (CONAC) will be located in the storage-sampler building. The RSM instrument will be used to monitor the sulfur content of the clean coal to ensure that an acceptable sulfur level is maintained at maximum Btu recovery. With information from the continuous sulfur readout, plant operators will be able to adjust the specific gravity of the heavy media to avoid both excessive coal loss from overwashing and possible penalties if SO_2 limitations are exceeded because of underwashing.

CONAC, to be installed at Paradise, will provide element analysis for carbon hydrogen sulfur, nitrogen, silicon, iron, aluminum, calcium, chlorine, sodium, and potassium. Additional sensors will provide moisture and coal mass flow rate information. Derived quantities, such as heating value, total ash content, and fouling and slagging indexes will also be displayed.[135]

We shall next describe work by Reynolds et al.[136] to illustrate the use of ^{252}Cf-source in coal analysis. In the measurements by Reynolds et al.,[136] prompt gamma photons emitted by nuclei in the coal have been measured using several detectors, including sodium-iodide and germanium-lithium. See Figure 33 for schematic of setup used to optimize prompt neutron activation of coal. Several coal types, including bituminous, subbituminous lignite, and anthracite were crushed to various top sizes and analyzed carefully by traditional wet chemistry techniques at two or three different laboratories. The elements (sulfur, hydrogen, carbon, aluminum, silicon, iron, calcium, sodium, nitrogen, and chlorine) were determined by prompt neutron activations and the quantities compared with those of the wet chemical analyses.

The measured neutron-capture gamma-ray spectra are affected by numerous parameters, all of which must be understood to design an optimum system. Spectral data were obtained to provide information about the following parameters:

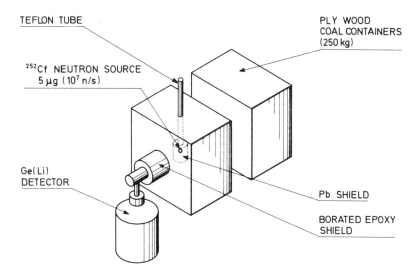

FIGURE 33. Schematic of setup used to optimize prompt neutron activation analysis of coal with ^{252}Cf source (after Reynolds et al.,[136] courtesy of EPRI, Palo Alto, Calif., with permission).

1. Coal thickness
2. Coal bulk density
3. Source-to-detector distance
4. Conveyor belt vs. bin geometry
5. Neutron source shielding
6. Detector shielding
7. Coal composition
8. Detector type and resolution

Measurements were made using high-sulfur Pittsburgh 8 and low sulfur Wyoming bituminous coals.

Using a high resolution Ge(Li) spectroscopy system, more than 100 gamma-ray peaks were identified from 14 elements, including those elements important to coal analysis — hydrogen, carbon, silicon, aluminum, sulfur, iron, nitrogen, sodium, potassium, titanium, and chlorine. To get consistent results for the peak areas, it was found necessary to develop a fast-running, relatively simple code designed specifically to analyze coal PNAA spectra. Measurements were also made with larger, more efficient NaI detectors. These spectra were enhanced and unfolded with the SAI MAZE code, to investigate the potential of specialized applications where assay of only a few of the more important coal elements is required.

Using the optimum detector shieldings of 5 cm borated epoxy with a Pb/Cu lining and the optimum source shield of 5 cm Pb, high resolution gamma-ray spectra were taken with a single Ge(Li) detector of 21% efficiency. Typical spectra for Pittsburgh #8 coal sample is shown in Figure 34. Ge(Li) spectra with good counting statistics were obtained at coal thicknesses of 25 and 40 cm for Pittsburgh coal, at low and high density. Numerous additional spectra were taken with short data collection times to study effects such as source-to-detector distance, and detector and source shielding. The measurements were repeated with Wyoming coal samples. There are some general observations to be made on comparing spectra from the two different coals. Both

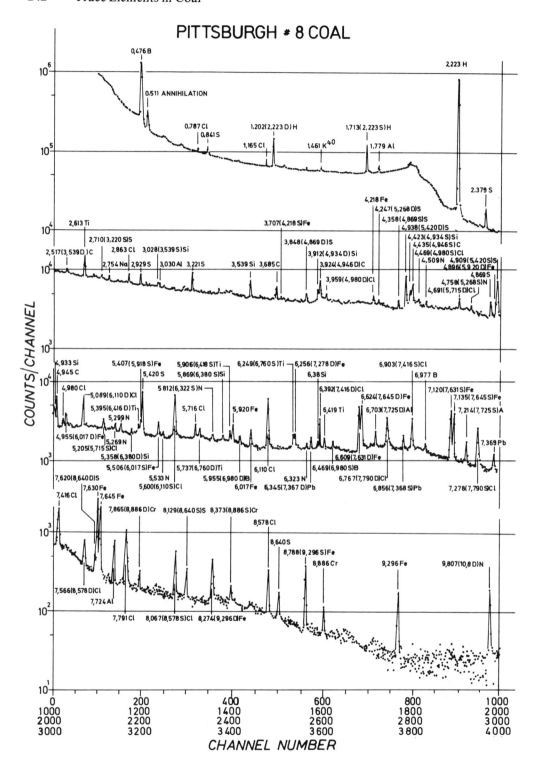

FIGURE 34. Spectrum of gamma rays detected with Ge(Li) detector after irradiation of Pittsburgh #8 coal with thermal neutrons (after Reynolds et al.,[136]) courtesy of EPRI, Palo Alto, Calif., with permission).

spectra are dominated by the 2.22 MeV line from hydrogen in the coal. The Doppler broadened peak at 0.467 MeV is from neutron capture in boron in the borated epoxy neutron shield around the detector. Except for the naturally occurring ^{208}Tl line (daughter product in the decay chain of ^{232}Th) at 2.61 MeV and K^{40} line at 1.46 MeV, the remaining peaks are due to neutron capture in elements in coal and to a much lesser extent to capture in other materials such as the Pb source shield and aluminum framework. For the most part, the same lines appear in the two spectra, although with varying intensities. For example, the Pittsburgh coal has about 7 times more sulfur than Wyoming, resulting in larger sulfur peaks, e.g., 5.42 MeV in Pittsburgh. The more numerous intense peaks between 4 and 8 MeV in Pittsburgh are due to neutron capture in chlorine, which our measurements show to be 8 times more abundant in Pittsburgh. Chlorine has an unusually large neutron capture cross section, making its presence obvious even though the weight percent in Pittsburgh is estimated to be only 0.08%.

The first steps in determining the elemental composition of coal from the complex Ge(Li) spectra in Figure 34 are to locate the significant peaks, determine the peak energy, and identify each peak with an element in the coal, using information in the Nuclear Data Tables or other gamma-ray catalogs. A more difficult task is to accurately determine the area (number of counts) in each peak. Because of the complexity, computerized techniques are required.

Table 29 summarizes the information on prompt neutron activation gamma-rays observed by Reynolds et al.[136] in Pittsburg #8 coal. The deduced information on coal composition of Pittsburg #8 coal is shown in Table 30 together with standard laboratory analysis data.

C. Fast Neutron Activation Analysis

Fast neutron activation has successfully been applied to Si, O, and N determination in coal and coal ash.

Determination of silicon was described by Block et al.[137,138] For the determination of silicon in coal and coal ash, the samples were after crushing and homogenization, packed into polyethylene rabbits with inner dimensions of 7 × 22 mm. Coal samples with low ash content (<20%) were packed as such; samples with a high ash content (>20%) were mixed with an equal amount of spectrally pure graphite. In both cases the total sample weight amounted to about 1.5 g. After ashing of the coal, about 130 to 300 mg of ash was mixed with graphite and also packed into polyethylene rabbits. Silicon standards were prepared from silica powder mixed with graphite.

The 14 MeV neutrons used were produced with a Sames type-I neutron generator (150 keV, 2.5 mA). Silicon was determined after a single irradiation of 15 sec at a neutron flux of 6×10^8 n·cm^{-2}s^{-1}, producing the isotope ^{28}Al by (n,p) reaction. The ^{28}Al activity was measured for 2 min, 75 sec after irradiation. Gamma spectrometry of ^{28}Al was performed with a 7.6 cm diameter × 7.6 cm NaI(Tl) detector, coupled to a preamplified and two single-channel analyzers.

Table 31 gives Si concentrations determined in NBS coal and fly ash standard reference materials. They are compared with the reference value given by the NBS for the coal material. The results are in agreement.

Oxygen determination in coal and coal ashes by fast neutron activation was done by Block and Dams,[138] Volborth et al.,[139-141] and Hamrin et al.[142] The procedure used by Block and Dams[138] is similar to the one they use for silicon determination except the activity of ^{16}N formed by (n,p) reaction on ^{16}O was measured for 15 sec after a decay period of 5 sec.

The method used by Volborth et al.[139-141] consisted of 20 sec irradiation and 20 sec

Table 29
PROMPT NEUTRON ACTIVATION GAMMA RAYS OBSERVED IN PITTSBURGH #8 COAL BY REYNOLDS ET AL.[136]

Carbon					
1.261	F	1.613	F	5.715	FSD
3.684	FSD	1.726	F	6.110	FSD
4.946	FSD	4.218	FS	7.416	FSD
		5.920	FSD	7.790	FSD
		6.017	FSD	8.578	FSD
Aluminum		7.278	D		
1.78	F[b]	7.631	FSD	Potassium	
2.960	F	7.645	FSD	0.770	F
4.133	F	9.296	FSD	1.460	F[a]
7.723	FSD				
		Nitrogen		Sodium	
Sulfur		4.508	FSD	0.870	F
0.841	F	5.268	FSD	2.754	F[b]
2.379	F	5.298	FSD	3.590	F
2.930	F	5.532	FS	3.982	F
3.220	FSD	5.560	F		
4.869	FSD	6.322	FS	Calcium	
5.420	FSD	10.83	D	1.942	F
7.800	FS			6.420	FSD
8.640	FS	Titanium			
		1.381	F	Thallium	
Silicon		6.418	FSD	2.614	F[a]
2.092	F	6.760	SD		
3.539	FSD			Boron	
4.934	FSD	Chlorine		0.477	F
6.380	FSD	0.787	F	6.977	FSD
		1.165	F		
Chromium		1.952	F	Lead	
8.884	FSD	1.959	F	7.368	FSD
		2.863	F		
Iron		3.823	F		
0.898	F	4.980	FSD		

Note: The following abbreviations are used: F, full energy peak; S, single escape peak; and D, double escape peak.

[a] Natural radioactivity (^{40}K, ^{208}Tl).
[b] Activation gamma-ray.

counting of 2 samples, a standard and the unknown, simultaneously. A special feature of their system was the alternate switching of samples into opposite transfer channels to correct for electronic drift and bias. The nuclear reaction ^{16}O (n,p) ^{16}N, and counting of the resulting 6.1 and 7.1 MeV gamma radiation with a half-life of 7.3 sec were used. The coal samples were packed into plastic containers (rabbits) and were sealed. Moisture at 105°C was determined on all samples, and the addition or loss of weight occurring while the samples were being packed was also determined by keeping a small sample exposed to ambient conditions for the duration of exposure. Some dried coal powders were very hygroscopic, gaining from 0.2 to 0.6% in weight during the 10 to 15 min of the packing procedure. Oxygen was determined at the time of packing. Oxygen was also determined on the dried samples (105°C). The precision of this method was ±0.2% (absolute) or better. A simple computer program computed the oxygen percentage in "as received" and in dried samples, taking the determined moisture as H_2O into account.

Table 30
COMPOSITION BY PNAA AND
LABORATORY ANALYSIS FOR
PITTSBURGH # 8 COAL[136]

	Lab. analysis (%)	PNAA[a] (%)
Carbon	72.0	72.0
Silicon	1.8	1.7
Aluminum	1.3	2.1
Iron	1.3	1.3
Sulfur	2.9	2.8
Nitrogen	1.4	1.4
Hydrogen	5.3	3.8
Titanium	0.030	0.065
Sodium	0.026	0.14
Potassium	0.11	0.57
Calcium	0.096	—
Chlorine	—	0.079
Magnesium	0.054	—
Oxygen (by difference)	13.6	14.1

[a] PNAA is prompt neutron activation analysis, normalized to carbon.

Table 31
DETERMINATION OF SILICON
CONCENTRATION (%) IN NBS REFERENCE
MATERIALS BY FAST NEUTRON
ACTIVATION ANALYSIS[138]

Element	Material	Block and Dams[138]	NBS
Si	NBS coal	3.5 ± 0.8	3.2
Si	NBS fly ash	23 ± 6	21 ± 2

In the measurement by Volborth et al.,[141] oxygen in 33 coals ranging from lignite to low-volatile bituminous coal had been determined and results were compared with "oxygen by difference". Considerable discrepancies were observed. Better stoichiometric results were obtained if oxygen in coal ash, in wet coal, and in the dried coal was determined. This permits the estimation of the true material balances in fuels using data of the ultimate and the proximate coal analysis. The oxygen determination provides the coal chemist with an accurate basis and can be used to rank coal. The summation of the percent of carbon, nitrogen, hydrogen, sulfur, and oxygen becomes more meaningful, some errors can be detected, and the state of completeness of coal analysis can be evaluated.[141]

Nitrogen and oxygen in U.S. coals were determined by Hamrin et al.[142] using fast neutron activation analysis. Samples were low-temperature ashed in a low-pressure, low-temperature, oxygen plasma using a six-chamber apparatus. Prior to ashing, approximately 40 g of each coal was heated at 378 K for 24 hr to remove moisture. Coal was then placed in 6 90-mm Pyrex® ashing dishes. The dishes were shaken at the beginning and end of the day to expose unashed coal to the plasma. Ashing was usually completed in 72 to 96 hr, and samples were then weighed and ashed for another 24 to 48 hr to ensure complete ashing.

The asher was operated at a low-power level to keep the temperature down. Approximately 24 W per chamber was used with an oxygen flow of 6670 mm^3/sec (400 mℓ/min) at a pressure of 133 Pa (1 torr). Under these conditions the estimated ashing temperature was less than 323 K.

The experimental method used by Hamrin et al.[142] is described in the following manner: oxygen and nitrogen were determined by 14 MeV neutron activation analysis employing a Kaman A-1250 neutron generator. Both the oxygen and nitrogen analyses employ a multiscaling technique, based on counting the 6.13 and 7.12 MeV spectral region for gamma rays emitted by 7.1 sec half-life, ^{16}N produced by the ^{16}O (n,p) ^{16}N reaction, and the 0.511 MeV annihilation gamma rays associated with the decay of ^{13}N produced by the ^{14}N(n,2n)^{13}N reaction. A 10.2 cm × 10.2 cm NaI(Tl) well-type scintillation detector was used in the oxygen determinations, while two 7.6 cm × 7.6 cm NaI(Tl) detectors operated in the coincidence mode were used to measure the annihilation radiation from ^{13}N decay for the nitrogen determinations.

The high-energy gamma radiation from the decay of ^{16}N was essentially free from spectral interferences for oxygen determinations in coal samples, although it should be noted that fluorine could cause a primary interference, if the F to O ratio exceeds 0.02. Interferences to the nitrogen determinations due to ^{31}P(n,2n)^{30}P, ^{12}C(p,γ)^{13}N, ^{63}Cu(n2n)^{62}Cu, ^{13}C(p,n)^{13}N, ^{16}O(p,α)^{13}N, ^{19}F(n,2n)^{18}F and ^{39}K(n,2n)^{38}K reactions were evaluated and the small corrections, where applicable, were applied to the data. The largest of these is due to the ^{16}O(p,α)^{13}N recoil proton-induced reaction which can be as much 4.6%, relative. The recoil proton interferences depend on factors such as packaging geometry, the hydrogen content of the sample, and the bulk and composition of the packaging material.

Sample sizes were generally in the range of 1.0 to 2.0 g and were packaged under an inert atmosphere of helium. The compound N-1-napthylacetamide (7.56% N) was used as a primary standard for the nitrogen determinations. The H, O, and C contents of this standard are similar to those of coal and help to minimize the problems associated with proton-induced interference reactions.

For 6 U.S. coals studied by Hamrin et al.,[142] total oxygen ranged from 9.4 to 28.7% and total nitrogen varied from 0.72 to 1.61%. To obtain values of organic oxygen and nitrogen, either a low-temperature ashing method or an acid-treatment method was suitable for bituminous coals. The mean difference of the experimentally determined values (O_{dmmf})LTA-(O_{dmff})AT = −0.82, s = 0.51, was found to be statistically significant at the 95% confidence level, but the comparable difference for nitrogen was not. By the LTA method, oxygen and nitrogen on the dmmf basis for bituminous coals showed no statistically significant difference with calculated dmmf values. Nitrogen was detected in all the LTAs varying from 0.38 to 1.67%. Formation of insoluble CaF$_2$ in the acid-treatment method caused an interference in the nitrogen determination due the ^{19}F(n,2n)^{18}F reaction, but was correctable.

In order to work with more success on the problems of coal processing, underground gasification, and to combat the methane danger, one needs a technique capable of rapid determination of hydrogen content in coal seams. According to Morstin and Woznizk,[143] some neutron methods, commonly applied in soil-moisture measurements and in petroleum well-logging, are also useful for hydrogen detection in coal. If coal moisture is known from other measurements, determining the total hydrogen content in coal can provide some useful information concerning coal quality and appropriation danger of methane, etc.

Registering the stationary epithermal neutron flux seems to be most reasonable for these purposes. Its dependence on hydrogen concentration is, even for relatively short source-to-detector distances of about 20 cm, sufficiently strong to obtain high accuracy

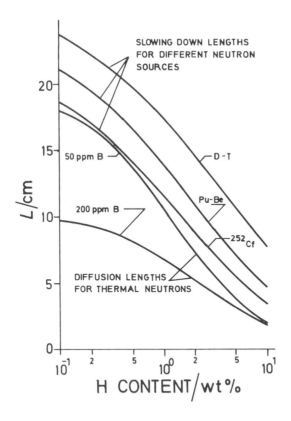

FIGURE 35. Slowing-down and thermal neutron diffusion lengths of coal (ϱ = 1.4 g/cm) vs. its total hydrogen content (after Morstin and Woznick[143]).

in determinations, unless a strong bore-hole effect takes place. The latter is a real problem and should be minimized by carrying out the measurements, if possible, in open holes of very small diameter and optimizing the source-to-detector distance.

Morstin and Woznick[143] have calculated the representative parameter, slowing-down length L_s for different neutron sources, which is plotted in Figure 35, vs. total hydrogen concentration in coal. Practically no matrix effects connected with the changes in ash content are observable. To simplify the problem of coal density variations, it was assumed that the pure carbon matrix of apparent density ϱ_o = 1.4 g/cm³ is saturated with hydrogen.

The high sensitivity of neutron absorption properties of coal restricts the feasibility of hydrogen determination with the use of thermal-neutron detection despite the strong dependence of the thermal neutron flux on the hydrogen content for very short source-to-detector distances. The situation improves slightly with dual detection systems. According to Morstin and Woznick[143] pulsed neutron logging can be applied in coal wells for determining the total hydrogen concentration in intersected coal seams.

VII. OTHER NUCLEAR METHODS

In this chapter we shall discuss analytical methods based on the application of nuclear phenomena to the study of coal. Neutron activation analysis was treated separately because of its widespread use. Other nuclear methods will be discussed only as much as they have been applied to coal analysis.

A. Charged Particle Activation

The physical principles of charged particle activation are the same as in neutron activation analysis, except for the nuclear reaction producing the radioactive isotopes. There are only a few reports on using charged particle activation to study coal.

For example, Schlyer et al.[144,145] have irradiated coal samples with ^{3}He ions to produce ^{11}C, ^{13}N, and ^{18}F via the following nuclear reactions: ^{12}C(^{3}He, ^{4}He)^{11}C; ^{12}C(^{3}He,d)^{13}N; ^{16}O(^{3}He,p)^{18}F. By comparing the ^{11}C to ^{18}F production, a direct comparison of the amounts of carbon and oxygen present in the coal samples can be obtained. The ^{13}N production provides an internal check for the amount of carbon present. In preference to calculating the yield of ^{11}C and ^{18}F on an absolute basis, organic standards with known oxygen-to-carbon ratios can be irradiated; this provides a direct correlation between the ^{18}F/^{11}C produced and the ^{16}O/^{12}C originally present in the sample. By using this ratio method, the necessity for accurately known reaction cross sections and beam fluxes is eliminated. If the standards and samples can be irradiated simultaneously, corrections for differences in saturation can be eliminated.

Here is the description of the measurements as performed by Schlyer et al.[145] All the samples were powdered under an inert atmosphere and dried at 60°C under vacuum for 16 hr to remove any moisture, which would result in an erroneous oxygen content. The samples were then placed into 1-mil aluminum packets such that each packet contained a sufficient thickness of the sample to stop the 10.0 MeV ^{3}He ions (the range of the ions at this energy is about 100 μm). Irradiations were performed on the Brookhaven National Laboratory 60-in. cyclotron. The beam energy was chosen to eliminate as many side reactions as possible and yet be sufficiently above threshold to produce an easily measurable quantity of activity in a reasonable amount of time. From examination of the excitation functions, it is evident that by keeping the energy at 10 MeV there should be no interfering isotopes produced. Since the beam characteristics are most stable at 56 and 33 MeV, the beam is degraded from 33 MeV to the desired energy by using 127 mg/cm² Al foils. By placing the samples (ten in this case) on a wheel so that as the wheel rotated each sample was exposed to the beam successively, the standards and coal samples could be irradiated essentially simultaneously. The wheel was operated at 60 rpm, and the duration of the irradiations was adequate to ensure that each sample was exposed to the beam about 300 times.

After the irradiation and an initial cooling-off period (~25 min to allow for decay of very short-lived isotopes produced during irradiation), the samples were transferred to counting vials for the assay of radioactivity. Since all the isotopes produced are positron emitters, the samples were counted in a NaI(Tl) well counter. The vials were counted at regular intervals for approximately 400 min. The data were punched onto paper tape that could be fed into a PDP-11 minicomputer. A computer program was written to sort as many as ten samples according to the start times and perform a least squares determination of the decay curve. The decay curve is analyzed for all three isotopes (^{13}N, $t_{1/2}$ = 10 min; ^{11}C, $t_{1/2}$ = 20 min; ^{18}F, $t_{1/2}$ = 110 min), or for only the two longer-lived components (^{11}C, ^{18}F) by waiting a sufficiently long time (120 min) for the ^{13}N to decay to insignificant levels. A gamma-ray spectrum was taken using a Ge(Li) detector and a 4096 multichannel analyzer to determine if any gamma-ray emitting isotopes were present. The only radiation apparent in the spectrum was that of the 511 keV gamma due to the positron annihilation. When irradiations were conducted with degrading foils, such that the energy of the ^{3}He impinging on the samples was greater than 10 MeV, there was a trace amount of the 478 keV gamma from ^{4}Be formed in the ^{12}C(^{3}He,2α)^{7}Be and ^{16}O(^{3}He,3α)^{7}Be nuclear reactions.

Figure 36 shows plot of ^{18}F to ^{11}C concentrations ratios for five organic standards as determined by Schlyer and Wolf[145] using charged particle activation analysis. Table 32 shows the characteristics of standards used.

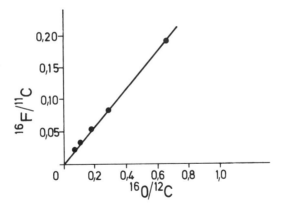

FIGURE 36. Plot of $^{16}F/^{11}C$ as determined by charged particle activation analysis vs. the $^{16}O/^{12}C$ of five organic standards (after Schlyer and Wolf[145]).

Table 32
STANDARDS FOR O/C CONCENTRATION RATIOS AS USED BY SCHLYER AND WOLF[145]

Standard	Atomic ratio (O/C)	O (wt %)	C (wt %)
1,5-diphenylcarbohydrazide	0.077	7	64
8-hydroxyquinoline	0.111	11	74
1,4-naphthoquinone	0.200	20	76
Benzoic acid	0.286	26	69
Adipic acid	0.667	44	49

Table 33
OXYGEN VALUES FOR DIFFERENT COAL SAMPLES AS DETERMINED BY SCHLYER AND WOLF[145]

Coal sample	C (wt%)	Actual O/C atomic ratio	Actual total oxygen (wt%)
PSOC 135	88.9	0.051 ± 0.003	6.0
PSOC 190	69.1	0.209 ± 0.004	19.2
PSOC 197	74.5	0.168 ± 0.004	16.7
N.D. lignite	58.1	0.307 ± 0.008	23.8
Penn. anthracite	86.2	0.043 ± 0.001	4.9
Penn. bituminous	80.2	0.097 ± 0.002	10.4
N. Mex. subbituminous	60.9	0.312 ± 0.010	25.3

Oxygen values determined for various samples are shown in Table 33. An added feature of this technique is the ability to vary the depth to which the charged particles penetrate by varying the energy of the particles (from about 0.1 μm to about 1 mm). By determining the oxygen concentration as a function of energy, it is possible to get an oxygen depth profile of the coal.

The chief advantages of charged-particle activation analysis according to Schlyer and Wolf[145] (1979) are

1. The samples are easily prepared for irradiation.
2. The O to C ratio is obtained directly from the analysis of the radioactive products and can be reported either as a ratio or as the absolute amount of oxygen present.
3. A large number of samples can be run during a given period of time, since the irradiation is short and the counting and irradiation procedures are separate.
4. The entire counting and analysis procedure may be easily automated.
5. The short irradiation times and ease of sample preparation keep the cost per sample very low.
6. Without modification of the procedures presently used in this laboratory, approximately 100 samples per day could be run including all operations.
7. Modifications of this technique have also been used by the authors on other oxygen-containing minerals and compounds where oxygen analysis is otherwise difficult or indirect.

B. Alpha, Beta, and Gamma Counting

Coal and coal ashes contain small concentrations of radionuclides which are members of one of the radioactive series shown in Table 34. Alpha and beta particles emitted by these nuclei can be used for the determination of concentration. Alpha spectroscopy can be used for the determination of ^{238}U, ^{234}U, ^{210}Po, and ^{230}Th, and some other radioisotopes, whereas beta counting is used for the determination of ^{210}Pb. Gamma spectroscopy is also used to measure the natural radioactivity of coal and as a measurement of uranium concentration in coal.

Here we shall describe the analytical procedures with alpha and beta counting as used by Styron et al.[146,147]

For uranium determination, Styron et al.[147] used the following procedure. The determination of uranium-234 and uranium-238 in samples involved a nitric acid-hydrofluoric acid dissolution, coprecipitation of uranium with iron hydroxide, an ether extraction, separation of uranium by anion exchange, and electrodeposition of the uranium followed by alpha pulse height analysis.

After uranium-232 tracer was added, organic matter was removed from the sample by heating at 600°C for 4 hr. Subsequently the sample was decomposed by a nitric acid-hydrofluoric acid digestion, and an iron hydroxide precipitation was performed to coprecipitate the uranium. The hydroxide precipitate was dissolved in 8 M hydrochloric acid, which was extracted with isopropyl ether to remove the bulk of the iron present. The 8 M hydrochloric acid solution was passed through an anion exchange resin column. Uranium, polonium, and bismuth were adsorbed onto the resin, while thorium and radium passed through the column. Plutonium and any unextracted iron were also retained by the resin, but were eluted with 6 M HCl containing hydrogen iodide. The iodide ion reduced plutonium (IV) to plutonium (III) and reduced the iron (III) to iron (II), and neither of these ions were retained by the ion exchange resin in 6 M HCl. The uranium was eluted from the column with 1.0 M HCl, while any zinc adsorbed on the column remained. The uranium was electrodeposited onto a stainless steel slide, and the slide was counted by alpha pulse height analysis using a silicon surface barrier detector for at least 1000 min.

The analytical procedure for thorium-230 in coal, soil, and vegetation as described by Styron et al.[147] involved a nitric acid-hydrofluoric acid dissolution, coprecipitation of thorium with iron hydroxide, separation of the thorium by anion exchange, electrodeposition of the thorium on stainless steel slide, and an alpha pulse height analysis with a minimum counting time of 1000 min.

Table 34
NAMES AND SYMBOLS FOR NUCLIDES OF RADIOACTIVE SERIES

(4n + 2)-series (uranium series)			(4n)-series (thorium series)			(4n + 3)-series (actinium series)		
Name	Old symbol	Symbol of nuclide	Name	Old symbol	Symbol of nuclide	Name	Old symbol	Symbol of nuclide
Uranium I	U I	^{238}U	Thorium	Th	^{232}Th	Actinouranium	Ac U	^{235}U
Uranium X$_1$	U X$_1$	^{234}Th	Mesothorium 1	Ms Th$_1$	^{228}Ra	Uranium Y	U Y	^{231}Th
Uranium Z, uranium X$_2$	U Z, UX$_2$	^{234}Pa	Mesothorium 2	Ms Th$_2$	^{228}Ac	Protoactinium	Pa	^{231}Pa
Uranium II	U II	^{234}U	Radiothorium	RdTh	^{228}Th	Actinium	Ac	^{227}Ac
Ionium	Io	^{230}Th	Thorium X	Th X	^{224}Ra	Radioactinium	RdAc	^{227}Th
Radium	Ra	^{226}Ra	Thoron	Tn	^{220}Rn	Actinium K	Ac K	^{223}Fr
Radon	Rn	^{222}Rn	Thorium A	Th A	^{216}Po	Actinium X	Ac X	^{223}Ra
Radium A	Ra A	^{218}Po	Thorium B	Th B	^{212}Pb	Antinon	An	^{219}Rn
Radium B	Ra B	^{214}Pb	Thorium C	Th C	^{212}Bi	Actinium A	Ac A	^{215}Po
Radium C	Ra C	^{214}Bi	Thorium C'	Th C'	^{212}Po	Actinium B	Ac B	^{211}Pb
Radium C'	Ra C'	^{214}Po	Thorium C"	Th C"	^{208}Tl	Actinium C	Ac C	^{211}Bi
Radium C"	Ra C"	^{210}Tl	Thorium D	Th D	^{208}Pb	Actinium C'	Ac C'	^{211}Po
Radium D	Ra D	^{210}Pb				Actinium C"	Ac C"	^{207}Tl
Radium E	Ra E	^{210}Bi				Actinium D	Ac D	^{207}Pb
Radium F (polonium)	Ra F	^{210}Po						
Radium G	Ra G	^{206}Pb						

A tracer, thorium-229, was used in evaluating chemical recovery of thorium. After the thorium-229 tracer was added to a sample, it was ashed to remove organic matter. The sample was decomposed by a nitric acid-hydrofluoric acid digestion with the addition of heat and stirring. Boric acid was added to complex the fluoride ion. Iron hydroxide was used to remove the thorium by coprecipitation. The hydroxide precipitate was dissolved in 8 M nitric acid and passed through an anion exchange resin column, onto which thorium was adsorbed. After 8 M nitric acid rinsing, the thorium was eluted from the column with 9 M hydrochloric acid. Electrodeposition of the thorium onto a highly polished, 0.75-in. diameter stainless steel disc was performed in an ammonium sulfate solution. The disc, or slide, was counted for a minimum of 1000 min in an alpha pulse height analyzer using a silicon surface-barrier detector.

The analytical procedure for polonium-210 in coal as described by Styron et al.[147] involved a hydrochloric acid leach of a raw coal sample, spontaneous deposition of the polonium onto a stainless steel slide, followed by an alpha pulse height analysis with a minimum counting time of 1000 min.

A tracer of polonium-208/209 was used to evaluate the chemical recovery of polonium-210. After polonium-208/209 tracer was added to the coal sample, the sample was dried at 30 to 40°C. The dried sample was then leached with 2 M HCl and filtered. The filtrate was transferred to a prepared electrodeposition cell. Polonium was isolated by deposition at 1.5 A current on a 0.75-in. stainless steel disc in the presence of ascorbic acid to complex the iron present in the sample. The disc was air dried and counted for a minimum of 1000 min in an alpha pulse height analyzer using a silicon surface-barrier detector.

Polonium was extracted from samples other than coal with a combination of nitric, hydrofluoric, and hydrochloric acids in the presence of polonium-208 tracer. For samples of soil, cement, and limestone, iron hydroxide precipitation removed the nitric acid and soluble fluorides. The iron was extracted with isopropyl ether from a solution adjusted to 8 M with hydrochloric acid. Instead of the iron hydroxide precipitation, vegetation samples were treated with 30% hydrogen peroxide and wet ashed as necessary to completely rid the sample of organic matter. Nitric and hydrofluoric acids were removed by heating. The precipitate was then dissolved, and the solution was adjusted to 1 M with hydrochloric acid from which the polonium was isolated by spontaneous deposition on a 1.88-in. diameter stainless steel disc. The polonium was then dissolved from the disc with nitric acid, and the iron was extracted with isopropyl ether, followed by electrodeposition onto a 0.75-in. diameter stainless steel disc for determination by alpha pulse height spectroscopy.

According to Styron et al.,[147] determination of lead-210 in samples involved dissolution with nitric and hydrofluoric acids, coprecipitation with iron hydroxide, lead nitrate precipitation, separation of lead by anion exchange, and mounting the lead on a 2-in. diameter stainless steel slide for counting.

After stable lead tracer was added, organic matter was removed from the sample by heating in a muffle furnace. The sample was further decomposed by a nitric acid-hydrofluoric acid digestion. Lead was coprecipitated with iron hydroxide and this precipitate was dissolved in dilute nitric acid. The nitric acid concentration of this solution was increased to 75%, precipitating out the nitrates of lead; the alkaline earth elements; and some aluminum, iron, and silicon. This precipitate was dissolved in distilled water and reprecipitated as the hydroxide which was readily soluble in 1.5 M HCl. The 1.5 M HCl solution was passed through an anion exchange resin column. Lead and bismuth were adsorbed on the column, while the other elements that may have been present passed through. Lead was eluted with distilled water, leaving the bismuth on the column. Atomic absorption analysis of this solution determined the lead recov-

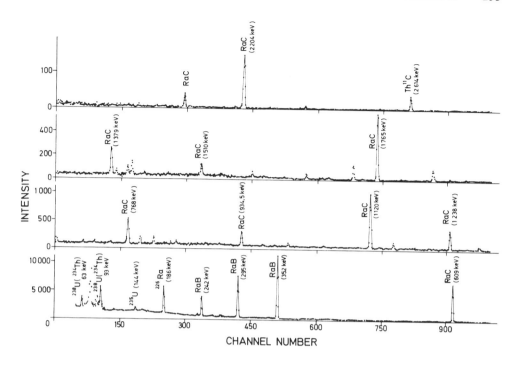

FIGURE 37. Gamma spectrum from a lignite sample obtained with Ge(Li) detector.

ery. The sample was mounted on a 2-in diameter stainless steel slide for beta counting of the ingrowth of bismuth-210. Data reduction by computer analysis determined lead-210 concentration from counting data.[147]

Gamma spectroscopy applied to coal analysis is often used to determine uranium concentration in coal and coal ash samples. Figure 37 shows the gamma spectrum from a lignite sample which contained several hundred parts per million of uranium. The lines present in the spectrum are due to nuclides from U and Th series.

C. Mössbauer Spectroscopy

Since its discovery in 1958, the Mössbauer Effect has been widely used by researchers in nuclear and solid-state physics. It is now being accepted as a standard research method in many laboratories. A variety of books describing the technique is available (see for example Wertheim,[148] May,[149] and Gibb,[150]). Owing to the great importance of iron in the crust of the earth and its abundance in rock-forming minerals, a considerable amount of work has been carried out using the Mössbauer isotope ^{57}Fe.

The most common use of the Mössbauer effects in minerology and geology has been in the determination of the oxidation states of iron in various minerals. The study of the Mössbauer spectral area also gives valuable information on the abundance of the different minerals in rocks. Application of the Mössbauer effect to poorly crystallized materials has acquired enormous relevance owing to the unique sensitivity of the method. Process development and control for removal of pyrite (FeS_2) from coal requires techniques for the accurate measurements of the FeS_2 content. The use of Mössbauer Spectroscopy is a simple, effective and relatively inexpensive monitor of pyrite in coal. In particular, Mössbauer Spectroscopy provides convincing evidence that the standard wet chemical technique for the determination of pyrite in coal (ASTM D 2492-68, reapproved 1974) can be inaccurate by up to 20%.[151]

The ASTM method is based upon the assumption that pyritic iron in coal is insoluble in hydrochloric acid but soluble in nitric acid, and that the iron soluble in hydrochloric acid is also soluble in nitric acid. It follows that the difference Fe(HNO₃)-Fe(HCl) should be equivalent to the pyritic iron. This then gives the pyrite sulfur in the coal. According to Levinson and Jacobs,[151] this procedure involves three potential problems:

1. It neglects the possibility that some (extremely fine particle) FeS_2 in coal might be removed by the HCl.
2. It is possible that some of the pyrite particles could be completely surrounded by the coal (even after crushing in the test sample) and might therefore not be removed by the nitric acid.
3. Many coals contain other iron-bearing minerals. Some of these might dissolve in hydrochloric acid but be only partly soluble in nitric acid.

The characteristic of Mössbauer spectroscopy when applied to the measurements of pyrite in coal are [151]

1. The technique is sensitive only to Fe in coal and will usually differentiate between various iron-bearing impurities in coal, e.g., between FeS_2 and siderite, magnetite, ferrous sulfate, etc.
2. The measurement is unaffected by the presence or absence of organic sulfur.
3. Carbon (coal) surrounding the FeS_2 does not affect the measurement.
4. MS of pyrite in coal is unaffected by any poor crystallinity of the pyrite.

Let us describe the work by Montano.[152] The Mössbauer spectrometer used in this work was a conventional constant acceleration spectrometer. A 25-mC ^{57}Co source was used. In this source the ^{57}Co is transformed by electron capture into the 137 keV state of ^{57m}Fe. After 10^{-8} sec, this level decays with the gamma emission of 123 keV to the 14.4 keV level. This transition is used in the Mössbauer effect. After $\sim 10^{-7}$ sec, the 14.4 keV level decays through conversion or gamma emission of 14.4 keV to the ground state. The detector in this experiment was a Kr-10% CO_2 proportional counter.

Samples from major West Virginia coal seams were selected for this study. The time interval between sampling and preparation varied from 2 months to 2 years, but most of the samples reported here were prepared within 3 months of collection. Grinding, screening, and riffling of samples to obtain 50 g of 60 U.S. mesh coal for analysis were mounted in lucite containers that were consequently sealed. The average surface densities of the samples were 150 to 250 mg/cm². Several samples were analyzed from the same seam in order to check for consistency of the results. The Mössbauer spectra were analyzed using a nonlinear least-square fit program and assuming Lorentzian lineshapes. Equal intensities for the pyrite peaks were assumed and introduced as a constraint in the fit program.

Figure 38 shows Mössbauer spectra for two different coal samples. On the top of the figure is a spectrum of high pyrite coal. The two lines responsible for the FeS_2 spectrum can be clearly identified. Two other lines, identified as those of Fe^{2+}, are also observed. The figure on the bottom shows a Mössbauer spectra of a low-pyrite coal. Here there is more iron sulfate present relative to pyrite, probably owing to moisture in the coal that oxidizes some of the pyrite, producing iron sulfates.

It has been suggested that a part of the sulfates found in the low-temperature ash could have been formed during the ashing process. In order to examine such a possibility, Montano[152] has carried out Mössbauer measurements on samples after low-temperature ashing. In the measured spectra, he could detect the presence of a new phase $Fe^{3+}(Fe_2(SO_4)_3)$. The iron oxidation took place at the expense of the sulfate and the

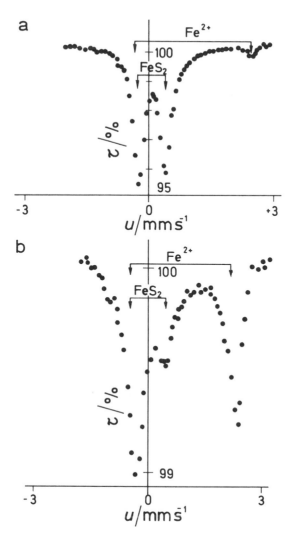

FIGURE 38. Mössbauer spectra for two different coal samples: (a) high pyrite coal, and (b) low pyrite coal (after Montano[152]).

pyrite. The appearance of a quadrupole splitting can be understood if the Fe^{3+} is formed on the surface of FeS_2 and $Fe(SO)_4$, or appears in its hydrate form. This experiment can be considered as a definite proof of the production of new mineral species in the ashing process. Similar results were observed with the other samples. In the high-temperature ash, all the iron compounds were converted to Fe_2O_3 for all the samples studied.

D. Nuclear Magnetic Resonance (NMR) Spectroscopy

The applications of nuclear magnetic resonance to coal studies can be numerous. NMR can be used to study the chemical identification of molecular fragments CH_3, or -OH or $-NH_2$ in molecules via the resonant frequency of absorption of the nucleus within the fragment. NMR active nuclei in coals include ^1H, ^{13}C, and ^{15}N. The area under the absorption peak associated with a particular nucleus in a given arrangement can be used to provide a quantitative analysis of that fragment. The type of resolution needed for this qualitative and quantitative analysis of molecular fragments, combined with current knowledge of the dispersion of resonant frequencies (chemical shifts) of ^{1}H and ^{13}C in solid coals indicates that with present state-of-the-art experiments, the following may be identified in solid coals: (1) total aromatic carbon, (2) total aliphatic carbon, (3) carbonyl carbon, (4) carbon adjacent to oxygen in aliphatic others, (5) aromatic plus hydroxyl hydrogen, and (6) aliphatic hydrogen. Considerations involving the total nitrogen in coals, the natural abundance of the same isotope ^{15}N, and the sensitivity to detection of this nuclide indicate that detection should be possible at fields of above 7 T.

The situation in coal-derived liquids is slightly different. The liquids have less of a dispersion of chemical fragments than do solid coals. Use of selective enhancement of nuclei in specific molecular environments via J-cross polarization (JCP) might distinguish nitrogen in primary, secondary, and tertiary amines, and between these and aromatic nitrogen in pyridine-type structures.

NMR can be used to monitor self-diffusion constants of small molecules such as H_2O, CH_3OH, Pyridine, Benzene, and hydrogenated naphthalenes in coals.

NMR can be used to probe dimensions of pores and voids in solid coals via the shielding anisotropies of ^{13}C in doped molecules of selected sizes, e.g., CO_2, C_6H_6, and $C_{10}H_{16}$.

Let us describe some of the work published in the literature. In an interesting study done by Retcofsky et al.,[31,153] the distributions of hydrogen and carbon among various organic structures in solvent extracts of selected coals were determined by high-resolution proton and carbon-13 magnetic resonance spectrometry. Structural parameters including the aromaticity, the degree of aromatic ring substitution, and the average size of the condensed aromatic ring system have been deduced for each extract using the nuclear magnetic resonance data in conjunction with the elemental analysis of the material. The potential of two other magnetic resonance techniques, proton-enhanced nuclear induction spectroscopy and proton-decoupled high-resolution carbon-13 magnetic resonance in coal research, is discussed. The results of the investigation are in accord with commonly held views of coal metamorphism; they do not, however, support recent reports challenging the classical view that coals are highly aromatic materials.

Gerstein and Pembleton[154] have applied pulsed nuclear magnetic resonance spectrometry for nondestructive determination of hydrogen in coal. Initial value of free induction decays are shown to yield a rapid, nondestructive quantitative measure of proton in coals provided: (1) an appropriate $\ln[A(t)]$ vs. time plot is made; (2) the zero of time is chosen to be the center of the pulse; (3) the pulse width for a 90-degree pulse is not longer than $T_2^*/4$; (4) recovery time of the receiver-dc amplifier system is of the order of the pulse width; (5) both coal sample and calibration sample are exposed to a uniform rf field; and (6) the calibration sample is measured under the same experimental conditions as is the coal sample.

A new technique, Computer-Assisted Molecular Structure Construction (CAMSC), has been developed by Oka[155] for the structural elucidation of coal-derived compounds. It utilizes elemental analysis, n.m.r. spectra, and molecular weight data to determine the allowable combinations of functional groups constituting the structure. The results compare well with those obtained by other methods and in some cases provide additional information.

In the measurements done by Alemany et al.,[156] the proton and carbon spectra of the butylated Illinois #6 fractions were examined in chloroform-d or pyridine-d_5. The Fourier-transform proton spectra were recorded at 22°C, the probe temperature, in 5 mm tubes at 270 MHz with a Bruker HS-270 spectrometer. Neither filtration, the removal of oxygen, nor variations in the solvent or in the temperature up to 50°C had a discernible impact on the quality of the spectra. The spectra recorded after several months were not different from the spectra recorded initially.

The Fourier-transform, proton-decoupled carbon spectra were recorded at 32°C, the probe temperature, in 20 mm bulbs suspended in deuterium oxide at 15.09 MHz with a Transform Technology 14 spectrometer. Both the proton and the carbon n.m.r. spectra are dominated by the resonances of the added butyl groups. Thus, their interpretation requires information concerning the chemical shifts for the hydrogen and carbon atoms of butyl groups in different environments. The shifts for the hydrogen atoms are quite well known. The shifts for the carbon atoms in many butyl groups were collected from recent articles. To complement these data, the authors have prepared other compounds including cis-9,10-dibutyl-9,10-dihydroanthracene, 1-butylnaphthalene, benzyl butyl ether, 1- and 2-naphthyl butyl ether, and N-butylquinolinium and N-butylisoquinolinium iodide.

The data for about 100 compounds indicate that the resonances of the α, β, γ, and δ carbon atoms are dispersed over 66, 13, 9, and 3 ppm, respectively. The extent of C-butylation compared to O-butylation was assessed by examination of the C_γ resonances which appear at δ 20.0 ± 0.7 for butyl ethers (12 compounds) and at δ 23.4 ± 0.8 for butyl groups in alkanes, alkenes, and arenes (30 compounds). The extent of C-butylation on quaternary, tertiary, and aromatic carbon atoms was assessed by analysis of the C_β resonances. The data for 22 compounds indicate that the C_β signal for a butyl group bonded to a quaternary carbon atom is at δ 26.7 ± 0.3, while the C_β signal for a butyl group bonded to a tertiary carbon atom is at δ 30.0 ± 0.8. The related resonances of butyl aryl ethers appear at δ 31.5 ± 0.2 and those of butyl alkyl ethers appear at δ 32.5 ± 0.6. The C signals for monobutylarenes occur at 33.4 ± 0.6. Finally, the C_α resonances of butyl aryl ethers appear at δ 67.8 ± 0.2, whereas this signal is downfield of δ 70 for butyl alkyl ethers.

Definite information concerning the structures of the alkylated products can be obtained from the carbon n.m.r. spectra. Moreover, many aspects of the interpretation can be tested by the scrutiny of the related proton spectra.[156]

E. Electron Spin Resonance (ESR) Spectroscopy

Although electron spin resonance (ESR) is not a nuclear technique, strictly speaking, it should be mentioned here because of its usefulness in the study of coal structure. Electron spin resonance is an important tool in the exploration of the microenvironment where there are large molecules in their natural state without chemical decomposition or physical deformation. Furthermore, solubilization of the substance is unnecessary, thus allowing spectra to be taken *in situ*.

Yen[157] has performed a number of ESR studies on fossil fuels, mainly in crude oils. Information from such studies might help one to understand the mechanisms of biogenesis, transformation, and metamorphism of fossil remains in nature.

The study of free radicals in coal by ESR spectroscopy has been studied by several research groups. Recently it has been also applied to the study of liquefaction of coal. Free radicals are generally accepted as playing an important role in the liquefaction of coal. Petrakis and Grandy[159] tried to determine the relation between the nature and properties of free radicals and process variables such as residence time in a reactor, heating rates to achieve highest temperature and the role of various solvents and gases in stabilizing free radicals in lower-molecular-weight moieties. The free radical concentration after thermal treatments correlates well with the propensity of a given solvent to donate its hydrogen.

In the measurement by Petrakis and Grandy,[159] the samples for ESR measurements were prepared by placing approximately 0.3 g of coal, mixed in a 1:1 ratio by weight with the appropriate solvent, in a 2-mm i.d. ESR tube. The tube and sample were then put in a 6-mm i.d. stainless-steel bomb reactor, purged repeatedly with either H_2 or N_2 gas and pressurized to about 9.3 MPa. A tube furnace was then placed around the reactor and heating was begun. Temperature control and measurement were accomplished by several thermocouples attached to the outside wall of the reactor. Previous calibration runs have shown that there is a 2- to 3-min. delay in the sample reaching the temperature of the outside wall of the reactor vessel. Operating pressure after heating to the desired temperature was adjusted to 12 MPa. After heating for the specified period of time, the reactor was cooled to ambient temperatures in about 3 min by air jets. The system was then depressurized and the sample tube removed from the reactor and sealed in air. At least one duplicate sample was prepared for each set of experimental conditions. The significance of any variable on an ESR parameter was determined by constructing an analysis of variance table and using the proper f test.

The ESR instruments used to measure the g-values, linewidths and concentrations of radicals in the samples were a pair of Varian V-4500® spectrometers with single and dual cavity capabilities.

The Wyodak subbituminous coal was selected to determine the most appropriate range of experimental conditions. In experiments performed without solvent, a slight difference in radical concentrations (less than a factor of 2) was found between samples run in H_2 (29×10^{18} spins per gram) vs. N_2 (45×10^{18} spins per gram). Samples treated in a vacuum with the same heating rate, temperature, and residence time had a radical concentration twice that of the N_2-treated samples (90×10^{18} spins per gram). These samples were heated to 440°C at 13°C min^{-1} and held at that temperature for 2 hr. No differences in g-values (2.0028) or linewidths (5.1 vs. 5.5 G) owing to the N_2 or N_2 gas treatments were evident.

The Wyodak coal was then reacted with tetralin or naphthalene in H_2 or N_2 gas under the same experimental conditions described above. Linewidths and g-values were the same within experimental error, for all 4 combinations of gases and solvents, i.e., about 5 G and 2.0027 to 2.0028, respectively. The radical concentration of tetralin-treated Wyodak coal appears to be invariant with respect to the gas used and is about 25 to 30×10^{18} spins per gram.[159,172]

VIII. MASS SPECTROSCOPY

Mass spectroscopy has been in use for almost half a century, and it is an accepted method in many laboratories. The mass spectrometer is actually an instrument that determines the mass to charge ratio (m/e) of gaseous ions. These ions are generally positive ions that are produced within the same vapor (which is contained in a vacuum system). They are sorted out according to their m/e ratio, and measured. A mass spectrometer can be said to consist of four parts: sample introduction system, ion source, mass analyzer, and ion detector. There are numerous variants to each of these four entities, and to discuss them fully is beyond the scope of this book.

There are different ways of introducing the sample into the instrument. The simplest method of sample introduction is as a gas. Such samples can be admitted through a metered glass manifold. Liquid samples are also easily introduced: a micropipet or similar device can be used to inject the sample into the system, whereupon it immediately vaporizes. Some solids of low melting point can be vaporized through heating in an introduction chamber; whereas others have to be treated by spark discharge technique which introduces the samples directly into the chamber. The material to be analyzed is molded into electrodes, placed in the chamber, and a high intensity radio frequency spark applied.

Spark source mass spectrometry (SSMS) has been applied to coal analysis by many researchers. In a SSMS analysis, an ion beam of the substance being investigated is produced in a vacuum by igniting a spark between two conductors with a pulsed, high-frequency potential of 50 kV. During this process, the electrode substance is evaporated and ionized. The ions produced are accelerated by a constant potential of 25 kV through the source slits into an electrostatic radial field which functions as an energy filter. As the ions pass through the magnetic field, the ion beam is split by deflection according to the mass-to-charge ratio. These charged particles impinge in focus on an ion-detector (photographic plate or electron multiplier) to form the mass spectrum. Elemental identification and abundance measurements can be made from the position and the relative intensity of the lines when the total ion beam current is known. This total ion current is measured by a monitor located just ahead of the magnetic analyzer.

SSMS is a rapid, multielement technique with minimal matrix effects. It is capable of simultaneously determining all elements of interest. SSMS complements INAA by providing analytical capability for several elements (Cd, Cu, Hg, Ni, Pb, and Zn) not readily determined by INAA; in addition, it provides a check for INAA results for the remaining elements (As, Co, Cr, Mn, Mo, Se). SSMS permits analysis without sample dissolution, thereby minimizing sample preparation time and reducing the probability of contamination or constituent loss. Determinations may be made for many elements at concentrations as low as 1 ppb; also semiquantitative determinations as low as 100 ppb for some elements can be performed by electrical scanning. This technique allows detection of all elements simultaneously during an electrical scan, including interstitial gases with minimal spectral overlap, matrix effects, or interelement effects. Also, this method exhibits linear response for ionic species of any element with the ion intensity being proportional to the concentration of that element in the sample.[158]

Although detection limits with this method for most elements in coal and fly ash are in the part-per-billion range, accuracy, they may be only ±50%, varying with the concentration of interferents, as well as with data interpretation. For increased precision, isotope-dilution spark source mass spectrometry (IDSSMS) can be applied to the analysis of many elements to give ±3% precision.[160,161,173] It is normally advisable to dissolve samples for IDSSMS to assure isotope equilibrium during analysis. SSMS analyzes extremely small samples, and for certain applications it offers distinct advantages over other techniques. The analysis of such small samples, however, tends to enhance inhomogeneity effects.

Sharkey et al.[162-164] analyzed sample pairs of respirable-range mine dusts and prepared coal dusts (3.3 to 3.5 μm) by a spark source mass spectrometry for 64 trace elements to determine differences in composition. The coal samples used represented 8 coal seams from the following states: Pennsylvania, West Virginia, Virginia, and Utah. Respirable mine dust samples were obtained from personal sample filters and were collected during actual mining operations, while prepared coal dust samples were obtained from crushing, riffling, and separating procedures. Results indicated that for a majority of the sample pairs a number of elements, including Ag, Cd, Cu, Cr, Rb, Ca, Cl, P, and Br, showed higher concentrations in the mine dust collected during mining operations than was found in the corresponding respirable range coal dust prepared in the laboratory.

For additional SSMS applications to coal analysis see the paper by Brown et al.[158] In their work, coal and its combustion by-products were examined to assess environmental impact of potentially toxic trace element emissions. From analysis of a whole coal, fly ash, and bottom ash, the fraction of trace elements volatilized and lost in the stack gas and that retained in particulates within the plant was determined.

In the work by Carter et al.[165,166] spark source (SSMS) and thermal emission (TEMS) mass spectrometry are used to determine parts per billion to parts per million quantities of elements in energy sources such as coal, fuel oil, and gasoline. Toxic metals — cadmium, mercury, lead, and zinc — may be determined by SSMS with an estimated precision of ±5%, and metals which ionize thermally may be determined by TEMS with an estimated precision of ±1% using the isotope dilution technique. An environmental study of the trace element balance from a coal-fired steam plant was done by SSMS using isotope dilution to determine the toxic metals and a general scan technique for 15 other elements using chemically determined iron as an internal standard. In addition, isotope dilution procedures for the analyses of lead in gasoline and uranium in coal and fly ash by TEMS are presented in their paper.

St. John et al.[167] have used field ionization (FIMS) and field desorption mass spectrometry for the characterization of coal liquefaction and fuel products. These include the capability of determining molecular weight profiles up to 1000 atomic mass unit (amu) with amu resolution and the ability to obtain and record molecular weight profiles as a function of sample temperature during a temperature-programed evaporation of the analyzed sample. Combined with appropriate liquid chromatographic separation techniques or certain fast and quantitative derivatization procedures to separate coal liquefaction products into families of compounds, advanced FIMS offers an unprecedented, precise, and meaningful analytical methodology for the characterization of coal products. The field ionization technique described in this paper is not yet perfect, and it requires some further development in the areas of instrumentation, sample pretreatment, and data handling. However, there is sufficient evidence that this technique can provide the basis for one of the most comprehensive analytical methodologies ever available to coal research. Extensive experimental data are included.

Hayatsu et al,[168] have used the following coals in their study: a Wyoming lignite (Decker Coal Co., Sheridan, Wy. composition: 67.3% C, 4.8% H, 1.3% N, 1.2% S, 25.4% O by difference), a high-volatile Illinois bituminous (Peabody Coal Co., Seam #2, composition: 77.8% C, 5.4% H, 1.4% N, 2.1% S, 13.3% O by difference) and a Pennsylvania anthracite (Penn State PSOC #85, composition: 91.3% C, 3.9% H, 0.6% N, 1.1% S, 3.1% O by difference). Prior to the oxidation experiments, the coals were leached by heating with concentrated hydrochloric acid to minimize problems in isolating the resulting carboxylic acids and to remove trapped volatiles. For investigations of the trapped organic compounds, freshly ground raw coals were used.

Hayatsu et al.[168] used a modified Bendix Model 12 time-of-flight mass spectrometer (TOFMS) to obtain low resolution mass spectral data. Samples are admitted to the TOFMS via either vacuum lines, a variable-temperature solid inlet, or two gas chromatograph interfaces. One GC-MS consists of a temperature-programed wall-coated open tubular (WCOT) column connected directly to the source of the TOFMS. Total ion current of the TOFMS was used for detection with coincident mass spectra providing identification. The other GC-MS used a Perkin-Elmer 3920B gas chromatograph interfaced to the TOFMS with a continually variable split between a flame-ionization detector and the source of the mass spectrometer.

High-resolution mass spectral data were obtained on an AEI MS-902 using a PDP-8 computer for data reduction.

Solvent extraction is described by Hayatsu et al.[168] in this way: about 25 g of raw coal was extracted by refluxing with 150 mℓ of benzene-methanol (3:1) for 24 to 48 hr. Yields as percentage of the raw coal weights were 4.3 for lignite, 6.3 for bituminous, and 0.04 for anthracite coal. The lignite and bituminous coal extracts were separated into 4 and 3 fractions, respectively, on a neutral alumina column using hexane-benzene solvent mixtures with an increasing percentage of benzene.

Aqueous sodium dichromate oxidation is described by Hayatsu et al. in this way: a

3-g sample of coal was heated at 250°C for 40 hr with 120 mℓ of 0.4 M $Na_2Cr_2O_7$, in an autoclave agitated with a magnedash stirrer. The autoclave was cooled to 60 to 80°C, emptied, and washed 3 times with 20 mℓ of hot water. The unreacted coal and the hydrated Cr_2O_3 which were formed in the reaction were removed by filtration.

The clear, alkaline filtrate was extracted with benzene-ether (1:1) to remove nonacidic compounds, and then acidified with cold concentrated HCl in an ice bath. The acidic solution was distilled off under reduced pressure at <40°C. The resulting mushy residue was extracted 3 times with 30 mℓ methanol-ether (3:1) and twice with 20 mℓ chloroform. The combined extracts were dried over Na_2SO_4, and then evaporated at room temperature under an N_2 stream. The resulting acids were dissolved in methanol-ether (3:1), and esterified with diazomethane. All compounds were identified and measured by a TOF variable-temperature solid inlet, high-resolution mass spectrometry and GC-TOFMS.

Essentially 100% of the lignite and bituminous coals were converted to volatile compounds and compounds soluble in either water or organic solvents, but only 62 to 67% anthracite was so converted. The weights of the soluble compounds were 51%, 70%, and 55% of the original coal weights for lignite, bituminous, and anthracite coals, respectively. The anthracite also yielded 17% of a humic acid-type material (of high molecular weight soluble only in alkaline aqueous solution).

Photochemical oxidation is described by Hayatsu et al.[168] in this way: a 1-g sample of coal was suspended in 50 mℓ of 10% aqueous HCl in a 300 mℓ quartz flask. The flask was irradiated with ultraviolet light from a high-pressure mercury lamp for 8 days while air was bubbled through the HCl solution. The temperature rose to 45 to 50°C during the reaction. After filtration, the nonreacted residue was washed successively with hot water, methanol, and chloroform. The filtrate and washings were evaporated under reduced pressure below 40°C. The resulting acids were esterified with diazomethane. Almost no photo-oxidation was observed for anthracite, while lignite and bituminous coal were oxidized in yields of 25 and 30%, respectively. If the unreacted residue was further oxidized by the same procedure, 22 to 24% of the residue was oxidized, again with the same distribution of products.

Hydrogen peroxide-acetic acid oxidation is described by Hayatsu et al.[168] in this way: a 2-g sample of coal which had been solvent extracted previously was stirred in 250 mℓ of glacial acetic acid at 40°C; 4 20 mℓ aliquots of 30% aqueous H_2O_2 were added dropwise at 24 intervals. After 96 hr, the mixture was filtered and the filtrate evaporated under vacuum. The resulting solids were taken up in methanol and methylated with diazomethane. The conversion of the solid coals was 80% for lignite and 77% for the bituminous coal with a weight yield for soluble acids of 96 and 127%, respectively.

Characterization of coal products by mass spectrometry is also described in a paper by Lumpkin and Azel.[169] They have used mass spectrometry to analyze coal molecules produced by solvent refining, liquefaction, pyrolysis, and extraction. In such a way, accumulated knowledge helps in understanding the organic chemistry of coal and is vital for the development of coal liquefaction processes and the further upgrading of the liquefaction products. Their paper describes some of the mass spectrometry procedures and their uses in the analysis of coal-derivative products. The following fractions were analyzed—naphtha boiling range; saturated high boiling range; aromatic and polar fractions of higher boiling range.

Another use of mass spectrometry was reported by Kershaw and Barrass.[170] They have studied the mechanisms of coal hydrogenation by the use of deuteration. In their study, hydrogenation and deuteration of the coal (50 g; New Wakefield Colliery, Transvaal, chemical analysis: 79.1% C, 5.4% H, 2.1% N, 2.3% S, 11.1% O) was carried out in a 1-ℓ rotating autoclave, using stannous chloride (1% tin by mass of coal) as catalyst at 450°C for 1 hr. The autoclave was charged with a mixture of helium and deuterium (run 1) or hydrogen (run 2). The ratio of deuterium or hydrogen to helium was 1:1.9, and the pressure was 25 MPa at 450°C. The final pressure at room temperature showed that an excess of deuterium or hydrogen was present during the reaction. Every care was taken to see that the conditions of the two runs were as identical as possible.

After the autoclave had cooled, the gases (gas yield 23% coal chemical) from the autoclave, in run 1, were released through a liquid-nitrogen trap to the atmosphere. The trap, which condensed the gases formed during deuteration, was removed from the liquid nitrogen and slowly allowed to warm up. The gases released from the trap were bled into a quadrupole mass spectrometer and the mass spectra recorded. Methane, 50% of the gas, on a hydrogen (deuterium) and air-free basis as determined by gas chromatography), which was the predominant gas formed on hydrogenation, has a much lower boiling point than the other gases. Thus, by controlling the temperature of the trap, predominantly methane was allowed into the mass spectrometer. As the other gases present do not have strong fragment ions where methane and its deuterated isomers have their major ions, the mass spectra of only methane and its deuterated isomers were recorded, within the range m/e 14 to 20. From the known fragmentation patterns of methane and the deuteromethanes, the relative amounts of each of the deuteromethanes were calculated as 13% CH_4, 21% CH_3D, 37% CH_2D_2, 22% CHD_3, and 7% CD_4. With a high pressure of deuterium, neither hydrogen abstraction by methyl, methylene, or methine radicals from other molecules nor hydrogen formation from the coal should occur to a significant extent. Therefore, the nondeuterated methane was presumably present in the coal while the deuteromethanes came from differently substituted carbon atoms in the coal molecule. The source of methane, under the conditions of hydrogenation as used in run 1, was 13% CH_4, 21% CH_3, 37% CH_2-, 22% >CH- and 7% >C<. The other alkanes (ethane to pentane) were also monitored by mass spectrometry and the spectra showed that all but two of the hydrogen atoms of these alkanes were replaced by deuterium in significant percentages. The complexity of the fragmentation patterns and variety of possibly isometric deuterium substitution makes quantitative work impossible.[170]

IX. COMPARISON OF DIFFERENT METHODS

Many interlaboratory analysis comparisons have been done with the aim of helping analysts to be more critical about the method and procedure they use. The conclusions published after such interlaboratory comparisons are of questionable value and should not always be taken very seriously. We shall mention here only some of such publications when different methods are used for coal and coal ash analysis. Some of the work in this area has already been noted.[38,52]

We should mention again the program initiated by U.S. Environmental Protection Agency, results of which were published by von Lehmden et al.[38] In this study, 9 laboratories were asked to determine the concentration of 28 elements in portions of the same coal and ash samples. The analytical methods employed were neutron activation analysis, atomic absorption, optical spectrometry, anodic stripping voltametry, spark-source mass spectrometry, and X-ray fluorescence. The determinations from the different laboratories were evaluated to assess the comparability of the various methods as applied to these matrices. Table 35 shows the results for the coal samples. The following results were obtained from this study. Only definitive concentrations were used for these conclusions (no less-than values were considered).

Table 35
COAL ANALYSIS FOR TRACE ELEMENTS — COMPARISON OF METHODS[38]

Analytical method (ppm, by weight)

Element	SSMS[a] (1)	SSMS[a] (3)	SSMS[a] (6)	OES[b] (1)	OES[b] (3)	NAA[c] (2)	NAA[c] (3)	NAA[c] (4)	NAA[c] (5)	NAA[c] (6)	AAS (3)
Hg	>2	>2	>0.10			>0.2		>0.02	0.03		0.0511
Be	0.4		0.4					>40			
Cd	6	>1	0.7	>1	>0.1		>3			1	
As	2	2	0.25	>30	>10	>1	1.4	1.6		6.0	
V	10		7.7	>100	>50	7.0	5.5	7		5.0	
Mn	20	3	1.9	10	10	7.6	4.8	6.7			
Ni	40	4	6.0	>10	20			>20			
Sb	0.6		0.04	>30	>10	0.14	0.2	0.4			
Cr	>30	7	12	>10	>30	3.4	5.0	4.8			
Zn	>100	5	6.6	>100	>50			>100			
Cu	10	9	4.5	10	10			0.4			
Pb	>4	4	1.8	>30	>10						
Se	>15	>8	0.1			1.0	5.0	2.0	1.5		
B	15	5	14	10	7						
F	>2	4	60								
Li	0.3		2.8	>300	10						
Ag	>2		>0.1	>1	>1			>2			
Sn	3		0.19	>30	>10						
Fe	2.000	2.000	1.800	2.000	3.000	2.400	2.700	3.140		8.000	
Sr	100	50	46	>30		160		120		80	
Na	600	100	660	300	500	800	870	840		800	
K	100	50	200	150	20			280		100	
Ca	10.000	10.000	5.800	8.000	10.000		2.200				
Si	6.000	10.000	10.000	3.000	20.000		5.500	7.070			
Mg	2.000	700	2.000	600	100	2.600		920		1.000	
Ba	400	30	110	500	200		220	430		>2.0	

Note: Abbreviations used in table are: SSMS, spark source mass spectrometry; OES, optical emission spectrometry; NAA, neutron activation analysis; and AAS, atomic absorption spectrometry.

[a] Analysis on sample direct. [b] DC arc on sample direct. [c] Instrumental NAA.

Table 36
SENSITIVITIES OF DIFFERENT METHODS FOR COAL ANALYSIS AS REPORTED BY DULKA AND RISBY[171]

	NAA[a] (g)	SSMS (ng)	CIMS[b] (g)	ICPAES (µg/mℓ)	NFAAS	XRFS (µg)	ASV
Ag	10^{-10}—10^{-9}	0.2		0.004	0.001 ng/mℓ	1.2	0.25 ppb
As	10^{-10}—10^{-9}	0.06				0.11	
B							
Ba	10^{-10}—10^{-9}	0.2		0.001	6×10^{-12}g	0.12	
Be		0.008		0.005	3×10^{-14}g		
Cd	10^{-9}—10^{-8}	0.3		0.002	0.03 ng/mℓ	0.40	0.005 ng/mℓ
Co	10^{-10}—10^{-9}	0.05	1×10^{-11}	0.003	2×10^{-12}g	0.05	
Cr	10^{-8}—10^{-7}	0.05	1×10^{-11}	0.001	1.2×10^{-12}	0.00006	
Cs	10^{-9}—10^{-8}	0.1				0.15	
Cu	10^{-10}—10^{-9}	0.08	1×10^{-11}	0.001	6×10^{-13}g	0.00002	0.005 ng/mℓ
Ga	10^{-10}—10^{-9}	0.09		0.014	1×10^{-12}g	0.01	0.4 ng/mℓ
Hg	10^{-10}— 10^{-9}	0.6		0.2	8×10^{-11}g	0.24	$4.0 \times 10^{-9} M$
La	10^{-11}—10^{-10}	0.1	5×10^{-11}	0.003	0.1 ng/mℓ	0.12	
Li		0.0006					
Mn	10^{-12}—10^{-11}	0.05	1×10^{-11}	0.0007	2×10^{-13}g	0.00015	
Ni	10^{-8}—10^{-7}		1×10^{-11}	0.006	4×10^{-12}g	0.06	0.1 g/mℓ
P						0.001	
Pb	10^{-7}—10^{-6}	0.3		0.008	0.002 ng/mℓ	0.0003	0.01 ng/mℓ
Rb						0.0075	
Sb	10^{-10}—10^{-11}						
Se							
Sn		0.3		0.3	2×10^{-12}g	3.9 ppm	2.0 ng/mℓ
U	10^{-10}—10^{-11}			0.08		0.00002	
V	10^{-11}—10^{-10}	0.04	1×10^{-11}	0.006	5×10^{-11}g		
Zn	10^{-8}—10^{-7}	0.1	1×10^{-11}	0.002	2×10^{-14}g	0.00004	0.04 ng/mℓ
Zr	10^{-8}—10^{-7}	0.1		0.005	5.0 µg/nℓ	0.00002	

[a] Sensitivity based on irradiation period of 0.5 T½ per 2 or 10 hr (whichever is less at a flux 1f 10^{13} neutrons cm^{-2}S^{-1}). Activity measured by Na/(Ti) gamma spectrometry.
[b] Chelation coupled with CIMS.

1. For at least eight trace elements in coal and ash, reported concentrations varied by more than one order of magnitude. For coal, these elements included Mn, Sb, Se, F, Li, Sn, K, and Ba. For ash, these elements were As, V, Zn, Se, Li, Ag, Sn, Na, and Mg.
2. Reported concentrations for three elements (Se, Li, and Sn), varied by more than an order of magnitude in both coal and ash matrices.
3. Agreement was within an order of magnitude in both matrices for only 9 of the 28 elements. These nine were Si, Ca, S, Sr, Fe, Cr, Ni, Be, and B.

This study showed the need for good standard reference material for coal and coal ashes; this was subsequently developed by U.S. National Bureau of Standards. The first standard reference materials to be developed were coal SRM #1632 and coal ash SRM #1633. Portions of these samples were sent to 85 laboratories for analysis. It was found that for many of the elements measured, there were surprisingly wide variations of concentrations reported by the participating laboratories, far outside the uncertainties usually quoted for the techniques used. For this reason, it was clear that the standards were badly needed so that laboratories could check their procedures.

Dulka and Risby[171] have tried to estimate sensitivities of different methods for coal analysis. Table 36 shows the sensitivities for neutron activation analysis (NAA), spark source mass spectrometry (SSMS), chemical ionization mass spectrometry (CIMS), inductively coupled plasma sources for atomic emission spectrometry (ICPAES), nonflame atomic absorption spectrometry (NFAAS), X-ray fluorescence spectrometry (XRFS), and stripping voltametry (ASV). It also should be noted however that NAA and XRF can analyze samples directly, whereas CIMS, ICPAES, NFAAS, and ASV, require the sample to be in solution, and SSMS requires the oxidation of the organic material.

The author of this book finds that the best discussion of this problem is as presented by Gluskoter.[10] The following comments summarize his observations and conclusions where an element has been determined by two or more methods:

Bromine — Average of INAA and XRF data; agreement between the methods was good for moderate to high values. For the low values, INAA data were preferentially used because the technique had better sensitivity.

Cadmium — Average of AA and optical emission — direct reading (OE-DR) data; where there was a choice of lower limits or a choice between lower limits and a real value, the AA value was usually chosen. In general, the agreement was good between the two techniques. The recently developed OE-DR procedure proved to be very effective.

Chlorine — Average of INAA, XRF, and ASTM data; in general, the XRF values were slightly higher and the INAA values were slightly lower than the ASTM data. Only XRF data were used for float-sink samples.

Chromium — Average of INAA, OE-DR, and optical emission—photographic detection (OE-P) data; agreement among the three methods was good. In a few instances, INAA appeared to have a high bias.

Cobalt — INAA, OE-P, and OE-Dr data were in excellent agreement. Results by these methods were averaged.

Copper — Average of AA, OE-P, and OE-DR data; in general, the agreement was good. The XRF data were excluded because of a consistently high bias.

Germanium — Average of OE-P and OE-DR data; in those cases where an uncertainty arose, the OE-DR results were usually chosen.

Iron — Average of INAA and XRF data; the agreement was only fair between the two methods. The INAA data tended to have a high bias in a number of samples. Only the XRF data were used for the float-sink samples.

Lead — Average of OE-P and AA data; in those cases where a choice of limits existed, the AA results were usually chosen. In general, the two sets of data were in good agreement. Since LTA (150°C) was used for AA, and since HTA (500°C) was used for OE-P, this confirms previous findings that Pb appears to be quantitatively retained in the high-temperature ash sample.

Manganese — Average of INAA and OE-P data with good agreement. Samples C-18820 through C-19000 are based on OE-P results only.

Molybdenum — Average of INAA, OE-P, and OE-DR data; the INAA data occasionally tended to have a high bias at the lower values. The agreement among the three techniques was only fair. The XRF data were not used since they were neither consistent nor comparable.

Nickel — Average of AA, OE-P, OE-DR, and XRF data; the agreement was generally good among the four techniques with XRF results occasionally being excluded for having a high bias.

Potassium — Average of INAA and XRF data; agreement was very good at moderate to high concentrations. At the lower concentrations INAA data were chosen because of greater sensitivity. Only the INAA data were used for float-sink samples.

Strontium — Average of OE-DR and INAA data; the methods agreed well in the

Table 37
ANALYTICAL PROCEDURES USED TO DETERMINE TRACE ELEMENT VALUES IN WHOLE COAL BY GLUSKOTER[10]

Element	Procedure
Rb, Cs, Ba, Ga, In, As, Sb, Se, I, Sc, Hf, Ta, W, La, Ce, Sm, Eu, Tb, Dy, Lu, Th, U, Yb, (Au)	INAA
Na, K, Br, Fe	INAA, XRF
Cl	INAA, XRF, ASTM
Mg, Ca, Al, Si, P, Ti	XRF
Be, Ge, Zr	OE-P, OE-DR
Cr, Co, Mo	OE-P, OE-DR, INAA
Ag, Sn	OE-P
Ni, Zn	OE-P, OE-DR, AA, XRF
Hg	NAA(Re)
Be	OE-DR
Pb	OE-P, AA
Sr	OE-DR, INAA
F	ISL
V	OE-P, OE-DR, XRF
Cu	OE-P, OE-DR, AA
Mn	OE-P, INAA
Cd	AA, OE-DR

low to intermediate concentrations; at higher concentrations, the INAA results were usually used.

Vanadium — Average of XRF, OE-DR, and OE-P data with only fair agreement. Some INAA results were obtained on several samples for confirmation.

Zinc — Average of AA, OE-P, OE-DR, and XRF data; the agreement among the four techniques was only fair owing to inhomogeneity of the samples for Zn. INAA results were not considered because of resolution problems.

Zirconium — Average of OE-P and OE-DR data with good agreement.

This is also summarized in the Table 37.

Instead of finishing this chapter with a suggestion as to which method to use, we shall leave this to the reader to decide for himself. The work of different research groups has been described, using in most cases their own words, and the reader should be able to decide which method is the most appropriate for the problem he intends to study. The only thing which really does count is to do good work.

REFERENCES

1. Averitt, P., Breger, I. A., Swanson, V. E., Zubovic, P., and Gluskoter, H., Minor elements in coal — a selected bibliography, U.S. Geol. Surv. Prof. Paper 800-D, 1972, 169.
2. Averitt, P., Hatch, J. R., Swanson, V. E., Breger, I. A., Coleman, S. L., Medlin, J. H., Zubovic, P. and Gluskoter, H. J., compilers. Minor and trace elements in coal — a selected bibliography of reports in English. U.S. Geol. Surv. Open File Report 76-481, 16 pp, 1976.
3. Freedman, R. W. and Sharkey, A. G., Jr., Recent advances in the analysis of responsible coal dust for free silica, trace elements, and organic constituents, *Ann. N.Y. Acad. Sci.,* 200, 7, 1972.

4. Gluskoter, H. J., Shimp, N. F., and Ruch, R. R., Coal analyses, trace elements, and mineral matter, in *Chemistry of Coal Utilization, 2nd Supplementary Volume,* Elliott, M. A., Ed., National Research Council, Wiley-Interscience, New York, chap. 7, in press.
5. Ignasiak, B. S., Ignasiak, T. M., and Berkowitz, N., Advances in coal analysis *Rev. Anal. Chem.,* 2, 278, 1975.
6. Konieczynski, J., Problems of microelement analysis in coal and its by-products, *Koks, Smola, Gaz,* 14, 13, 1969.
7. Hattman, E. A., Schultz, H., and McKinstry, W. E., Solid and gaseous fuels, *Anal. Chem.,* 49, 176R, 1977.
8. Babu, S. P., et al., Suitability of West Virginia coals to coal conversion processes, Coal-Geology Bulletin, No. 1, West Virginia Geological and Economic Survey, December 1973, 32.
9. Ford, C. T., Care, R. R., and Bosshart, R. E., Preliminary evaluation of the effect of coal cleaning on trace element removal, Trace Element Program, Report 3. Bituminous Coal Research, Inc., Monroeville, Pa., 1976, 115.
10. Gluskoter, H. J., Illinois State Geological Survey, Circular 499, Trace elements in coal: Occurrence and distribution, 1977, 154.
11. Karr, C., Jr., Ed., *Analytical Methods for Coal and Coal Products,* Academic Press, New York, 1978.
12. Pollock, E. N., Trace impurities in coal by wet chemical methods, *Adv. Chem. Ser.,* 141, 23, 1975.
13. Swanson, V. E. and Huffman, J. C., Guidelines for sample collecting and analytical methods used in the U.S. Geological Survey for Determining Chemical Composition of Coal, U.S. Geol. Surv. Circ. 735, 1976, 15.
14. Simon, F. O. and Huffman, C., Analytical methods used by the U.S. Geological Survey for determining the chemical composition of coal, *Am. Chem. Soc., Div. Pet. Chem., Prepr.,* 22, 580, 1977.
15. Wewerka, E. M., Williams, J. M., and Vanderborgh, N. E., Contaminants in coals and coal residues, 4th National Conference on Energy and the Environment, Los Alamos Scientific Laboratory, LA-UR 76-2197, 1976, 23.
16. Wewerka, E. M., Williams, J. M., and Wanek, P. L., Assessment and control of environmental contamination from trace elements in coal processing wastes, from ERDA Energy Res. Abstr. No. LA-UR-76-86, 1976, 7.
17. Yaverbaum, L., Fluidized bed combustion of coal and waste materials, Noyes Data Corp., Park Ridge, New Jersey, 1977.
18. Estep, P. A., Kovach, J. J., and Karr, C., Jr., Quantitative infrared multicomponent determinations of minerals occurring in coal, *Anal. Chem.,* 40, 358, 1968.
19. Estep, P. A., Kovach, J. J., Karr, C., Jr., Childers, E. E., and Hiser, A. L., Characterization of iron minerals in coal by low-frequency infrared spectroscopy, *Am. Chem. Soc., Div. Fuel Chem., Prepr.,* 13, 18, 1969.
20. Botoman, G. and Stith, D. A., Analysis of Ohio coals, *Ohio Div. Geol. Surv., Inf. Circ.,* 47, 1, 1978.
21. Gladfelter, W. L. and Dickerhoof, D. W., Determination of sulphur forms in hydrodesulphurized coal, *Fuel,* 55, 355, 1976.
22. Glass, G. B., Analysis and measured sections of 54 Wyoming coal samples (collected in 1974), Report of investigations, 11, from ERDA Energy Res. Abstr. No. 22521, Report No. NP-20935, 1975, 228.
23. Guttag, N. S. and Grimaldi, F. S., Fluorimetric determination of uranium in shales, lignites, and monazites after alkali carbonate separation. XV. Collected papers on methods of analysis for uranium and thorium, *U.S. Geol. Surv. Bull.,* 1006, 111, 1954.
24. Hatch, J. R., Affolter, R. H., and Farrow, R. A., Chemical analyses of coal and shale from the Wasatch Formation in core hole B-1, City of Buffalo, Johnson County, U.S. Geological Survey Open-file Report 78-901, 1978, 28.
25. King, J., Maries, M. B., and Crossley, H. E., Formulae for the calculation of coal analyses to a basis of coal substance free of mineral matter, *J. Soc. Chem. Ind. London,* 57, 277, 1936.
26. Knudson, C. L., Schiller, J. E., and Rudd, A. L., Temperature effects on coal liquefaction, rates of depolymerization and product quality as determined by gel permeation chromatography, *Am. Chem. Soc., Div. Fuel Chem., Prepr.,* 22, 1977; *Prod. of Liq. Fuels Ill.,* American Chemical Society, Washington, D.C., 1977, 49.
27. Mahajan, P. O., Tomita, A., Nelson, J. R., and Walker, P. L., Differential scanning calorimetry studies on coal. II. Hydrogenation of coals, *Fuel,* 56, 33, 1977.
28. Ode, W. H., Coal analysis and mineral matter, in *Chemistry of Coal Utilization,* Lowry, H. H., Ed., John Wiley & Sons, New York, 1962, 202.
29. Parr, S. W., *The Analysis of Fuel, Gas, Water and Lubricants,* McGraw-Hill, New York, 1932, 49.
30. Ruschev, D. and Bekyarova, E., Determination of the rate of extraction of germanium component groups of solid fuels during chemical analysis, *Chem. Anal.,* 53, 569, 1971.

31. Retcofsky, H. L., Investigation of the chemical structure of coal by nuclear magnetic resonance and infrared spectrometry, *Appl. Spectrosc.*, 31, 116, 1977.
32. Senftle, F. E., Tanner, A. B., Philbin, P. W., Boynton, G. R., and Schram, C. W., In situ analysis of coal using a californium-252-lithium drifted germanium borehole sonde, U.S. Geol. Survey, Reston, Va., *Min. Eng.*, 30, 666, 1978.
33. Schweighardt, F. K., Retsofsky, H. L., and Friedel, R. A., Chromatographic and NMR analysis of coal liquefaction products, *Fuel*, 55, 131, 1976.
34. Scherbov, D. P., Plotnikova, R. N., Astaf'eva, I. N., Fluorimetric determination of germanium in mineral raw material, *Zavod. Lab.*, 36, 528, 1970.
35. Slates, R. V., Methods for analysis of trace elements in coal, coal fly ash, soil and plant samples, from ERDA Energy Res. Abstr. 1977, Abstr. No. 25780, Report, No. DP-1421, 1976, 21.
36. Swanson, V. E., Medlin, J. H., Hatch, J. R., Coleman, S. L., Wood, G. H., Jr., Woodruff, S. D., and Hildebrand, R. T., Collection, chemical analysis, and evaluation of coal samples in 1975, U.S. Geol. Survey, Open file report 76-468, 1976, 503.
37. Szonntagh, J., Farady, L., and Janosi, A., Chromatographic determination of uranium in the ash of Hungarian coals, *Magy. Kem. Foly*, 61, 312, 1955.
38. VonLehman, D. J., Jungers, R. H., and Lee, R. E., Jr., Determination of trace elements in coal, fly ash, fuel oil, and gasoline, preliminary comparison of selected analytical techniques, *Anal. Chem.*, 46, 239, 1974.
39. Watt, J. S. and Gravitis, V. L., Analysis of coal or coke, Ger. Offen. Pat. No. 2648434, 1977.
40. Wedge, W. K., Bhatia, D. M., and Rueff, A. W., Chemical analysis of selected Missouri coal and some statistical implications, Missouri Department of Natural Resources of Investigations, 60, 1976, 36.
41. Weyesser, J. L. G., Coal analysis and their relationship to combustion characteristics, Illinois Engineering Experiment Station Circular Series No. 43, University of Illinois, 1942, 72.
42. Montgomery, W. J., ASTM approach to the standardization of new techniques for coal analysis, *Am. Chem. Soc., Div. Fuel Chem., Prepr.*, 22, 1977.
43. Bernas, B., A new method for decomposition and comprehensive analysis of silicates by atomic absorption spectrometry, *Anal. Chem.*, 40, 1682, 1968.
44. Rantola, R. T. T. and Loring, D. H., New low-cost Teflon decomposition vessel, *Atomic Absorption Newsletter*, 12, 97, 1973.
45. Buckley, D. E. and Cranston, R. E., Atomic absorption analyses of 18 elements from a single decomposition of aluminosilicate, *Chem. Geol.*, 7, 273, 1971.
46. Dreher, G. B. and Schleicher, J. A., Trace elements in coal by optical emission spectroscopy, *Adv. Chem. Ser.*, 141, 35, 1975.
47. Hamrin, C. E., Catalytic activity of coal mineral matter. Annual Report, Jan. 1976 - March 1977, (Report FE-2233-3) National Technical Information Service, Springfield, Va., 1977.
48. Peterson, M. J. and Zink, J. B., A semiquantitative spectrochemical method for analysis of coal ash, *U.S. Bur. Mines, Rep. Invest.*, 6496, 15, 1963.
49. Peterson, M. J. and Zink, J. B., Semiquantitative spectrochemical method for analysis of coal ash, U.S. Bureau of Mines, College Park, Md., Report BM-RI-6496, 1964.
50. Filby, R. H., Shah, K. R., Hunt, M. L., Khalil, R. S., and Sauther, C. A., Solvent refined coal (SRC) Process, Trace elements NTIS, Report FE-496-T-17, National Technical Information Service, Springfield, Va., 1978.
51. Filby, R. H., Shah, K. R., and Sautter, C. A., Trace elements in the solvent refined coal process, U.S. Environm. Prot. Agency, Off. Res. Dev., Report No. EPA/600/7-78/063, 1978, 266.
52. Ondov, J. M., Zoller, W. H., Olmez, I., Aras, N. K., Gordon, G. E., Rancitelli, L. A., Abel, K. H., Filby, R. H., Shah, K. R., and Ragaini, R. C., Elemental concentrations in the National Bureau of Standards, Environ. Coal and Fly Ash Standard Reference Materials, *Anal. Chem.*, 47, 1102, 1975.
53. Lyon, W. S., Nuclear methods in coal combustion research, from Energy Res. Abstr. No. 13719, 1978, Report, Conf-771072-2, 1977, 9.
54. Volborth, A., Miller, G. E., Garner, C. K., and Jarabek, P. A., Oxygen determination and stoichiometry of some coals, *Am. Chem. Soc., Div. Fuel Chem., Prepr.*, 22, 1977; Symp. on the New Tech. in Coal Anal., presented at 174th Am. Chem. Soc. Natl. Meet., 9-20, 1977, 9.
55. Neavel, R. C. and Keller, J. E., Estimation of sulphur content in coal from titration of calorimeter bomb washings, *Fuel*, 58, 402, 1979.
56. Abernethy, R. F. and Gibson, F. H., Rare elements in coal, U.S. Bureau of Mines, Washington, D.C., Report BM-IC-8163, 1963.
57. Kinson, K. and Belcher, C. B., *Fuel*, 54, 205, 1975.
58. Cavallaro, J. A., Deubrouck, A. W., Gibbon, G. A., Hattman, E. A., and Schultz, H., A washability and analytical evaluation of potential pollution from trace elements in coal, *Anal. Methods Coal Prod.*, 1, 435, 1978.

59. Ruch, R. R., Cahill, R. A., Frost, J. K., Camp, L. R., and Gluskoter, H. J., Trace elements in coals of the United States determined by activation analysis and other techniques, *Am. Nucl. Soc. Trans.*, 21, 107, 1975.
60. Ruch, R. R., Cahill, R. A., Frost, J. K., Camp, L. R., and Gluskoter, H. J., Survey of trace elements in coal and coal-related materials by neutron activation analysis, *J. Radioanal. Chem.*, 38, 415, 1977.
61. Ruch, R. R., Gluskoter, H. J., and Shimp, N. F., Distribution of trace elements in coal, *Environ. Prot. Technol. Ser.*, EPA-650-2-74-118, 49, 1974.
62. Ruch, R. R., Gluskoter, J. H., and Shimp, N. F., Occurrence and distribution of potentially volatile trace elements in coal: a final report, *Environ. Geol. Notes*, 72, Ill. Geol. Surv., 1974, 96.
63. Scott, T. W., Chu, K. C., and Venugopalan, M., ESR studies of plasma treated coal, Phys. Chem. Lab. West. Illinois Univ. Macomb Ill. Chem. Ind., London, 17, 739, 1976.
64. Govindaraju, K., Mevelle, G., and Chouard, C., Automated optical emission spectrochemical bulk analysis of silicate rocks with microwave plasma excitation, *Anal. Chem.*, 48, 1325, 1976.
65. Hubbard, D. P., Annual Reports on Analytical Atomic Spectroscopy, Vol. 1, Society for Analytical Chemistry, London, 1971.
66. Schlesinger, M. D. and Schultz, H., Analysis for mercury in coal, Bureau of Mines, Technical Progress Report 43 (September, 1971).
67. Davison, R. L., Natusch, D. F. S., Wallace, J. R., and Evans, C. A., Trace elements in flyash-dependence of concentration on particle size, *Environ. Sci. Technol.*, 8, 1107, 1974.
68. Pollock, E. N., Trace impurities in coal, *Am. Chem. Soc., Div. Fuel Chem., Prepr.*, 18, 92, 1973.
69. Pollock, E. N. and West, S. J., The Determination of Antimony at Submicrogram Levels by Atomic Absorption Spectrophotometry, *At. Absorpt. Newsl.*, 11, 104, 1972.
70. Muter, R. B. and Nice, L. L., Major and minor constituents in siliceous materials by atomic absorption spectroscopy, *Adv. Chem. Ser.*, 141, 57, 1975.
71. Lyon, W. S., Lindberg, S. E., Emery, J. F., Carter, J. A., Ferguson, N. M., Van Hook, R. I., and Raridon, R. J., Analytical determination and statistical relationships of fourty-one elements in coal from three-coal fired steam plants, in *Nuclear Activation Techniques in the Life Science*, International Atomic Energy Agency, Vienna, 1978.
72. Gladfelter, W. L. and Dickerhoof, D. W., Use of atomic absorption spectrometry for iron determinations in coals, *Fuel*, 55, 360, 1976.
73. Gluskoter, H. J. and Lindhal, P. C., Cadmium - mode of occurrence in Illinois coal, *Science*, 181, 264, 1973.
74. Martin, T. D., Kopp, J. F., and Ediger, R. D., Determining selenium in water, wastewater, sediment, and sludge by flameless atomic absorption spectroscopy, At. Absorpt. Newsl., 14, 109, 1975.
75. Aruscavage, P., The Determination of Arsenic, Antimony and Selenium in Coals by Atomic Absorption Spectrometry With a Graphite Tube Atomizer, *U.S. Geol. Survey J. Research*, 5, 1977.
76. Huffman, C., Jr., Rahill, R. L., Shaw, V. E., and Norton, D. R., Determination of Mercury in Geologic Materials by Flameless Atomic Absorption Spectrometry, U.S. Geol. Survey, Prof. Paper 800-C, C203-C207, 1972.
77. Coleman, W. M., Szabo, P., Wooten, D. L., Dorn, H. C., and Taylor, L. T., Minor and trace metal analysis of solvent refined coal by flameless atomic absorption, *Fuel*, 56, 195, 1976.
78. Coleman, W. M., Perfetti, P., Dorn, H. C., and Taylor, L. T., Trace element distribution in various solvent refined coal fractions as a function of the feed coal, *Fuel*, 57, 612, 1978.
79. Fowkes, W. W., Wilhelm, T. D., and Kube, R, W., Sample preparation and electron microprobe examination of lignite ash deposits, *Metallography*, 2, 209, 1969.
80. Nandi, B. N., Montgomery, D. S., and Martin, E., The study of coal by a scanning electron microscope and electron probe, *Am. Chem. Soc., Div. Fuel Chem., Prepr.*, 14, 81, 1970.
81. Augustyn, D., Iley, M., and Marsh, H., Optical and scanning electron microscope study of brown coals, *Fuel*, 55, 25, 1976.
82. Solomon, P. R. and Manzione, A. V., New method for sulphur concentration measurements in coal and char, *Fuel*, 56, 393, 1977.
83. Finkelman, R. B., Determination of trace element sites in the Waynesburg coal by SEM analysis of accessory minerals, *Scanning Electron Microsc.*, 1, 143, 1978.
84. Finkelman, R. B. and Stanton, R. W., Identification and significance of accessory minerals from a bituminous coal, *Fuel*, 57, 763, 1978.
85. Valković, V., *Trace Elements*, Taylor and Francis, Ltd., International Scientific Publishers, 1975.
86. Valković, V., Nuclear Microanalysis, Garland Pub. Co., New York, 1977.
87. Valković, V., Analysis of Biological Materials for Trace Elements Using X-ray Emission Spectroscopy, CRC Press, Boca Raton, Fla., 1980.
88. Kuhn, J. K., Harfst, W. F., and Shimp, N. F., X-ray fluorescence analysis of whole coal, *Am. Chem. Soc., Div. Fuel Chem., Prepr.*, 18, 72, 1973.
89. Kuhn, J. K., Harfst, W. F., and Shimp, N. F., X-ray fluorescence analysis of whole coal, in Babu, S. P., Ed., American Chemical Society, Washington, D.C.; Advances in Chemistry, 141, 66, 1975.

90. Kuhn, J. K. and Henderson, L. R., Survey of emission x-ray analysis of coal, *Am. Chem. Soc., Div. Fuel Chem., Prepr.,* 22, 68, 1977.
91. Wheeler, B. D. and Jacobus, N. C., Qualitative analysis on coal and coal ash by energy dispersive x-ray fluorescence, EG&G ORTEC, Oak Ridge, Tenn., Report, January 1980.
92. Burnstien, F., Particle Size and Mineralogical Effects in Mining Applications, 11th Annu. Conf. Applications of X-Ray Analysis, Denver Research Institute, University of Denver, 1962.
93. Campbell, W. J. and Thatcher, D., *Advances in X-ray Analysis,* Vol. 2, University of Denver, Plenum Press, New York, 1958.
94. Wheeler, B. D., 1967; as referenced in Wheeler B. D. and Jacobus N.C.: EG&G/ORTEC, Oak Ridge, Tenn., Report, 1980.
95. Wheeler, B. D., 1969; as referenced in Wheeler B. D. and Jacobus N.C.; EG&G/ORTEC, Oak Ridge, Tenn., Report, 1980.
96. Liebhafsky, H. G., Wilson, E. H., and Zemany, P. D., *X-Ray Absorption and Emission in Analytical Chemistry,* John Wiley & Sons, New York, 1966.
97. Hasler, M. F. and Kemp, J. W., Suggested Practices for Spectrochemical Compositions, ASTM, E2 SM2-3, Philadelphia, Pa., 1957.
98. Anderman, G. and Allen, J. D., X-ray emission analysis of finished cements, Anal. Chem., 33, 1695, 1961.
99. Spacek, F., Application of x-ray spectral analysis to the size measurement and evaluation of microconcretions in coal, *Freiberg. Forschungsh. A.,* A553, 259, 1976.
100. Lloyd, W. G. and Francis, H. E., Determination of sulfur in whole coal by x-ray fluorescence spectrometry, Proc. of ERDA Symp. on x- and Gamma-Ray Sources and Appl., Univ. of Michigan, Ann Arbor, May 19-20, 1976. (CONF-760539).
101. Cooper, J. A., Sheeler, B. D., Wolfe, G. J., Bartell, D. M., and Schlafke, D. B., Determination of sulfur ash, and trace element content of coal, coke and fly ash using multielement tube-excited x-ray fluorescence analysis, *Adv. X-ray Anal.,* 000077, 20, 421, 1977.
102. Cooper, J. A., Schlafke, D. B., and Wheeler, B. D., Energy dispersive x-ray fluorescence analysis of coal, ores and geological materials, EXT. Abstr., Int. Symp. Anal. Chem. Explor. Min. Process. Mater. 148, 1977.
103. Cooper, J. A., Wheeler, B. D., Wolfe, G. J., Bartell, D. M., and Schlafke, D. B., Determination of sulfur, ash and trace element content of coal, coke and fly ash using multielement tube-excited x-ray fluorescence analysis, Proc. ERDA Symp. X-Gamma Ray Sources Appl., 1976, 169.
104. Cooper, J. A., Schlafke, D. B., and Wheeler, B. D., Energy dispersive x-ray fluorescence analysis of coal, ores, and geological materials. Ext. Abstr. - Int. Symp. Anal. Chem. Explor. Min. Process. Mater., 147-155, 1977.
105. Cooper, J. A., Sheeler, B. D., Wolfe, G. J., Bartell, D. M., and Schlafke, D. B., Determination of sulfur, ash, and trace element content of coal, coke, and fly ash using multielement tube-excited x-ray fluorescence analysis, *Adv. X-Ray Anal.,* 20, 431, 1977.
106. De Kalb, E. L. and Fassel, V. A., Energy dispersive x-ray fluorescence analysis of coal, Proc. of ERDA Symp. on X-ray and Gamma Ray Sources and Appl., Univ. of Mich., Ann Arbor, CONF-760539, 1976.
107. Prather, J. W., Tarrer, A. R., and Guinn, J. A., X-ray fluorescence of trace metals in solvent refined coal, *Am. Chem. Soc., Div. Fuel Chem., Prepr.,* 22, 72, 1977.
108. Orlić, I., Pavlić, M., Rendić, D., Marijanović, P., Valković, V., Budnar, M., and Cindro, L., Determination of trace elements in coal by X-ray emission spectroscopy. Paper presented at Int. Congr. Analytical Techniques in Environmental Chemistry, Barcelona, 1981.
109. Lyon, W. S., Ed., *Trace Element Measurements at the Coal-fired Steam Plant,* CRC Press, Boca Raton, Fla., 1977.
110. Lyon, W. S., *Trace Element Measurement at the Coal-fired Steam Plant,* Blackwell Scientific, Oxford, 1977, 144.
111. Sparks, C. J., Jr., Cavin, O. B., Harris, L. A., and Ogle, J. C., Simple quantitative x-ray fluorescent analysis for trace elements, in *Proc. of the Seventh Annual Conference on Trace Substances in Environmental Health,* Hemphill, D. D., Ed., 1973, 361.
112. Sparks, C. J., Jr., Ogle, J. C., Dunn, H. W., and Hulett, L. D., Development of high sensitivity x-ray fluorescence for analyses of trace toxic elements, in *Ecology and Analysis of Trace Contaminants,* Fulkerson, W., Shults, W. D. and Van Hook, E. L., Eds., ORNL-NSF-EATC-6, Oak Ridge, Tennessee, 1974, 243.
113. Miklavžič, U., 1981, private communication.
114. Valković, V., Liebert, R. B., Zabel, T., Larson, H. T., Miljanić, D., Wheeler, R. M., and Phillips, G. C., Trace element analysis using proton induced X-ray emission spectroscopy, *Nucl. Instrum. Methods,* 114, 573, 1974.

115. Simms, P. C., Rickey, F. A., and Mueller, K. A., Multielemental analysis using proton induced photon emission, *Am. Chem. Soc., Div. Fuel Chem., Prepr.,* 22, 1977; Symp. on New Tech. in Coal Anal., Chicago, Aug. 29-Sep. 2, 1977.
116. Cronch, S. W., Ehmann, W. D., Laumer, H. W., and Gabbard, F., Elemental analysis of coal by proton-induced X-ray emission analysis, Proc. ERDA Symp. X- and Gamma-Ray Sources and Applications, Ann Arbor, 1976, 190.
117. Budnar, M., Kregar, M., Miklovžič, U., Ramsak, V., Rauniker, M., Rupnik, Z., and Valkovič, V., Some improvements in scattering chamber for PIXE, *Nucl. Instrum. Methods,* 180, 1980.
118. Chen, J. R., Kneis, H., Martin, B., Bobiling, R., Traxel, K., Chao, E. C. T., and Minkin, J. A., Trace elemental analysis of bituminous coals using the Heidelberg proton microprobe, 2nd PIXE Conf., Lund, Sweden, 1980.
119. Uemura, Y. J., Kuno, Y., Koyama, H., Yamazaki, T., and Kienle, P., *Nucl. Instr. Meth.,* 153, 573, 1978.
120. Ondov, J. M., Zoller, W. H., Olmez, I., Aras, N. K., Gordon, G. E., Rancitelli, L. A., Abel, K. H., Shah, R. H., and Ragaini, R. C., Four laboratory comparative instrumental nuclear analysis, the NBS coal and ash standard reference materials, Natl. Bur. Stand. Spec. Publ. 422, 1976; Accuracy in Trace Analysis Sampling, Sample Hand. Anal., Proc. of the Mater. Res. Symp., Gaithersburg, 1974, 211.
121. Sheibley, D. W., Trace elements analysis of coal by neutron activation, *Am. Chem. Soc., Div. Fuel Chem., Prepr.,* 18, 59, 1973.
122. Sheibley, D. W., Trace elements by instrumental neutron activation analysis for pollution monitoring, *Adv. Chem. Ser.,* 141, 98, 1975.
123. Abel, K. H. and Rancitelli, L. A., Major, minor, and trace element composition of coal and fly ash, as determined by instrumental neutron activation analysis, *Adv. Chem. Ser.,* 141, 118, 1975.
124. Frost, J. K., Santoliquido, P. M., Camp, L. R., and Ruch, R. R., Trace elements in coal by neutron activation analysis with radiochemical separations, *Adv. Chem. Ser.,* 141, 84, 1975.
125. Lyon, W. S. and Emery, J. F., Neutron activation analysis applied to the study of elements entering and leaving a coal-fired steam plant, *Int. J. Environ. Anal. Chem.,* 4, 125, 1975.
126. Millard, H. T., Recent application of neutron activation analysis to coal, *Am. Chem. Soc., Div. Fuel Chem., Prepr.,* 22, 1977; 174th Symp. on New Tech. in Coal Anal., 1977, 64.
127. Rhodes, J. R., Neutron-gamma techniques for on-stream analysis of coal, *Am. Chem. Soc., Div. Fuel Chem., Prepr.,* 22, 1977; Symp. on New Tech. in Coal Anal. presented at 174th Am. Chem. Soc. Natl. Meet., 1977, 21.
128. Rhodes, J. R., Daglish, J. C., and Clayton, C. G., A coal monitor with low dependence on ash composition, Radioisotope Instruments in Industry and Geophysics, Proc. Symp. Warsaw, 1965, International Atomic Energy Agency, Vienna, 1966, 447.
129. Greenwood, M. R., X-ray fluorescence analysis applied to small samples, Report LBL-6451, 1978.
130. Nargolwalla et al., ref. 11 in Greenwood, M. R., X-ray fluorescence analysis applied to small samples, Report LBL-6451, 1978.
131. Senttle, F. E., Macy, R. J., and Mikesell, J. L., Determination of the optimum-size californium-252 neutron source for borehole capture gamma-ray analysis, *Nucl. Instrum. Methods,* 158, 293, 1979.
132. Loska and Gorski, ref. 13 in Greenwood.[129]
133. Mott and Spooner, ref. 14 in Greenwood[129]
134. Weaver, J. N., Neutron activation analysis of trace elements in coal, fly ash and fuel oils, *Anal. Methods Coal Proc.,* 1, 377, 1978.
135. Yeager, K., R & D Status Report, Coal Combustion Systems Division, EPRI Journal, p. 32, 1981.
136. Reynolds, G., Bozorgmanesh, H., Elias, E., Gozani, T., Maung, T., and Orphan, V., Nuclear Assay of Coal, FP-989, EPRI Research Project 983-1, Final Report, January, 1979.
137. Block, C., Dams, R., and Hoste, J., Chemical composition of coal and fly ash, *Meas. Detect. Control Environ. Pollut. Proc. Int. Symp.,* International Atomic Energy Agency, Vienna, 1976, 101.
138. Block, C. and Dams, R., Determination of trace elements in coal by instrumental neutron activation analysis, *Anal. Chim. Acta,* 68, 11, 1974.
139. Volborth, A., Miller, G. E., Garner, C. K., and Jerabek, P. A., Material balance in coal. I. Material balance and oxygen stoichiometry of six coals from Wyoming, *Fuel,* 56, 1976.
140. Volborth, A., Miller, G. E., Garner, C. K., and Jerabek, P. A., Oxygen in coal ash, a simplified approach to the analysis of ash and mineral matter in coal, *Fuel,* 56, 1976.
141. Volborth, A., Miller, G. E., Garner, C. K., and Jerabek, P. A., Material balance in coal. II. Oxygen determination and stoichiometry of 33 coals, *Fuel,* 57, 49, 1978.
142. Hamrin, C. E., Johannes, A. H., James, W. D., Sun, G. H., and Ehmann, W. D., Determination of oxygen and nitrogen in coal by instrumental neutron activation analysis, *Fuel,* 58, 48, 1979.
143. Morstin, K. and Woznick, J., Hydrogen in coal, the feasibility of determination by neutron methods, IAEA-SM-216/33, Vienna, 119, 1976.

144. Schlyer, D. J., Ruth, T. J., and Wolf, A. P., Oxygen content of selected coals as determined by charged particle activation analysis, *Fuel*, 58, 208, 1979.
145. Schlyer, D. J. and Wolf, A. P., A study of coal oxidation by charged particle activation analysis, 4th Int. Conf. on Nuclear Methods in Environmental and Energy Research, Columbia, Missouri, (BLN 27607), 1980.
146. Styron, C. E., Preliminary assessment of the impact of radionuclides in Western coal and health and environment, in Technology for Energy Conservation, National Technical Information Services Report MLM-2497/OP, 1978.
147. Styron, C. E., Casella, V. R., Farmer, B. M., Hopkins, L. C., Jenkins, P. H., Phillips, C. A., and Robinson, B., Assessment of the radiological impact of coal utilization, NTIS Report MLM-2514, UC-90a, 1979.
148. Wertheim, G., *Mössbauer Effect, Principles and Applications*, Academic Press, New York, 1965.
149. May, L., *An introduction to Mossbauer Spectroscopy*, Plenum Press, 1971.
150. Gibb, T. C., *Principles of Mössbauer Spectroscopy*, Chapman and Hall, London, 1976.
151. Levinson, L. M. and Jacobs, I. S., Mössbauer spectroscopic measurement of pyrite in coal, *Fuel*, 56, 453, 1977.
152. Montano, P. A., Mössbauer spectroscopy of iron compounds found in West Virginia coals, *Fuel*, 56, 397, 1977.
153. Retcofsky, H. L., Hough, M., and Clarkson, R. B., American Chemical Society, Fuel Div., Prepr. 24, 83, 1979.
154. Gerstein, B. C. and Pembleton, R. G., Pulsed nuclear magnetic resonance spectrometry for nondestructive determination of hydrogen in coal, *Anal. Chem.*, 49, 75, 1977.
155. Oka, M., Chang, H.-C., and Gavalas, G. R., *Fuel*, 56, 3, 1977.
156. Alemany, L. B., King, S. R., Stock, L. M., Proton and carbon n.m.r. spectra of butylated coal, *Fuel*, 57, 738, 1978.
157. Yen, T. F., in *Trace Substances in Environmental Health*, Vol. 6, Hemphill, D. D., Ed., Univ. of Missouri, 1973, 347.
158. Brown, R., Jacobs, M. L., and Taylor, H. E., A survey of the most recent applications of spark source mass spectrometry, *Am. Lab.*, 4, 29, 1972.
159. Petrakis, L. and Grandy, D. W., Free radicals in coals and coal conversion. II. Effect of liquefaction processing conditions on the formation and quenching of coal free radicals, *Fuel*, 59, 227, 1980.
160. Bolton, N. E., Fulkerson, W., Van Hook, R. I., Lyon, W. S., Andren, A. W., Carter, J. A., Emery, J. F., Feldman, C., Hulett, L. D., Dunn, H. W., Sparks, C. J. Jr., Ogle, J. C., and Mills, M. T., Trace element measurements at the coal-fired Allen steam plant, Oak Ridge National Laboratories, Oak Ridge, Tenn., Progress report, February 1973-July 1973.
161. Bolton, N. E., Van Hook, R. I., Fulkerson, W., Emery, J. F., Lyon, W. S., Andren, A. W., and Carter, J. A., Trace Element Measurements at the Coal-Fired Allen Steam Plant, Progress Report, June 1971-January 1973, ORNL/NSF/EP 43, Oak Ridge National Laboratory, Oak Ridge, Tenn.
162. Sharkey, A. G., Jr., Kessler, T., and Friedel, R. A., Trace elements in coal dust by spark-source mass spectrometry, *Trace elements in Fuel*, Baby, S. P., Ed.; Advances in Chemistry Series 141, Am. Chem. Soc., Div. Fuel Chem. 166th Meet., Chicago, Ill., 48-56, Aug. 30, 1973.
163. Sharkey, A. G., Jr., Kessler, T., and Friedel, R. A., Trace elements in coal dust by spark-source mass spectrometry, *Adv. Chem. Ser.*, 141, 48, 1975.
164. Sharkey, A. G., Schultz, J. L., Schmidt, C. E., and Friedel, R. A., Mass spectrometric analysis of stream from coal gasification and liquefaction processes, PERC/RI-75/5, U.S. Energy Research and Development Administration, 1975.
165. Carter, J. A., Walker, R. I., and Sites, J. R., Trace impurities in fuels by isotope dilution mass spectrometry, *Adv. Chem. Ser.*, 141, 78, 1975.
166. Carter, J. A., Donohue, D. L., and Franklin, J. C., Trace metal analysis in coal by multielement isotope dilution spark source mass spectrometry, *Am. Chem. Soc., Div. Fuel Chem., Prepr.*, 22, 1977.
167. St. John, G. A., Buttrill, S. E., Jr., and Anbar, M., *Am. Chem. Soc. Div. Fuel Chem. Prepr.*, 22, 5, 1977; Symp. on the Org. Chem. of Coal, presented at 174th Am. Chem. Soc. Natl. Meet., Chicago, Ill., August 29-Sep. 2, 1977, 141.
168. Hayatsu, R., Winans, R. E., Scott, R. G., Moore, L. P., and Studier, M. H., Trapped organic compounds and aromatic units in coal, *Fuel*, 57, 541, 1978.
169. Lumpkin, H. E. and Azel, T., Characterization of coal products by mass spectrometry, Am. Chem. Soc., Div. Fuel Chem., Prepr. 22, 1977; Symp. on the Org. Chem. of Coal, Aug. 29-Sep. 2, 1977, 135.
170. Kershaw, J. R. and Barrass, G., Deuterium studies of coal hydrogenation, *Fuel*, 56, 455, 1977.
171. Dulka, J. J. and Risby, T. H., Ultratrace metals in some environmental and biological systems, *Anal. Chem.*, 48, 640A, 1976.

172. Petrakis, L. and Grandy, D. W., *Anal. Chem.*, 50, 303, 1978.
173. Bolton, N. E., Carter, J. A., Emery, J. F., Feldman, C., Fulkerson, W., Hulett, L. D., and Lyon, W. S., Trace element mass balance around a coal-fired steam plant, *Adv. Chem. Ser.*, 141, 175, 1975.
174. EG&G ORTEC, Oak Ridge, Tennessee.
175. Electric Power Research Institute (EPRI), Palo Alto, Calif.

INDEX

A

Absorption spectrophotometry, 173
Acetylene, 68
Acid bomb combustion, 178—179
Acid mist, 36
Acid rains, 103, 158
Airborne radioactive materials, 137
AIRDOS computer code, 138
Air pollution, 36, 104—116, 131, 133, 160
Alpha counting, 250—253
Alumina, 16, 123
    extraction, 48—51
Aluminum, 4, 23, 31, 120, 127, 129, 133—134, 188, 230, 240—241, 252
    hydrochloric acid leach, 49—50
    lime-sinter process, 49—52
    sulfuric acid leach, 49—50
    sulfurous acid process, 49—50
Aluminum oxide, 15, 47
Aluminum silicate, 108
Alunite, 48
Amines, 59—60
Ammonia, 60
Analytical methods, see also specific topics, 173—273
    alpha, beta, and gamma counting, 250—253
    charged-particle activation, 248—250
    classical chemical methods, 181—185
    comparison, 262—265
    electron microscopy, 197—202
    electron spin resonance (ESR) spectroscopy, 257—258
    general considerations, 173—181
    instrumental, 173
    mass spectroscopy, 258—262
    Mössbauer spectroscopy, 253—255
    neutron activation, 228—247
    nuclear magnetic resonance (NMR) spectroscopy, 255—257
    optical, 185—197
    other nuclear methods, 247—258
    sample preparation, 177—179
    standards, 179—181
    X-ray analysis, 202—228
Anhydrite, 16
Anodic stripping voltammetry, 262, 265
Anorthite, 16
Anthracite coal, 17, 60, 260—261
Antimony, 3—4, 16, 26, 114—115, 120, 127, 156, 163, 188, 191, 195, 205, 230, 236
Argon, 133
Arsenic, 3—4, 8, 10, 16, 27, 114, 117, 120, 125, 127, 129, 156, 159, 162—163, 178, 188, 191, 195, 213, 236
Ash, 2, 11, 14—35, 40, 71, 75, 115, 123, 132—133, 140, 149, 154, 177, 186, 193, 199, 213—214, 243
    calculated, 25
    combustion residues, 1
    content, 18, 23—24, 54, 57, 64, 73
    determination, 21—25
    disposal, 130
    fouling, 20—21
    fusibility, 39
    fusion, 14—15, 18—21
    production, 47
    recycling complex, 47
    utilization, 31—35
Ashing, 178—179, 187, 245
Atmospheric deposition, 116—120
Atmospheric gases, 1
Atomic absorption, 6, 149, 174, 186, 262
Atomic absorption analysis, 11, 178, 181, 194—195, 252
Atomic absorption prcedures, 147
Atomic absorption spectrometry, 176, 194—195, 230
Atomic absorption spectrophotometry, 72, 114, 174
Atomic absorption spectroscopy, 8, 112, 173, 177, 185

B

Backscattering of electromagnetic radiation, 21—22
Barium, 4, 19, 27, 115, 120, 125, 127, 163, 188, 201
Battelle hydrothermal coal process, 41, 45
Bauxite, 49
Bayer process, 49
Benzene, 68, 260—261
Bernas dissolution procedure, 177—178
Beryllium, 3—4, 19, 26, 53, 112, 115, 120, 125, 127, 129, 177, 181, 188, 197
Beta counting, 250—253
Bi-Gas process, 78
Biological aspects of coal use, 156—163
Bismuth, 16, 27, 120, 141, 188, 191, 250, 252—253
Bituminous coal, 16—17, 40, 43, 60, 67, 115, 125, 147, 246, 260—261
Bituminous fly ash, 30
Bituminous-type ash, 14
Boilers, 1, 18
Boiling-water reactors, 134, 136
Bone dose, 138—140
Boron, 3—4, 19, 26, 35, 120, 125, 127, 159, 162—163, 178, 188, 243
Bottom ash, 1, 3—6, 10—11, 25—29, 46—47, 103, 108, 112, 114—116, 135, 140—143, 149, 259
Bremsstrahlung radiation, 213, 222

Bromine, 113, 163, 236, 265
Bronchitis, 161
Brown coal, 67

C

Cadmium, 3—4, 26, 114—115, 117, 120, 124, 127, 129, 133, 156, 159—160, 162, 178, 181, 185, 188, 193, 196, 236, 265
Calcite, 14, 16, 25, 57, 133, 179, 203
Calcium, 3—4, 21, 23, 31, 71, 120, 127, 129, 180, 188, 240
Calcium carbonate, 73
Calcium oxide, 14—15, 25, 43, 47, 51
Calcium sulfate, 65
Californium, see ^{252}Cf
Calorific value, 127, 129, 132, 182
Calorimetric methods of trace element analysis, 173
Calsinter process, 52
Cancer, 133—134, 161
Carbon, 3, 14, 30, 44, 48, 58, 68—69, 75, 132, 162, 182, 240—241, 245, 248, 256—257
Carbonates, 21, 115, 201
Carbon crucible, 185
Carbon dioxide, 79, 147, 184
Carbonization, 4—5
Carbon monoxide, 2, 48, 79
Carbon rod atomizer (CRA), 196—197
Carcinogenesis, 159
Cerium, 16, 188, 230
Cesium, 26, 188, 236
^{252}Cf, 237, 239
^{252}Cf-source, 240
  neutron activation using, 237—243
Chalcophiles, 13
Chalcophilic elements, 14
Char, 68, 78, 147, 149, 153—154
Charged-particle activation analysis, 248—250
  chief advantages of, 250
Chemical composition, 3
Chemical ionization mass spectrometry, 265
Chemical methods of trace element analysis, 173—174
  classical, 181—185
Chlorides, 112
Chlorination method, 41
Chlorine, 4, 20—21, 23, 26, 54, 163, 240—241, 243, 265
Chlorophyll, 163
Chromium, 3—4, 18—19, 53, 114, 117, 120, 124, 127, 159, 162, 181, 188, 201, 265
Classical chemical methods of trace element analysis, 181—185
Clay, 25, 50—51, 115, 200—201
Clay material, 135
Clay mineral, 16, 112
Coal
  ash, see Ash
  burning process, 1—3
  cleaning, 120—129, 146

combustion, 1—5, 10—12, 42, 52, 103, 105, 112, 119—121, 129, 146, 157, 159, 175—176
conversion, 4, 44, 67—69, 72—73, 77, 121, 146, 148, 153, 158
extraction, 157
gasification, 41, 68, 74—80, 146—147, 149, 153—154, 163
hydrogenation, see Hydrogenation
liquefaction, see Liquefaction of coal
processing, 157, 246
pyrolysis, 68
rank, see Rank of coal
seams, 237, 246
slag, see Slag
utilization, 1—102, 125
  ash, 14—35
  biological aspects, 156—163
  combustion, 1—5
  desulfurization, 35—46
  health aspects, 156—163
  material balance in power plant, 5—14
  new technologies, 64—80
  trace element recovery, 46—64
  waste, 131—133, 174—175
Coal ash, see Ash
Coal-fired power-generation plants, residues from, 1
Coal-fired power plants, trace contaminants from, 103—120
Coalification, 16
Coal-uranium breeders, 61, 63—64
Cobalt, 4, 19, 53, 117, 120, 133, 159, 163, 181, 188, 265
Coffinite, 135
Coke, 68, 115
Coking, 4
Collector ash, 115
Colorimetric techniques of trace element analysis, 174, 177
Colorimetry, 112
Combustion, see Coal combustion
  process, 67
  system, 11
Compton scattering, 220
Computer-Assisted Molecular Structure Construction (CAMSC), 256
Computer tomograph (CT) scanner, 1—2
Continuous nuclear analysis of coal (CONAC), 237, 239—240
Copper, 3—4, 16, 18—19, 26—27, 49, 71, 115, 117, 120, 124, 127, 129, 133, 159—160, 163, 188, 237, 256
Coquimbite, 203
Coronary disease, 161
Cyclone furnaces, 1, 18, 20

D

Desulfurization, 1, 35—46, 52, 108—109, 121, 146
  acid-base neutralization, 38

nucleophilic displacement, 38—39
oxidation, 38
reduction, 38
solvent partition, 38
thermal decomposition, 38
Deuteration, 261—262
Deuterium, 262
Direct toxicity, 159
Dispersion coefficient, 143
Disposal of waste, 129—131
Distillation of tar, 5
Distribution of trace elements, 3—4, 11, 26, 73, 147
Dolomite, 16, 65, 67
Dry ashing, 187, 191

E

Effluent limitations, 134
Electrodeposition, 250, 252
Electron microprobe, 198—199
  analysis, 46, 115
Electron microscopy, 57, 197—202
Electron spin resonance (ESR) spectroscopy, 257—258
Electrostatic precipitation, 106, 113—114, 118, 135, 141
Emission factors for trace elements, 106
Emission of elements, calculation of, 105
Emission rates of trace elements, 11—12
Emission spectrochemical analysis, 190
Emission spectroscopy, 71, 112, 175, 181, 185
Emphysema, 161
Enrichment factors, 113
Enrichment values, 6
Environmental considerations, 103—172
  air pollution, 105—116
  atmospheric deposition, 116—120
  biological aspects of coal use, 156—163
  coal cleaning, 120—129
  health aspects of coal use, 156—163
  new technologies of coal utilization, 146—156
  radioactivity, 133—146
  soil pollution, 116—120
  trace contaminants from coal-fired power plant, 103—120
  waste management, 129—133
Environmental pollution, 119
Erbium, 26
ESR, see Electron spin resonance spectroscopy
Estimated emission of trace elements, 107
Exinite, 227

F

Fast neutron activation analysis, 243—247
Feldspars, 201
Ferric oxide, 14—15, 20, 25, 47
FGD sludge, 108, 122

Field ionization mass spectrometry, 260
Fischer-Tropsch technology, 69—70
Fission products, 135
Flame atomic absorption, 191, 194
Flameless atomic absorption, 174, 194—195
Flameless atomic absorption spectroscopy, 154, 195—196
Flocculation, 59
Flue gas, 3—5, 8, 10—11, 36, 43, 105, 113—114, 122—123, 156
Flue-gas desulfurization (FGD), see Desulfurization
Fluidized-bed boiler, 66, 129
Fluidized bed combustion, 11, 41—42, 63, 65, 67, 146
Fluidized bed combustion boilers, 129
Fluidized bed reactor, 45
Fluoride, 176, 185
Fluorimetry, 177
Fluorine, 4, 11, 26—27, 103, 120, 124—125, 129, 133, 149, 158, 163, 175, 181, 185, 246
Fluorosis, 158
Fly ash, 1, 3—6, 8, 10—14, 21, 25—32, 43, 46, 51—52, 54—55, 103, 105, 108—119, 131, 135—136, 138, 140—143, 149, 157, 159, 162—163, 177, 185, 191, 198, 219—220, 222, 230—231, 233—234, 243, 259—260
  utilization, 31—35
Free radicals, 257
Fuel cycle, 157
Fuel fraction, 24
Fuel oils, 108, 115, 122

G

Gadolinium, 26
Gallium, 3, 19, 120, 188, 236
  recovery, 52—54
Gamma counting, 250—253
Gamma-ray spectrometry, 231
Gamma spectroscopy, 135, 250, 253
Gas chromatography, 40
Gasification, see Coal gasification
  processes, 68
Ge(Li) detector, 229—231, 235—236, 241—243, 248
Germanium, 3, 19, 49, 120, 125, 127, 163, 188, 191, 265
  recovery, 52—54
Gravimetric procedure, 21
Gypsum, 14, 16, 52, 57

H

Hafnium, 26—27
Halogens, 108, 134
Health aspects of coal use, 156—163
Heating rate, 54
Heat of combustion, 2, 183
Hematite, 29

High-temperature ash and ashing (HTA), 178—180, 187, 191, 205, 255
Holmes-Streetford Process, 42—43
Hydrocarbons, 158
Hydrochloric acid, 254, 260
Hydrogasification, 68
Hydrogen, 2, 38, 45, 48, 69, 75—76, 79, 153, 182, 240—241, 243, 245—247, 256—257, 262
Hydrogenation, 69, 71, 261—262
Hydrogenolysis, 71
Hydrogen sulfide, 36, 41, 78
Hydroxides, 115

I

Illite, 16, 115, 203
Imbalances, 9—10
Indium, 188
Inductively coupled plasma sources for atomic emission spectrometry, 265
Inertinite, 227
Inhalation pathways, 160—161
Inlet fly ash, 10
Inlet precipitator ash, 3
In situ gasification, 79—80, 246
Instrumental neutron activation analysis (INAA), 229—230, 259
Instrumental photon activation analysis (IPAA), 230
Instrumental techniques of trace element analysis, 173
Iodine, 26—27, 133, 139, 184
Ion microprobe mass spectrometry, 115
Ion-selective electrode, 149, 177
Iron, 4, 16, 22—24, 31, 44, 48, 52, 58, 71, 76, 117, 120, 123, 127, 129, 133—134, 160, 180, 188, 194, 213, 224, 230, 240—241, 250, 252—254, 265
Iron sulfate, 254
Iron sulfide, 71, 199
Isotope-dilution spark source mass spectrometry, 259
Isotopes used for coal analysis, 238

J

Jarosite, 14, 57

K

Kaolin, 43
Kaolinite, 14, 16, 29, 48, 57, 71, 203
Kidney dose, 138—140

L

Land-disposal techniques, 130

Lanthanum, 19, 188, 230
Lead, 3—4, 11, 19, 27, 49, 71, 114, 117, 120, 124, 127, 129, 135, 139—141, 143—144, 156, 158—159, 162, 178, 181, 188, 193, 252, 260, 265
Light-water reactors, 136
Lignite, 17, 43, 56—57, 60—61, 63, 147, 149—153, 199, 203, 220—221, 253, 260—261
Lignite ash, 16, 21, 59
Lignite fly ash, 30, 33
Lignite-type ash, 14
Lime, 20, 43, 122, 129
Limestone, 43, 52, 65, 67, 122, 129
Liquefaction of coal, 40—41, 68—74, 78, 146, 154—155, 257, 261
Liquefaction processes, 73
Lithium, 3, 19, 26—27, 120, 163, 188, 194
Lithophiles, 12
Lithophilic elements, 14
Liver dose, 138—140
Low-temperature ash and ashing (LTA), 179, 186, 191, 200, 203, 205, 245, 254
Lung dose, 138—140
Lurgi process, 77, 146

M

Macerals, 73, 198—199, 227
Magnesium, 4, 31, 60, 120, 127, 188, 230
Magnesium oxide, 14—15, 43, 47
Magnetic desulfurization, 44
Magnetic filtration, 43
Magnetic separation, 39, 41—43, 46, 54—55
Magnetic susceptibility, 48
Magnetite, 29, 200
Magneto-hydrodynamics (MHD), 64
Manganese, 3—4, 19, 26, 49, 115, 120, 124—125, 127, 129, 133—134, 160, 163, 181, 188, 265
Marcasite, 48, 133
Mass absorption coefficient, 24—25, 216, 218—219
Mass balances, 5—6, 8, 10—11, 63, 106
Mass spectrometry, 174, 261
Mass spectroscopy, 258—262
Material balance in power plant, 5—14, 105
Matrix effects, 191, 195, 215, 247, 259
Maximum individual dose commitments, 136, 138—139
Melting behavior of ash, 20
Mercury, 3—4, 6, 8, 11, 103, 112, 116—117, 120, 124—125, 127, 129, 147, 149, 156, 158—159, 162, 178, 181, 195, 205, 236, 260
Metakaolinite, 16
Metamorphism, 256
Methanation, 75—76, 78
Methane, 68, 75, 153, 246
Meyers Process, 40, 44, 46
MHD technology, 64—65
Microorganisms, 158

Mineral matter, 16, 22, 39—40, 71, 73, 124, 132, 179, 185, 198
Minerals, 73, 133, 198, 200, 207
Minimum detectable concentration, 209, 211
Minimum detection limit (MDL), 224, 226
Moisture content, 22, 246
Molybdenosis, 158
Molybdenum, 4, 19, 35, 49, 58, 60, 62, 120, 127, 158—160, 162—163, 181, 188, 227, 265
Montmorillonite, 14
Mössbauer Effect, 253
Mössbauer spectroscopy, 253—255
Mullite, 16, 25, 29

N

NaI detector, 235—236, 241
NaI(Tl) detector, 231, 243, 246
NaI(Tl) well counter, 248
National Bureau of Standards, see NBS standards
NBS reference materials, see NBS standards
NBS standards, 181—184, 219, 230, 264
Neodymium, 16, 188
Neptunium, 236—237
Neutron activation, 182, 186, 228
Neutron activation analysis, 8, 71—72, 108, 112, 114, 147, 173—174, 177—178, 181, 228—247, 262, 264
    fast neutron activation analysis, 243—247
    neutron activation using ^{252}Cf source, 237—243
    neutron activation with thermal neutrons, 228—237
New technologies of coal utilization, 64—80
    environmental impact of, 146—156
Nickel, 3—4, 18—19, 53, 76, 114, 117, 120, 124, 127, 129, 133—134, 156, 159, 163, 181, 188, 200, 237, 265
Niobium, 16, 120, 188
Nitrates, 115
Nitric acid, 183, 194—195, 236, 252, 254
Nitrogen, 2, 69, 160, 162, 182—183, 240—241, 245—246, 256
NMR, see Nuclear magnetic resonance spectroscopy
Noble gases, 133—134, 139
Nonflame atomic absorption spectrometry, 265
Nuclear magnetic resonance (NMR) spectroscopy, 255—257
Nuclear power plants, 133—146
Nuclear techniques, 21

O

Oil, 73, 77, 149
On-site disposal areas, criteria for selection of, 130
Optical emission, 173—174, 178
    results, 192
Optical emission spectroscopy, 174, 177, 186, 230
Optical methods of trace element analysis, 185—197
    atomic absorption spectroscopy, 191—197
    emission spectroscopy, 185—191
Optical microscope, 199
Optical spectrometry, 262
Organic materials, 73
Organometallic complexes, 71
Organometallic compounds, 148
Outlet fly ash, 10
Outlet precipitator ash, 3
Oxides, 115
Oxygen, 2, 69, 75, 184—185, 237, 243, 245—246, 248—249, 257
Oxygen bomb combustion, 191

P

Particle size, 3
Particle size-intensity relationship, 209
Pathways of trace contaminants, 104
Phosphates, 115
Phosphorus, 3, 49, 175, 177, 188
Phosphorus pentoxide, 15, 47
Physical synergism, 159
Physiological irritation, 159
Pitches, 4
Platinum crucibles, 178, 185
Plutonium, 250
Polarography, 174, 185
Pollution control, 48
    equipment, 1
Polonium, 133, 135, 139—141, 143—144, 250, 252
Polychlorinated biphenyls, 115
Polycyclic organic material, 115
Population dose commitments, 139
    percentage contributions of radionuclides to, 140
Potassium, 3, 20, 23, 31, 120, 188, 240—241, 265
Potassium oxide, 15, 47
Potentially hazardous elements in coal resource development, 161—162
Potentiometry, 173, 185
Pressurized-water reactors, 134, 136
Prompt neutron activation, 240—241, 243—244
Protactinium, 135
Proton-decoupled high-resolution carbon-13 magnetic resonance, 256
Proton-enhanced nuclear induction spectroscopy, 256
Proton induced X-ray emission (PIXE) spectroscopy, 220—228
Proton microprobe, 227
Proximate analysis, 181—182
Pulverized coal-fired boiler, 5, 18—21
Pulverized-coal-fired (PC-fired) furnaces, 1, 156
Pulverized coal-fired power plant, 6
Pulverized fly ash (PFA), see Fly ash
Pyrites, 2, 14, 16, 21, 25, 37, 39—40, 43—44, 48, 73, 124—125, 133, 179, 203, 253—254
Pyrolysis processes, 68

Q

Quad, 156
Quartz, 16, 25, 29, 57, 133, 200, 203,

R

Radicals, 258
Radioactive lignites, 62
Radioactive series, 250—251
Radioactive source excitation, 219—220
Radioactive sources, 219
Radioactivity, 133—146, 229, 248, 250
Radiochemical separations, 235
Radionuclides, 133—136, 138—144, 146, 230—235, 250
Radium, 135, 139—140, 146, 157, 250
Radon, 133—134, 136, 140—141, 144, 146
Rain, 118
Rank of coal, 39, 54, 132
Rare earth elements, 201
Recycling of waste, 129—130
Relative order of volatility, 3
Rhodium, 213
Rubidium, 16, 18, 188, 236

S

Salts, 115
Samarium, 26
Sample preparation for trace element analysis, 177—179
Scandium, 19, 26, 120, 188
Scanning electron microscope (SEM), 108, 110, 198—200
Scanning electron photomicrograph, 202
Scrubber sludge, 36
Secular equilibrium, 135
Selenium, 3—4, 11, 26—27, 114, 120, 125, 127, 129, 156, 158—159, 162, 178, 191, 195, 205, 236
Seracite, 115
Shale, 115
Siderophilic elements, 14
Si(Li) detector, 199, 204, 208, 215, 222
Silica, 16, 49, 51, 123
Silicates, 115
Silicon, 16, 23, 31, 120, 129, 133, 189, 199, 213, 237, 240—241, 243, 245, 252
Silicon dioxide, 15, 47
Silicon surface barrier detector, 250, 252
Silver, 4, 26, 117, 120, 189
Slag, 1, 8, 10—11, 47, 105, 133, 177
Sludge, 116, 131
Sodium, 3, 20—21, 31, 71, 120, 127, 163, 189, 199, 213, 230, 240—241
Sodium oxide, 15, 21, 43, 47
Soil pollution, 105, 116—120
Solid waste material, 149, 153
Solvent refined coal (SRC), 69—73, 76, 154, 156, 195
Spark source mass spectrometry, 115, 147, 149, 191, 259, 262, 264—265
Spark source mass spectroscopy, 8
Specific ion electrode method of analysis, 174, 176, 185
Spectrographic method results, 188—190
Spectrometer parameters, 208
Spectrometric method results, 189
Spectrophotometric system principle, 186
Spectrophotometry, 177
Spectroscopic parameters, 191
Sphalente, 202
Sphalerite, 16, 202
Spleen dose, 138—140
Standard additions, method of, 155
Standard reference material, 180—181, 224, 230, 243, 264
Standards in trace element analysis, 179—181
Steam-boiler furnace, 25—26
Stoker-fired boilers, 1, 18, 21
Strontium, 19, 26, 120, 163, 181, 189, 201, 230, 265—266
Subbituminous coal, 16—17, 29, 43, 154, 258
Sulfates, 115, 199
Sulfides, 115, 199—200
Sulfur, 2—4, 21—22, 36—37, 39—40, 43—46, 48, 54, 65, 69, 71, 73, 75—76, 78, 114, 120—123, 129, 132—133, 162, 182—183, 194, 199, 215, 239—241, 243, 245, 254
removal, 67, 78
Sulfur dioxide, 35—36, 65, 67, 78, 108, 122, 158, 160
emission, 127, 129
Sulfuric acid, 36, 48—50, 52—53, 55, 58—59, 61, 122—123, 160, 183, 194—195
Sulfur trioxide, 15, 36, 47
Sulfurous acid, 49—50
Syngas production, 78—79
Synthane process, 78
Szomolnokite, 203

T

Tars, 4, 46, 68, 78, 147, 149
distillation, 5
Tellurium, 115, 191, 196
Terbium, 26—27
Thallium, 3, 16, 114, 156, 189, 196
Thermal emission mass spectrometry, 259—260
Thermal neutrons, neutron activation with, 228—237
Thorium, 26, 31, 133—136, 138, 140—141, 157, 250, 252
Thulium, 26
Thyroid dose, 138—140
Time-of-flight mass spectrometer (TOFMS), 260
Tin, 19, 26, 120, 127, 189, 191

Titanium, 4, 23, 31, 49, 52, 71, 115, 117, 120, 123, 125, 127, 189, 200, 241
  recovery, 54—56
Titanium dioxide, 201
Titanium oxide, 15, 47
Tourmaline, 200
Toxicity, 163
Trace contaminants from coal-fired power plant, 103—120
Trace elements
  analysis, methods of, see Analytical methods
  mobilization, 103—104
  recovery, see also specific trace elements, 46—64
Transmission electron microscope (TEM), 197—198
Tritium, 133—134
Tube excited X-ray emission, 204—219
Tumors, 160
Tungsten, 26, 160, 189, 230

U

Ultimate analysis, 132, 181—182
Underground gasification, see *In situ* gasification
Uranite, 135
Uranium, 3—4, 26—27, 31, 49, 133—136, 138, 140—141, 143—144, 146, 177, 220—222, 227, 236, 250, 253, 260
  recovery, 56—64

V

Vanadium, 3—4, 16, 19, 71, 115, 120, 127, 129, 159—160, 163, 189, 230, 266
Vaporization, 2—3
Viscosity, 21
Vitrinite, 227
Volatile compounds, 2, 4, 8
Volatile matter, 14, 132
Volatilization, 105, 112, 195
  losses, 147, 178
Voltammetry, 173, 185

W

Waste management, 129—133
Wastewaters, 1
Water pollution, 105
Wet chemistry, 6, 173, 181, 185, 191, 240, 253
Whole body dose, 138—140

X

X-ray absorption, 21, 215
X-ray analysis, see also specific types, 202—228
X-ray backscattering, 23
X-ray diffraction, 16, 29, 57, 73, 202—203
  results, 206
X-ray diffractograms, 204—205
X-ray emission, 186
X-ray emission spectrometry, 203
X-ray emission spectroscopy, 178, 182, 203—228
  proton induced X-ray emission (PIXE) spectroscopy, 220—228
  radioactive source excitation, 219—220
  tube excited X-ray emission, 204—219
X-ray fluorescence, 6, 23—24, 112, 147, 174, 181, 208, 213, 215, 262
X-ray fluorescence spectrometry, 265
X-ray fluorescence spectroscopy, 24, 177
X-ray spectrogram, 202
X-ray spectrometry, 176

Y

Ytterbium, 19, 189, 230
Yttrium, 19, 189, 201

Z

Zinc, 3—4, 19, 26—27, 35, 49, 114—115, 117, 120, 127, 129, 133—134, 156, 159, 163, 181, 189, 200, 236, 250, 260, 266
Zirconium, 19, 120, 125, 127, 189, 200, 266